New Materials and Advanced Procedures of Obtaining and Processing

New Materials and Advanced Procedures of Obtaining and Processing

Editor

Andrei Victor Sandu

MDPI • Basel • Beijing • Wuhan • Barcelona • Belgrade • Manchester • Tokyo • Cluj • Tianjin

Editor
Andrei Victor Sandu
Faculty of Materials Science
and Engineering
Gheorghe Asachi Technical
University of Iasi
Iasi
Romania

Editorial Office
MDPI
St. Alban-Anlage 66
4052 Basel, Switzerland

This is a reprint of articles from the Special Issue published online in the open access journal *Applied Sciences* (ISSN 2076-3417) (available at: www.mdpi.com/journal/applsci/special_issues/new_advanced_materials).

For citation purposes, cite each article independently as indicated on the article page online and as indicated below:

LastName, A.A.; LastName, B.B.; LastName, C.C. Article Title. *Journal Name* **Year**, *Volume Number*, Page Range.

ISBN 978-3-0365-8243-6 (Hbk)
ISBN 978-3-0365-8242-9 (PDF)

Contents

applied sciences

MDPI

Editorial

New Materials and Advanced Procedures of Obtaining and Processing—Applied Sciences Insights

Andrei Victor Sandu [1,2,3]

[1] Faculty of Materials Science and Engineering, Gheorghe Asachi Technical University of Iasi, 61 D. Mangeron Blvd., 700050 Iasi, Romania; sav@tuiasi.ro
[2] Romanian Inventors Forum, St. P. Movila 3, 700089 Iasi, Romania
[3] National Institute for Research and Development in Environmental Protection INCDPM, Splaiul Independentei 294, 060031 Bucharest, Romania

Citation: Sandu, A.V. New Materials and Advanced Procedures of Obtaining and Processing—Applied Sciences Insights. *Appl. Sci.* **2023**, *13*, 3153. https://doi.org/10.3390/app13053153

Received: 2 February 2023
Accepted: 24 February 2023
Published: 1 March 2023

The emphasis of this Special Issue is on showcasing the most recent advancements in the field of materials and techniques used in a variety of applications, including medical and civil engineering.

The primary uses concern the materials used in environmental engineering, health care, dentistry, and civil engineering, as well as the methods used in the handling and processing of these materials.

Materials ranging from nano- to macro-sized, including alloys, ceramics, composites, biomaterials, polymers, and more, have been investigated in order to gain knowledge of both processes and materials.

In the first article [1], the properties of a novel material based on pillared graphene and the icosahedral clusters of boron B12 as a supercapacitor electrode material were examined using the first-principle density functional theory (DFT) approach. The efficiency of the new composite material is demonstrated by its specifically high quantum capacitance, specific charge density, and negative value for formation heat. It is demonstrated that as clusters are added, the density of electronic states rises, naturally increasing the electrode conductivity. We forecast that the effectiveness of current supercapacitors will be enhanced with the usage of a composite made of pillared graphene and boron.

The necessity to develop methods for drive shaft designs that are more reliable and can guarantee the development of frozen soils during a deposit investigation confirms the relevance of the research. The goal of study [2] was to justify the critical loads and stresses in hardened gear coatings acting under the intense wear of the contact surface (with a broken contact symmetry) in order to prolong the service life and improve the energy efficiency of the highly loaded drive gear teeth of core drilling pump transmission shafts. Additionally, the interaction between the eccentric shaft gear and transmission shaft gear teeth at different axial torques was studied. The drive transmission shaft gear and eccentric shaft gear, which define the energy consumption of the drill bit's depth stroke, were justified in terms of their effective power. This paper also suggests a way to use Legendre polynomials to support the technological and power parameters of the transmission shaft. The relationship between the variation in the load distribution factor and the contact spot deviation factor from the design axis and the contact stress base cycles was determined using a nomographic chart.

In another study [3], the linear polarization resistance method (LPR) was used to quantify the corrosion rate in pipelines utilizing a three-electrode corrosion setup. To investigate the sample's surface and the corrosion products, optical and SEM tests were carried out. With the dissolved oxygen content in the solution kept at 6 mg/L, how the concentration of NaCl affects the rate of corrosion at various pH levels, temperature ranges, and flow rates (6 ppm) is investigated. It was discovered that the corrosion rate varies between 1 and 10 mils per year, increases with flow velocity, and reaches its peak at

Reynolds numbers above 10,000. An increased fluid velocity demonstrates that corrosion is flow-insensitive.

In other research [4], TA3 alloys were implanted with Nb ions, and the impact on TA3's tribocorrosion behavior in Ringer's solution as well as biological corrosion behavior was comprehensively examined. The implanted samples had a smoother surface as a result of the sputtering and radiation damage, and the Nb ions predominated in the alloy as the solid solution element according to the surface microstructure and XRD data. The implantation of Nb ions can raise the corrosion potential of the samples, demonstrating greater thermodynamic stability according to electrochemical polarization tests. In a corrosive environment with wear behavior, the implanted samples show greater thermodynamic stability according to the results of tribocorrosion tests. The worn surface also had fewer pitting pits, indicating improved corrosion resistance. However, the partial softening of the surface and brittle passivation film caused the sample's abrasive wear and oxidation wear degree to increase.

The actual experiments and experimental processing results collected in order to construct a model for forecasting the surface roughness based on the optimization of cutting parameters are additional major contributions made in other publications [5]. The method of obtaining the regression equation of the surface roughness from a standard end-milling process on aluminum alloy 7136 in temper T76511, using two statistical methods of data analysis, adds novelty to the paper. This process used standard milling tools, standard milling parameters recommended by the tool manufacturer, a three-axis CNC machine, and a standard vice. For the suggested research area, this material is little-known and used to make extruded parts. The purpose of this study is to establish the surface roughness equation derived by milling aluminum alloy 7136 using Taguchi's experimental design once and the central composite design once. An effective mathematical model is created using the Taguchi method and the central composite design to forecast the ideal value of specific processing parameters. ANOVA analysis is used to compare the calculated and experimental surface roughness values. The Minitab application is used to examine the beginning features (surface roughness) and the controlled factors (cutting speed, depth of cut, and feed). The advantages and disadvantages of the two methods used were then analyzed and presented.

The movement precision and accuracy of a 3D printer's extruder system in relation to the print bed are examined in this work by utilizing the characteristics of 2D circular trajectories produced by simultaneous displacements on the x and y axes. The sampling of displacement evolutions, obtained with two non-contact optical sensors, is made possible using a computer-aided experimental setup. To assess and explain circular trajectory errors, some processing methods for displacement signals were proposed (e.g., open and closed curves fitting, the detection of recurrent periodical patterns in x and y-motions, low-pass numerical filtering, etc.). The description of these faults can attest to the 3D printer's proper operation for maintenance purposes and, particularly, for computer-aided accuracy corrections [6].

In the next research paper [7], a theoretical and experimental investigation of the vibrating beating phenomena is proposed. This phenomenon is produced when two rotating, unbalanced shafts are placed inside the headstock of a lathe and are connected by a flat friction belt. The work was conducted using a straightforward computer-assisted experimental setup for the capture, processing, and simulation of absolute vibration velocity signals. A horizontal geophone that is used as a sensor and mounted on a headstock produces the input signal. A vibration velocity signal was transformed into a vibration displacement signal via numerical integration (using a novel antiderivative calculus and signal correction method). This method allowed for the conversion of an absolute velocity vibration sensor into an absolute displacement vibration sensor. Numerical modeling made an important discovery that was fully supported by experiments on the evolution of the resultant vibration frequency (or combination frequency) of the beating vibration displacement signal. It was found that the combination frequency is modestly variable (tens of millihertz variance over the complete frequency range) and exhibits a periodic pattern, in contrast to certain previously reported research results. Depending on how the amplitudes and frequencies of the vibrations engaged in the beating relate to one another, this pattern has negative or positive peaks that are strategically placed in the nodes of the beating phenomenon. Other accomplishments on problems related to the description of the beating phenomena were also made. With an entire frequency range inaccuracy of less than 3 microhertz, research on a simulated signal demonstrated that the method employed for combination frequency measurements exhibited great theoretical accuracy. On the basis of a computer-aided analysis (curve fitting) on the free damped response, a study on the experimental measurement of the dynamic amplification factor of the combined vibration (5.824) due to the resonant behavior of the headstock and lathe on its foundation was also carried out. These accomplishments guarantee a better understanding of the conditions and demands for dynamic balancing as well as the phenomenon of vibration beating.

Regarding the biomaterials field, for a predictable result, the optimum biomaterial used in endodontics to seal radicular canals should have a number of characteristics, including biocompatibility, the start of ontogenesis and cementogenesis, ease of handling, enough time for manipulation, and an affordable price [8]. The root canal procedure can be followed with prosthetic repair for a flawless seal. The purpose of the study is to quantify the local response to the implantation of three biomaterials in the subcutaneous connective tissue of rabbits. The study concentrated on qualitative and quantitative analyses based on histopathological examinations, which were supported by the positive findings of the study's application of oral rehabilitation treatments and, in turn, resulted in an increase in patients' quality of life of 95% and produced the stomatognathic system's optimal functioning.

The utilization of diverse nanoparticles [9] as medication delivery systems to target and eliminate pathogenic bacteria may be a good solution for the prevention and treatment of severe illness, according to recent nanotechnology research findings. Antimicrobial medication encapsulation into nano-sized systems emerged in recent years as a viable substitute that improved treatment efficacy and reduced side effects. Both Ery-PLA and Ery-PLGA nanostructures exhibit a sustained drug release, according to the erythromycin release profile from PLA/PLGA. Ring-shaped, stiff, and spherical nanoparticles were visible in the morphology photos. Thermal analyses of the Ery-PLA and Ery-PLGA samples revealed that the pure medicine exhibits an endothermic peak with a melting temperature of around 150 °C. For thermographs of antibiotic-loaded PLA and PLGA nanoparticles, the antibiotic melting peak vanished, indicating the presence of erythromycin. This shows that the antibiotic is evenly distributed at the nanoscale across the host's polymer matrix. The chemical structure of drug-loaded polymer nanoparticles was not altered, as shown by the almost identical peaks in the FTIR spectra of the Ery-PLA and Ery-PLGA nano-architectures.

Another in vitro experiment [10] examined the impact of dentifrices containing nano-hydroxyapatite (n-HAp) on mineral deposition and dentinal tubule blockage. Ten human teeth dentin samples were placed in 40% citric acid for 30 s before being separated into four groups at random (three study groups and one control group). All samples from the control group displayed a full and wide opening of the dentinal tubules, but varying degrees of tubule closure by mineral depositions were observed in the study groups. Dentinal tubules were significantly occluded by toothpaste containing n-HAp, and mineral deposition on the dentin surface increased significantly.

Measurements of facial tissue thickness and estimated feature shapes are necessary for cranial reconstruction, which frequently serves as the last stage in medicolegal identification [11]. The purpose of this study is to develop a valid and repeatable method for estimating the maximum nose width (MNW) based on the maximum nasal aperture width (MAW) in an adult sample from Romania. An adult Romanian subject sample of 55 computer tomography (CT) images was chosen from a neurosurgery hospital's database. Using 3D systems Freeform Modelling Plus Software, two measurements of the MNW and the MAW comprised the craniometrics that were taken. A moderate relationship between the MAW and the MNW was found using correlation analysis. MAW and sex were found to form a statistically significant regression pattern using regression analysis. Based on MAW measurements taken on the skull, the preliminary results offer accurate forecasts of MNW for facial reconstruction.

Finally, a very important aspect of the applied sciences is the field of cultural heritage.

An analysis of three bronze socketed axes found in Romania's Neamţ County is presented in the paper [12]. In order to clarify the nature of the materials utilized and the production procedures, the surface structures, as well as those from the interface of the corrosion layer with the metal core of the basic alloy, were examined. In conjunction with X-ray spectrometry (EDX), optical and electron microscopy analyses revealed the type of deterioration that occurred throughout the depositional period as a result of chemical alterations and physical damage. A number of metallurgical methods employed were also identified, along with some finishing and ornamentation procedures that contributed to the determination of the objects' functionality.

Another study [13] proposed an interdisciplinary inquiry model of pottery that enables the scientific study of this group of objects. Examinations were conducted on 11 ceramic pieces from the Middle Bronze Age settlement of Piatra Neamţ-Lutărie in Eastern Romania, taking into account details about the color, production method, kind, size, usefulness, and category of the vessel, as well as information regarding ceramic paste inclusions. Optical microscopy (OM), scanning electron microscopy (SEM) with energy-dispersive X-ray analysis (EDX), and micro-Fourier-transform infrared spectroscopy (FTIR) were used to examine the samples. The results showed how various vessel categories functioned in a prehistoric population and provided valuable information about pottery manufacturing technology, such as raw material sources and fire temperatures.

In conclusion, this Special Issue managed to collect high-quality papers on various applications in applied sciences, and we hope to provide a solid state-of-the-art reference in this research area.

Institutional Review Board Statement: Not applicable.

Informed Consent Statement: Not applicable.

Data Availability Statement: Not applicable.

Acknowledgments: I would like to thank all researchers involved in the realization of this Special Issue who dedicated their valuable time and effort. Special thanks to the reviewers for their constructive comments.

Conflicts of Interest: The author declares no conflict of interest.

References

1. Kolosov, D.A.; Glukhova, O.E. Boron-Decorated Pillared Graphene as the Basic Element for Supercapacitors: An Ab Initio Study. *Appl. Sci.* **2021**, *11*, 3496. [CrossRef]
2. Savinkin, V.V.; Kolisnichenko, S.N.; Sandu, A.V.; Ivanova, O.V.; Vizureanu, P.; Zhumekenova, Z.Z. Investigation of the Strength Parameters of Drilling Pumps during the Formation of Contact Stresses in Gears. *Appl. Sci.* **2021**, *11*, 7076. [CrossRef]
3. Ravikumar, A.; Rostron, P.; Vahdati, N.; Shiryayev, O. Parametric Study of the Corrosion of API-5L-X65 QT Steel Using Potentiostat Based Measurements in a Flow Loop. *Appl. Sci.* **2021**, *11*, 444. [CrossRef]
4. Xue, C.; Zhang, P.; Wei, D.; Hu, H.; Li, F.; Yang, K. Corrosion and Tribocorrosion Behaviors for TA3 in Ringer's Solution after Implantation of Nb Ions. *Appl. Sci.* **2020**, *10*, 8329. [CrossRef]
5. Ţîţu, A.M.; Sandu, A.V.; Pop, A.B.; Ţîţu, Ş.; Frăţilă, D.N.; Ceocea, C.; Boroiu, A. Design of Experiment in the Milling Process of Aluminum Alloys in the Aerospace Industry. *Appl. Sci.* **2020**, *10*, 6951. [CrossRef]
6. Munteanu, A.; Chitariu, D.-F.; Horodinca, M.; Dumitras, C.-G.; Negoescu, F.; Savin, A.; Chifan, F. A Study on the Errors of 2D Circular Trajectories Generated on a 3D Printer. *Appl. Sci.* **2021**, *11*, 11695. [CrossRef]
7. Chitariu, D.-F.; Negoescu, F.; Horodinca, M.; Dumitras, C.-G.; Dogan, G.; Ilhan, M. An Experimental Approach on Beating in Vibration Due to Rotational Unbalance. *Appl. Sci.* **2020**, *10*, 6899. [CrossRef]
8. Aminov, L.; Sindilar, E.V.; Pasca, A.S.; Antohi, C.; Decolli, Y.; Stamatin, O.; Costin, L.I.; Bulancea, B.P.; Francu, L.; Mihalas, E.; et al. In Vivo Evaluation of Biocompatibility of Three Biomaterials Used in Endodontics for Prosthetic Purposes in Complex Rehabilitation Treatment. *Appl. Sci.* **2021**, *11*, 6519. [CrossRef]

9. Forna, N.; Damir, D.; Duceac, L.D.; Dabija, M.G.; Calin, G.; Ichim, D.L.; Gutu, C.; Grierosu, C.; Eva, L.; Ciuhodaru, M.I.; et al. Nano-Architectonics of Antibiotic-Loaded Polymer Particles as Vehicles for Active Molecules. *Appl. Sci.* **2022**, *12*, 1998. [CrossRef]
10. Bologa, E.; Stoleriu, S.; Iovan, G.; Ghiorghe, C.A.; Nica, I.; Andrian, S.; Amza, O.E. Effects of Dentifrices Containing Nanohydroxyapatite on Dentinal Tubule Occlusion—A Scanning Electron Microscopy and EDX Study. *Appl. Sci.* **2020**, *10*, 6513. [CrossRef]
11. Diac, M.M.; Earar, K.; Damian, S.I.; Knieling, A.; Iov, T.; Shrimpton, S.; Castaneyra-Ruiz, M.; Wilkinson, C.; Bulgaru Iliescu, D. Facial Reconstruction: Anthropometric Studies Regarding the Morphology of the Nose for Romanian Adult Population I: Nose Width. *Appl. Sci.* **2020**, *10*, 6479. [CrossRef]
12. Drob, A.; Vasilache, V.; Bolohan, N. The Interdisciplinary Approach of Some Middle Bronze Age Pottery from Eastern Romania. *Appl. Sci.* **2021**, *11*, 4885. [CrossRef]
13. Vasilache, V.; Diaconu, V.; Mircea, O.; Drob, A.; Sandu, I. The Archaeometallurgical Evaluation of Three Bronze Socketed Axes, Discovered in Eastern Romania. *Appl. Sci.* **2021**, *11*, 1811. [CrossRef]

applied sciences

MDPI

Article

Boron-Decorated Pillared Graphene as the Basic Element for Supercapacitors: An Ab Initio Study

Dmitry A. Kolosov [1] and Olga E. Glukhova [1,2,*]

1 Department of Physics, Saratov State University, Astrakhanskaya Street 83, 410012 Saratov, Russia; kolosovda@bk.ru
2 Laboratory of Biomedical Nanotechnology, I.M. Sechenov First Moscow State Medical University, Trubetskaya Street 8-2, 119991 Moscow, Russia
* Correspondence: glukhovaoe@info.sgu.ru

Abstract: In this work, using the first-principle density functional theory (DFT) method, we study the properties of a new material based on pillared graphene and the icosahedral clusters of boron B_{12} as a supercapacitor electrode material. The new composite material demonstrates a high specific quantum capacitance, specific charge density, and a negative value of heat of formation, which indicates its efficiency. It is shown that the density of electronic states increases during the addition of clusters, which predictably leads to an increase in the electrode conductivity. We predict that the use of a composite based on pillared graphene and boron will increase the efficiency of existing supercapacitors.

Keywords: pillared graphene; boron; specific quantum capacitance

Citation: Kolosov, D.A.; Glukhova, O.E. Boron-Decorated Pillared Graphene as the Basic Element for Supercapacitors: An Ab Initio Study. *Appl. Sci.* **2021**, *11*, 3496. https://doi.org/10.3390/app11083496

Academic Editor: Andrei Victor Sandu

Received: 19 March 2021
Accepted: 9 April 2021
Published: 13 April 2021

1. Introduction

Currently, the rapid development of electronics requires energy-intensive, safe, and efficient energy storage sources for portable and stationary applications. Lithium-ion batteries (LIBs) are usually used if high energy density and long-term operations are required. Supercapacitors (SCs) are applied when high power density with a short exposure time should be obtained [1]. The operation principles of these two sources are different: LIBs work on electrochemical Faraday processes, while SCs use an electrostatic charge storage process. Faraday processes are slower than electrostatic processes. This limits the maximum power density of batteries and their discharge/charge currents. SCs have a huge power density due to electrostatic interactions, but the energy density is much lower. Modern lithium batteries have a limited life span of tens of thousands of cycles and require sophisticated tracking systems to safely charge and discharge them [2,3]. In turn, the lifetime of SCs exceeds one million cycles, the charging method is quite simple and safe, and full discharge does not lead to degradation of the electrodes [4–6]. Thus, in terms of lifetime and ease of use, SCs have obvious advantages over batteries, but low-energy density is still the main limiting factor to their widespread application. The energy of a SC is contained in its capacitance of the double electric layer, which is formed at the media boundary and depends on the effective area of the electrodes [7]. In addition to the double-layer capacitance, there is also a quantum capacitance that directly affects the total electrode capacitance:

$$\frac{1}{C_{Total}} = \frac{1}{C_D} + \frac{1}{C_Q} \tag{1}$$

Quantum capacitance, first investigated in [8], is directly related to the density of states (DOS), so for materials with a low DOS near the Fermi level, the quantum capacitance will also be a low value. For metal electrodes (for example, electrodes of a classical dielectric capacitor) in which the DOS is high, the quantum capacitance will also be high.

Here, the quantum capacitance C_Q and the double-layer capacitance C_D contribute to the total capacitance C_{Total}, according to the law of addition of capacitors connected in series. From the expression above, it can be concluded that for increasing the efficiency of SCs, it is necessary to simultaneously increase both C_Q and C_D.

SCs are divided into two types of devices: EDLC and hybrid. In EDLC, both electrodes have the same structure and electrostatic mechanism of charge accumulation. In hybrid ones, electrostatic charge accumulation occurs on one of the electrodes, and the Faraday process occurs on the other [9]. Hereinafter, according to the SC, we denote the EDLC.

To increase C_D, the electrode material must have a sufficiently large effective area. Carbon materials, for example, widely used activated carbon, are suitable for this role [10–12]. Due to the developed surface, the electrolyte penetrates into the pores of the activated carbon and thereby creates a double-layer capacitance. The disadvantages of the activated carbon electrode are its low mechanical strength and electrical conductivity due to the amorphous structure and the gaps between the layers [13]. In addition, researchers [14,15] revealed a pattern of reduction in the total capacitance of carbon materials due to the low value of C_Q. It is possible to increase the C_Q value of carbon materials by adsorption of various compounds on their surfaces, as shown in theoretical [16–19] and experimental works [20,21]. Boron compounds are actively used for modification of such carbon structures as single-walled carbon nanotubes (SWCNTs) and graphene. The total capacitance of the synthesized SC in this case ranges from 32.2 F to 1544 F/g [22–25]. The carbon material by itself must have good electrical conductivity for efficient transfer of stored charge, high porosity for penetration of the electrolyte into the material cavities, and strong mechanical properties, since the flow of large discharge currents can lead to its overheating and destruction. These criteria are well met by a composite based on SWCNTs and graphene-pillared graphene (PGR). The use of this material in SC is currently being actively researched and discussed. Electrodes with a specific capacity of 145 to 200 F/g, excellent cyclicity, and high power density have already been synthesized [25–29], but a numerical experiment can indicate the most effective direction for researchers.

In our study, using the ab initio density functional theory (DFT) method performed in SIESTA software, we predicted methods to increase the efficiency of the SC electrode material on the base of pillared graphene and icosahedral boron clusters.

2. Materials and Methods

The SIESTA 4.1 software package [30,31] is widely used for geometry relaxation, for the search of the potential energy minima, and for the calculation of the carbon nanostructures electronic properties [32–34]. We used the density functional theory (DFT) basis set with generalized gradient approximation (GGA) and Perdew-Burke-Ernzerhof (PBE) parameterization, since these calculation parameters have been well-proven both in terms of computational accuracy and calculation duration. The force acting on each atom after relaxation was set to 0.03 eV/Å, and the energy limit was chosen to be 350 Ry. The Brillouin zone was sampled by a $4 \times 4 \times 1$ Monkhorst-Pack grid. The relaxation process was performed by the Broyden algorithm [35] and the Pulay corrections.

The expression for the $C_Q(V)$ calculation is given below [36]:

$$C_Q(V) = \frac{1}{mV} \int_0^V eD(E_F - eV)dV. \quad (2)$$

As can be seen from Equation (2), the quantum capacitance C_Q directly depends on the density of electronic states D, at an applied bias, the Fermi level E_F, and the bias V, calculated as a change in the Fermi level with a change in the object's charge and object mass m.

The considered model of pillared graphene consisted of SWCNT with a chirality index of (9,9) of 1.2 nm in diameter and two graphene sheets. The choice of SWCNT with such diameter was caused by its stability in pillared graphene framework [37].

3. Results and Discussion

Figure 1a shows a supercell of PRG containing 400 carbon atoms. The translation vectors of the supercell after geometric relaxation are 24.61 Å in the X direction and 21.44 Å in the Y direction. During the PGR supercell building, a SWCNT (9,9) with open edges was attached by chemical bonds to a graphene sheet in the area of the hole that was preliminary cut in the graphene surface. The resulting structure matched the experimental data of pillared graphene [38]. The heat of formation of the supercell is 1.2–142 kcal/mol·atom and does not depend on the CNT length. In the graphene-CNT contact area, six pairs of penta- and heptagons as well as three octagons were formed.

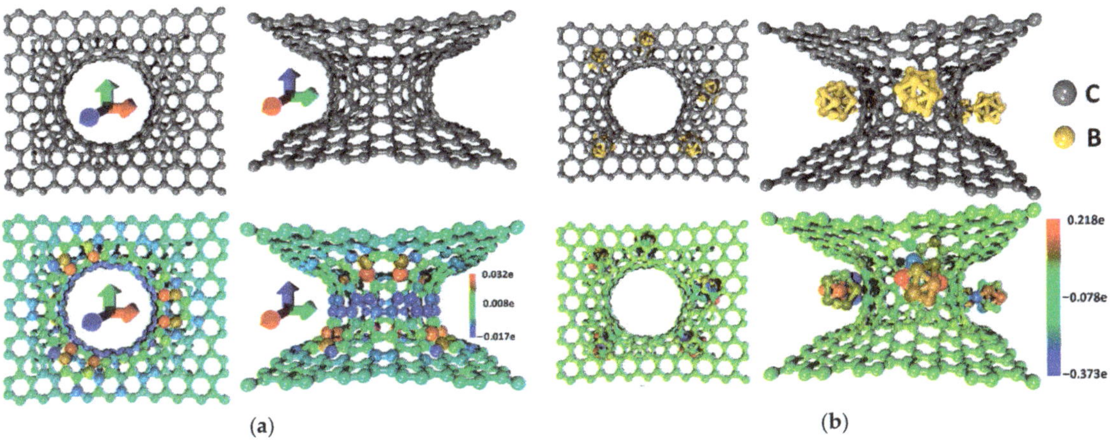

Figure 1. The PGR/B_{12} composite supercell. (**a**) The supercell of the pure PGR and the charge distribution over atoms in the XY and ZY planes; (**b**) the supercell of PGR with five boron clusters and the charge distribution over atoms in the XY and ZY planes. Red indicates a lack of electrons; blue indicates an excess of electrons.

The supercell in Figure 1b has five B_{12} clusters. B_{12} clusters are icosahedral clusters with 12 boron atoms. As can be seen from the figure, the B_{12} clusters formed bonds with the carbon near the defects in the area of the junction between the graphene sheets and SWCNTs. The length of the bonds varied in the range of 1.669 to 1.682 Å. The mass fraction of the boron ranged from 2.64% for one cluster and up to 11.92% for five clusters. The Mulliken charge distribution pattern between the PGR and B_{12} framework (Figure 1b) shows that the carbon framework received additional electrons from the B_{12} clusters. For the case of five clusters, the additional charge on the PGR is −0.287 electrons.

To calculate C_Q and predict changes in the material's conductivity, it was necessary to calculate the DOS curves for all considered cases of B_{12} concentration. Figure 2 shows the DOS curves for the pure PGR and for PGR decorated with one to five boron clusters. It can be seen that the growth in the B_{12} amount increases the DOS amplitude over the entire energy range by increasing the number of electronic states.

Earlier [39], we showed that B_{12} clusters in the modified CNT increase the DOS peak amplitude and the CNT conductivity. Here, we observed a similar effect, so we assumed that the resistance of the boron-decorated PGR-based electrode would also decrease.

To predict the effectiveness of the model as an electron SC material, we built C_Q curves in the voltage range from −3 to 3 V (Figure 3). In comparison with the pure PGR (black curve), the modified model had the characteristic peaks in the positive branch of the plot reaching ~1.266 kF/g for five B_{12} clusters. In accordance with Formula (2), an increase in DOS in the energy range from −3 to 3 eV (Figure 2) also leads to an increase in C_Q. Despite the increase in the electrode mass due to the addition of clusters, the C_Q continued to grow as the clusters introduced new electronic states. The details of all C_Q values are shown in Table 1.

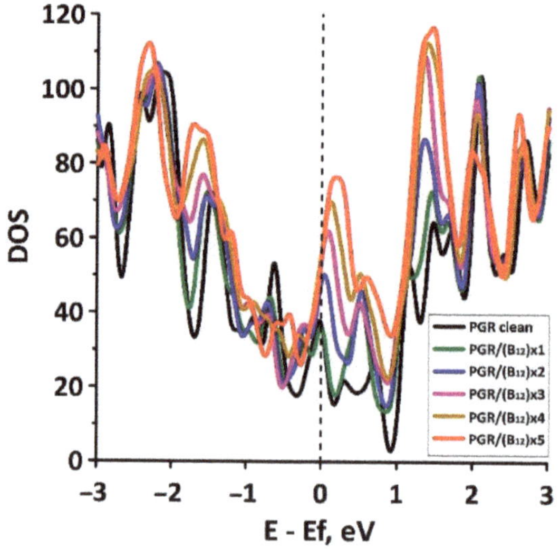

Figure 2. Density of states (DOS) curves of the considered model of electrode material.

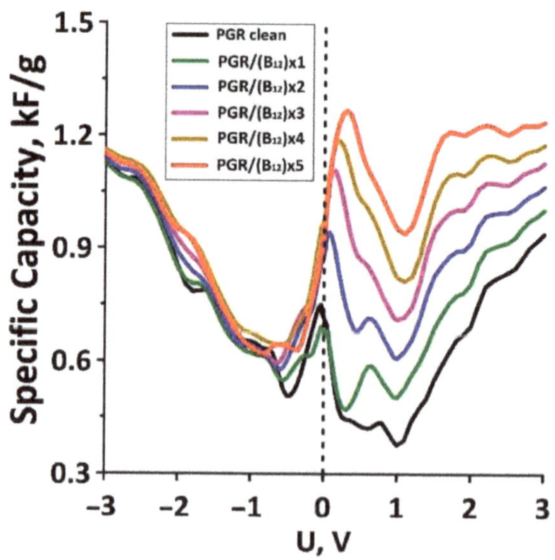

Figure 3. C_Q for the pure PGR and for PGR decorated with 1 to 5 boron clusters.

Table 1. Heat of formation, specific charge density, and specific capacitance for the pure PGR and for the PGR decorated with 1 to 5 boron clusters.

No.	ΔH, eV	Q_{SCD}, kC/g	C_Q, kF/g
Pure PGR	-	−2.384/1.748	0.749
PGR/(B_{12}) × 1	−0.00607	−2.359/2.036	0.691
PGR/(B_{12}) × 2	−0.00965	−2.466/2.418	0.943
PGR/(B_{12}) × 3	−0.01271	−2.532/2.757	1.105
PGR/(B_{12}) × 4	−0.01743	−2.608/3.030	1.187
PGR/(B_{12}) × 5	−0.02272	−2.534/3.317	1.266

In this study, we did not conduct a numerical estimate of the capacitance C_D. However, according to previous articles [26–29], it can be seen that the value of the total capacitance of SCs built on the basis of pillared graphene gives values in the range from 144.5 to 352 F/g, where the value C_D must be greater than these capacitance values. Thus, an increase in C_Q will lead to an increase in the total capacity of the SC.

Finally, to determine whether the electronic material for symmetric SC devices was ELCC or hybrid SC, specific charge density curves were calculated (Figure 4). Q_{SCD} were calculated by the method described in [40]:

$$Q_{SCD} = \int_0^V C_Q(V)dV \tag{3}$$

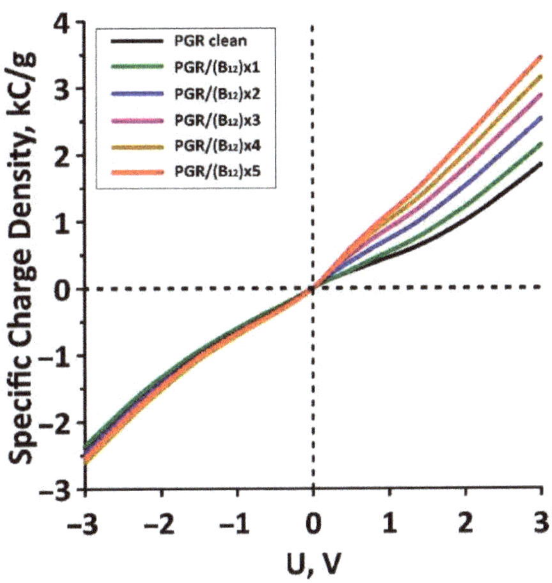

Figure 4. Q_{SCD} for the pure PGR and for the PGR decorated with 1 to 5 boron clusters.

The look of the curves in Figure 4 indicates the symmetrical behavior of the material with an increase in number of B_{12} clusters up to two. However, with further clusters up to five, the picture slightly changed, and the asymmetry of the negative and positive branches appeared. In the negative branch, the curves were almost constant, and this was directly related to the values of C_Q in the voltage range from −3 to 0 V. Here, the change in C_Q was not so pronounced. Specific charge density indicates how much charge, in terms of mass, the electrode material can store. The increase in charge capacity is explained as a direct result of the additional availability of states near the Fermi level in DOS. From the nature of the Q_{SCD} distribution, it follows that by changing the boron concentration, this electrode material can be used both in hybrid SCs and in symmetric SCs.

Heat of formation per atom (ΔH) for each added boron cluster, C_Q at the 0.1 V and Q_{SCD} for voltages −3 and 3 V, are shown in Table 1.

4. Conclusions

In the present paper, the objects of research were the models of PGR decorated with boron clusters. Based on the ab initio method in the SIESTA code, the specific quantum capacitance, electronic properties, and energy stability of PGR modified by clusters of boron-containing compounds were calculated. The calculation of the specific charge density demonstrated symmetry with respect to the positive and negative bias at a concentration of 5.13 wt% (two B_{12} clusters). With the increase in boron concentration, branch asymmetry was observed, which expanded the application of the electrode material for both symmetric and hybrid SC devices. Thus, the modification of PGR by B_{12} boron clusters can significantly improve the characteristics of the electrode material and expand its application in the electrode material SC.

Author Contributions: Conceptualization, O.E.G. and D.A.K.; methodology, O.E.G.; investigation, O.E.G. and D.A.K.; writing—original draft preparation, D.A.K.; writing—review and editing, O.E.G.; supervision, O.E.G.; funding acquisition, D.A.K. All authors have read and agreed to the published version of the manuscript.

Funding: This work was financed by the Ministry of Science and Higher Education of the Russian Federation within the framework of state support for the creation and development of World-Class Research Centers "Digital biodesign and personalized healthcare" No. 075-15-2020-926. The authors thank RFBR (project No. 19-32-90160) for supporting the DFT calculations.

Institutional Review Board Statement: Not applicable.

Informed Consent Statement: Not applicable.

Data Availability Statement: Not applicable.

Conflicts of Interest: The authors declare no conflict of interest.

References

1. Poonam Sharma, K.; Arora, A.; Tripathi, S.K. Review of supercapacitors: Materials and devices. *J. Energy Storage* **2019**, *21*, 801–825. [CrossRef]
2. Kim, H.-J.; Krishna, T.; Zeb, K.; Rajangam, V.; Gopi, C.V.V.M.; Sambasivam, S.; Raghavendra, K.V.G.; Obaidat, I.M. A Comprehensive Review of Li-Ion Battery Materials and Their Recycling Techniques. *Electronics* **2020**, *9*, 1161. [CrossRef]
3. Manthiram, A. A reflection on lithium-ion battery cathode chemistry. *Nat. Commun.* **2020**, *11*, 1550. [CrossRef] [PubMed]
4. Miller, J.R.; Simon, P. MATERIALS SCIENCE: Electrochemical Capacitors for Energy Management. *Science* **2008**, *321*, 651–652. [CrossRef]
5. Murray, D.B.; Hayes, J.G. Cycle Testing of Supercapacitors for Long-Life Robust Applications. *IEEE Trans. Power Electron.* **2015**, *30*, 2505–2516. [CrossRef]
6. Panda, P.K.; Grigoriev, A.; Mishra, Y.K.; Ahuja, R. Progress in Supercapacitors: Roles of Two Dimensional Nanotubular Materials. *Nanoscale Adv.* **2020**, *2*, 70–108. [CrossRef]
7. Chen, G.Z. Supercapacitor and supercapattery as emerging electrochemical energy stores. *Int. Mater. Rev.* **2016**, *62*, 173–202. [CrossRef]
8. Luryi, S. Quantum capacitance devices. *Appl. Phys. Lett.* **1988**, *52*, 501–503. [CrossRef]
9. Shao, Y.; El-Kady, M.F.; Sun, J.; Li, Y.; Zhang, Q.; Zhu, M.; Wang, H.; Dunn, B.; Kaner, R.B. Design and Mechanisms of Asymmetric Supercapacitors. *Chem. Rev.* **2018**, *118*, 9233–9280. [CrossRef]
10. Ahmed, S.; Ahmed, A.; Rafat, M. Supercapacitor performance of activated carbon derived from rotten carrot in aqueous, organic and ionic liquid based electrolytes. *J. Saudi Chem. Soc.* **2018**, *22*, 993–1002. [CrossRef]
11. Piñeiro-Prado, I.; Salinas-Torres, D.; Ruiz-Rosas, R.; Morallón, E.; Cazorla-Amorós, D. Design of Activated Carbon/Activated Carbon Asymmetric Capacitors. *Front. Mater.* **2016**, *3*, 16. [CrossRef]
12. Pina, A.C.; Amaya, A.; Marcuzzo, J.S.; Rodrigues, A.C.; Baldan, M.R.; Tancredi, N.; Cuña, A. Supercapacitor Electrode Based on Activated Carbon Wool Felt. *C J. Carbon Res.* **2018**, *4*, 24. [CrossRef]
13. Qing, Y.; Jiang, Y.; Lin, H.; Wang, L.; Liu, A.; Cao, Y.; Sheng, R.; Guo, Y.; Fan, C.; Zhang, S.; et al. Boosting the Supercapacitor Performance of Activated Carbon by Constructing Overall Conductive Networks Using Graphene Quantum Dots. *J. Mater. Chem. A* **2019**, *7*, 6021–6027. [CrossRef]
14. Paek, E.; Pak, A.J.; Hwang, G.S. A computational study of the interfacial structure and capacitance of graphene in [BMIM][PF6] ionic liquid. *J. Electrochem. Soc.* **2013**, *160*, A1–A10. [CrossRef]
15. Paek, E.; Pak, A.J.; Kweon, K.E.; Hwang, G.S. On the origin of the enhanced supercapacitor performance of nitrogen-doped graphene. *J. Phys. Chem. C* **2013**, *117*, 5610–5616. [CrossRef]
16. Hirunsit, P.; Liangruksa, M.; Khanchaitit, P. Electronic structures and quantum capacitance of monolayer and multilayer graphenes influenced by Al, B, N and P doping, and monovacancy: Theoretical study. *Carbon* **2016**, *108*, 7–20. [CrossRef]
17. Mousavi-Khoshdel, S.M.; Jahanbakhsh-bonab, P.; Targholi, E. Structural, electronic properties, and quantum capacitance of B, N and P-doped armchair carbon nanotubes. *Phys. Lett. A* **2016**, *380*, 3378–3383. [CrossRef]
18. Chen, L.; Li, X.; Ma, C.; Wang, M.; Zhou, J. Interaction and quantum capacitance of nitrogen/sulfur co-doped graphene: A theoretical calculation. *J. Phys. Chem. C* **2017**, *121*, 18344–18350. [CrossRef]

19. Sun, P.; Wang, R.; Wang, Q.; Wang, H.; Wang, X. Uniform MoS2 Nanolayer with Sulfur Vacancy on Carbon Nanotube Networks as Binder-free Electrodes for Asymmetrical supercapacitor. *Appl. Surf. Sci.* **2019**, *475*, 793–802. [CrossRef]
20. Miao, J.; Dong, X.; Xu, Y.; Zhai, Z.; Zhang, L.; Ren, B.; Liu, Z. Preparation and electrochemical performance of 1,4-naphthaquinone-modified carbon nanotubes as a supercapacitor material. *Org. Electron.* **2019**, *73*, 304–310. [CrossRef]
21. Wang, X.; Wu, D.; Song, X.; Du, W.; Zhao, X.; Zhang, D. Review on Carbon/Polyaniline Hybrids: Design and Synthesis for Supercapacitor. *Molecules* **2019**, *24*, 2263. [CrossRef]
22. Jain, R.; Wadekar, P.H.; Khose, R.V.; Pethsangave, D.A.; Some, S. MnO2@Polyaniline-CNT-boron-doped graphene as a freestanding binder-free electrode material for supercapacitor. *J. Mater. Sci. Mater. Electron.* **2020**, *31*, 8385–8393. [CrossRef]
23. Hao, H.; Xiaogang, S.; Wei, C.; Jie, W.; Xu, L.; Yapan, H.; Chengcheng, W.; Guodong, L. Electrochemical properties of supercapacitors using boron nitrogen double doped carbon nanotubes as conductive additive. *Nano* **2019**, *14*, 1950080. [CrossRef]
24. Maity, C.K.; Sahoo, S.; Verma, K.; Behera, A.K.; Nayak, G.C. Facile functionalization of boron nitride (BN) for the development of high-performance asymmetric supercapacitors. *New J. Chem.* **2020**, *44*, 8106–8119. [CrossRef]
25. Jung, D.; Muni, M.; Marin, G.; Ramachandran, R.; El-Kady, M.F.; Balandin, T.; Kaner, R.B.; Spokoyny, A.M. Enhancing cycling stability of tungsten oxide supercapacitor electrodes via a boron cluster-based molecular cross-linking approach. *J. Mater. Chem. A* **2020**, *8*, 18015–18023. [CrossRef]
26. Banda, H.; Périé, S.; Daffos, B.; Taberna, P.-L.; Dubois, L.; Crosnier, O.; Simon, P.; Lee, D.; Paëpe, G.D.; Duclairoir, F. Sparsely Pillared Graphene Materials for High Performance Supercapacitors: Improving Ion Transport and Storage Capacity. *ACS Nano* **2019**, *13*, 1443–1453. [CrossRef] [PubMed]
27. Wang, W.; Ozkan, M.; Ozkan, C.S. Ultrafast high energy supercapacitors based on pillared graphene nanostructures. *J. Mater. Chem. A* **2016**, *4*, 3356–3361. [CrossRef]
28. Ghosh, K.; Yue, C.Y.; Sk, M.M.; Jena, R.K. Development of 3D Urchin-Shaped Coaxial Manganese Dioxide@Polyaniline (MnO2@PANI) Composite and Self-Assembled 3D Pillared Graphene Foam for Asymmetric All-Solid-State Flexible Supercapacitor Application. *ACS Appl. Mater. Interfaces* **2017**, *9*, 15350–15363. [CrossRef] [PubMed]
29. Jiang, L.; Sheng, L.; Long, C.; Wei, T.; Fan, Z. Functional Pillared Graphene Frameworks for Ultrahigh Volumetric Performance Supercapacitors. *Adv. Energy Mater.* **2015**, *5*, 1500771. [CrossRef]
30. Ordejón, P.; Artecho, E.; Soler, J.M. Self-consistent order-N density-functional calculations for very large systems. *Phys. Rev. B* **1996**, *53*, 10441. [CrossRef]
31. Soler, J.M.; Artecho, E.; Gale, J.D.; Garcýa, A.; Junqera, J.; Ordejón, P.; Sanchez-Portal, D. The SIESTA method for ab initio order-N materials simulation. *J. Phys. Condens. Matter* **2002**, *14*, 2745. [CrossRef]
32. Naveh, Z.M.; Khajehdad, M.M.; Sarmazdeh, M.M. A theoretical study on the chirality detection of serine amino acid based on carbon nanotubes with and without Stone-Wales defects. *Struct. Chem.* **2019**, *31*, 455–464. [CrossRef]
33. Juarez, F.; Dominguez-Flores, F.; Quaino, P.M.; Santos, E.; Schmickler, W. Interactions of ions across carbon nanotubes. *Phys. Chem. Chem. Phys.* **2020**, *22*, 10603–10608. [CrossRef] [PubMed]
34. Chen, Y.R.; Lin, M.K.; Chan, D.H.; Lin, K.B.; Kaun, C.C. Ab Initio and Theoretical Study on Electron Transport through Polyene Junctions in between Carbon Nanotube Leads of Various Cuts. *Sci. Rep.* **2020**, *10*, 8033. [CrossRef] [PubMed]
35. Johnson, D.D. Modified Broyden's method for accelerating convergence in self-consistent calculations. *Phys. Rev. Lett.* **1988**, *38*, 12807. [CrossRef] [PubMed]
36. Shunaev, V.V.; Ushakov, A.V.; Glukhova, O.E. Increase of gamma-Fe2O3/CNT composite quantum capacitance by structural design for performance optimization of electrode materials. *Int. J. Quantum Chem.* **2020**, *120*, e26165. [CrossRef]
37. Slepchenkov, M.M.; Shmygin, D.S.; Zhang, G.; Glukhova, O.E. Controlling anisotropic electrical conductivity in porous graphene-nanotube thin films. *Carbon* **2020**, *165*, 139–149. [CrossRef]
38. Zhu, Y.; Li, L.; Zhang, C.; Casillas, G.; Sun, Z.; Yan, Z.; Ruan, G.; Peng, Z.; Raji, A.R.O.; Kittrell, C.; et al. A seamless three dimensional carbon nanotube graphene hybrid material. *Nat. Commun.* **2012**, *3*, 1–7. [CrossRef]
39. Kolosov, D.A.; Glukhova, O.E. A New Composite Material on the Base of Carbon Nanotubes and Boron Clusters B_{12} as the Base for High-Performance Supercapacitor Electrodes. *C J. Carbon Res.* **2021**, *7*, 26. [CrossRef]
40. Xu, Q.; Yang, G.; Fan, X.; Zheng, W. Improving the Quantum Capacitance of Graphene-Based Supercapacitors by the Doping and Co-Doping: First-Principles Calculations. *ACS Omega* **2019**, *4*, 13209–13217. [CrossRef]

applied sciences

MDPI

Article

Investigation of the Strength Parameters of Drilling Pumps during the Formation of Contact Stresses in Gears

Vitaliy Vladimirovich Savinkin [1,*], Sergey Nikolayevich Kolisnichenko [1], Andrei Victor Sandu [2,3,4,*], Olga Vladimirovna Ivanova [1], Petrica Vizureanu [2] and Zaure Zhetpysbayevna Zhumekenova [1]

1 Transport and Mechanical Engineering Department, M. Kozybayev North Kazakhstan University, 86, Pushkin Street, Petropavlovsk 150000, Kazakhstan; serega5@mail.ru (S.N.K.); itf.nkzu@mail.ru (O.V.I.); zhumekenova82@mail.ru (Z.Z.Z.)

2 Faculty of Materials Science and Engineering, Gheorghe Asachi Technical University of Iaşi, Boulevard D. Mangeron, No. 51, 700050 Iasi, Romania; peviz@tuiasi.ro

3 National Institute for Research and Development for Environmental Protection INCDPM, 294 Splaiul Independentei, 060031 Bucharest, Romania

4 Romanian Inventors Forum, Str. P. Movila 3, 700089 Iasi, Romania

* Correspondence: cavinkin7@mail.ru (V.V.S.); sav@tuiasi.ro (A.V.S.)

Abstract: The relevance of this research lies in the need to develop scientifically based methods for calculating and designing a transmission shaft with a hardened coating of increased strength and service life of a core drilling pump drive that can allow for a redistributing of resistance forces along the contact surfaces of the gear. This relevance is confirmed by the need to improve domestic methods for designing drive shafts of increased reliability which can ensure the development of frozen soils during deposits exploration. The purpose of the research is to increase the energy efficiency and service life of the high-loaded drive gear teeth of core drilling pump transmission shafts by justifying the critical loads and stresses in hardened gear coatings acting under intense wear of the contact surface with a broken contact symmetry. The criteria for the effective wear area with an uneven contact cross-section at the maximum bending moments of the transmission shaft of the drilling pump were justified and presented in the work. Additionally, the process of interaction of the transmission shaft gear teeth with the eccentric shaft gear at uneven axial torques was investigated. The effective power (Ng) of the gearing of the drive transmission shaft gear and the eccentric shaft gear, which characterizes the energy consumption of the drill bit depth stroke, was justified. This work also proposes a method of substantiating the technological and power parameters of the transmission shaft by using Legendre polynomials. A nomographic chart was developed for the determination of the dependence of the contact stress base cycles on the change in the load distribution factor and the contact spot deviation factor from the design axis λ.

Keywords: effective gearing power; drilling rig pump; service life; strength calculation; contact wear; transmission shaft; mathematical model; nomographic chart of basic cycles

Citation: Savinkin, V.V.; Kolisnichenko, S.N.; Sandu, A.V.; Ivanova, O.V.; Vizureanu, P.; Zhumekenova, Z.Z. Investigation of the Strength Parameters of Drilling Pumps during the Formation of Contact Stresses in Gears. *Appl. Sci.* **2021**, *11*, 7076. https://doi.org/10.3390/app11157076

Academic Editor: Ilaria Cacciotti

Received: 12 July 2021
Accepted: 28 July 2021
Published: 30 July 2021

1. Introduction

In modern mining and exploration drilling units, one of the most important installation units (and the main energy consumer) is the mud pumping unit (MPU).

The energy intensity of the drilling process significantly depends on the efficient operation of the transmission (drive) shaft of the pump, which is accompanied by large dynamic loads [1–3]. The range of changes in the pump cyclic load directly depends on the operating modes of the drill (start up - braking), the sum of the moments of forces that occur during different periods of drilling, and the deviation angle of the cutter at its entry. The energy efficiency of the pump operation largely depends on the dynamic, power, and energy characteristics of the transmission shaft, the kinematics of the gear and the design features of the drive.

Its technical condition and maximum permissible life indicators ensure the performance of technological operations that determine the efficiency and energy intensity of the well drilling process [3,4].

Studying the design and technological features of the pumping and the drilling unit, it was found that there are frequent failures along with its high margin of safety that can be associated with a sharp decrease in the supply pressure, *Ps*, and, as a result, in productivity. In some fields, intensive wear (i→max) of the cylinder-piston and plunger parts of the pump occurs during long-term core drilling at a milling cutter entry angle of up to 45°. This process can be explained by a sharp increase in the moments of resistance forces (Ff, ...) with an increase in the supply volume of a special clay-containing solution to the casing head when the milling cutter slows down the speed.

Since the main criterion for the performance of shafts and gears is their fatigue strength (endurance), for the manufacture of most shafts thermally improved medium-carbon steels 40, 45, and 50 are used. They are used for the manufacture of shafts of stationary machines and mechanisms. The billet made of such steels is subjected to an improving heat treatment (HRC $\leq$ 36) before machining. The shafts are turned on a lathe, followed by grinding of the seats and trunnions on a grinding machine.

If the endurance of the thermally improved shafts is unsatisfactory or if they have wear sections, the shafts made of these steels are subjected to surface quenching with high-frequency currents heating and low tempering in these places.

Thus, there is a need to form and systematize the technical and operational indicators of the drilling rig and establish the dependence of changes in the quality criteria of a hardened coating that characterize the efficiency and service life of its operation.

2. Technological Methods for Increasing Service Life of Drilling Pumps

Analytical studies of the achievements of domestic and foreign scientists Zi-Ming, et.al. [4]. Sagar in the field of service life of drilling rigs (DR) show that the factors are insufficiently substantiated and a causal relationship has not been fully established between the failure of the main pump assemblies of drilling rigs with dynamic processes occurring in the structural elements of the gearing of the transmission shaft gear [4,5]. The importance of the task of studying the service life of pumps lies not only in the justification of the unit performance criteria but also in understanding the physical phenomena of cause and effect relationships of fatigue processes with technological and operational factors (Table 1) [6,7].

Table 1. Factors influencing the determination of cause and effect relationships when justifying performance criteria.

Technical and Operational Parameter	Limits and Conditions of Variation	Variable Drilling Rig Output Parameters	Cause and Effect Relations
Supply pressure P_s, MPa Flow rate Q, L/min	5.0 ÷ 6.3 294 ÷ 384; 486 ÷ 594	Performance, *Q* Head pressure, *H*	$Q = S \cdot D \cdot \kappa \cdot \kappa v$ $N = (d1 - d2)/(f \cdot g) + V + p$ $\mathrm{E} = \sum \mathrm{E}_{z.kp} + \mathrm{E}_{HK}$
Zenith angle of the cutter entry, α	35 ÷ 60	Intensive wear, *I*	$\mathrm{N} = \frac{(\mathrm{P_H} \cdot \mathrm{Q_H}) - (\mathrm{M} \cdot \mathrm{n})}{45 \cdot \eta \cdot \eta}$ $i \rightarrow \min$, function of shape changing during wear $P_i(x) = \frac{1}{i2^i}\left[\left(x^2 - 1\right)^i\right]^{(i)}$
Electric motor power N, kW	30	moments of forces, M_k	$\mathrm{N}_M = \mathrm{N}_B/(\eta_T \cdot \eta_M)$ $M_{Mmin} = \frac{M_{Lmax} \cdot F_m}{F_d}$, $M_{Mmax} = \frac{M_{Lmax} \cdot F_{max}}{F_d}$
Uneven pressure at the pump outlet, %	No more than 12%	Average load, F_{mL}	$F_{mL} = \sqrt[3]{\frac{F_1^3 S1 + F_2^3 S2 + F_3^3 S3 + \ldots}{\sum S_i}}$

To understand the constituent elements of the equalities, we present their explanation. *S*—is the cross-sectional area of the piston; *D*—length of the stroke of the piston; *k*—speed of rotation of the shaft (rpm); *kv*—efficiency; *d1* is the pressure of the liquid in the intake tank, *d2* is in the receiving tank; f—is the density of the liquid; *g* — acceleration of gravity at a given density; V—the suction height of the solution; *p* — pressure loss; *Ed* —Drilling energy; *Egp* = The energy spent on the engagement of the pump gears; *Neff* —effective pump power; *Pr*—rated pressure; *Qr*—nominal flow rate; Mg—gear torque; ng—gear speed; ηg—gearbox efficiency; ηp—pump efficiency; N_S-power on the shaft, (*W*); η_t—transmission efficiency; η_p- the efficiency of the engine; M_{Mmin}—minimum rated torque required by the engine during the cycle; M_{Lmax}—maximum allowed primary; M_{Mmax}—the highest torque that the engine needs to reach during the cycle torque; F_{m}—average system load; F_{max}—maximum system load; F_{d}—dynamic load.

Works of leading researchers, such as Ladich, E. et al., are devoted to the problems of improving the reliability and service life of drilling equipment. Their contribution to the selection of promising equipment in terms of quality and reliability is reflected in the methodological basis of their resource approach to equipment maintenance and repair, as well as in their calculated equipment utilization rates.

The research of Vetter S. et al., is based on the solution of particular problems of increasing service life according to known technical and structural parameters of pump unit parts [8]. However, the causes of failures are not sufficiently studied, the influence of dynamic loads, and most importantly, the unevenness of their distribution over parts in different time intervals t_i (acceleration time, steady-state movement time and braking time) of equipment operation was not taken into account. The main problem of the known methods is their separate calculation of a separate part, based only on the physical and mechanical properties of this material and the applied point load to the static calculated "shoulder".

The service life of machine parts is determined by the following factors: the design of components and parts, the physical and mechanical properties of the friction surfaces, the working environment and operating conditions, and the quality of their manufacturing. Many factors that affect the service life and variability of the combinations of these factors in the operating conditions lead to the dispersion of the actual data on the failure of parts and components. It should be noted that it is quite difficult to solve some issues of increasing the service life of equipment due to the large number of tasks that require both extensive theoretical and practical research [8,9].

Researchers of machine service life [8–12] agree that measures to improve reliability can be implemented at three stages, namely during design, production, and in operation. The proposed methods for improving reliability are based on redundancy such as reducing the failure rate of the equipment, reducing the time of continuous operation, and reducing the average recovery time. However, the practice and reality of operating drilling units at a distance from the bases in harsh climatic and geographical conditions practically reduces the effectiveness of existing methods to zero.

Classical methods of power and energy calculation do not take into account the influence of the intensity and volume of wear on the change in the design geometry of the part. The degraded worn contact surface affects the dynamics of the redistribution of moments of forces and contact stresses σ_{-1} and, as a result, changes the structure of the relative position of the contact surfaces relative to the design axis of symmetry [13].

Thus, a scientific problem is formulated about the need to develop a unified methodology for assessing service life, taking into account the coefficients that determine the change in the structural relative position of the contact surfaces relative to the design axis of symmetry under dynamic loads. The developed methodology should have practical results in substantiating the structural parameters of the effective operation of contact surfaces. On the basis of the acquired knowledge, a design and technological solution that provides an adaptive connection with the changing design structure of the entire pumping unit should be developed.

To solve this problem, it is necessary to investigate how the load is distributed over the main drive shafts of the pump and the contact gears during operation.

3. New Method Development

Here we are presenting the development of methods and an algorithm for calculating the service life of a drilling pump, taking into account the wear of the contact area of the transmission and eccentric shafts.

Elastic deformation occurs up to known loading limits under the action of external forces applied to the transmission shaft of the pump and disappears when the load is removed. During the long-term operation of the pump, the physical and mechanical properties of the contact surface of the transmission shaft degrade, and the deformation does not disappear after the load is removed. In this case, the strength of the local section of the shaft is considered to be broken. The violation can manifest itself in the form of peeling of the contact layer, brittle destruction, hardening, etc. Thus, the violation of strength can be manifested not only by visible signs (fracture, microcracks), but also by the concentration of plastic deformations and the deviation of linear dimensions from the design axis of symmetry to the contact surfaces. This means that the study of the principles of load distribution over the transmission shaft of the pump and the reasons for the loss of service life should begin with the method of calculation for dangerous spots and permissible stresses.

A general diagram of the drilling pump unit was constructed (Figure 1) to determine the power and kinematic properties of the drilling pump drive. Schematically, the calculated sections that characterize the relationship between the structural elements and the nodes of the pumping unit are denoted from I to IV.

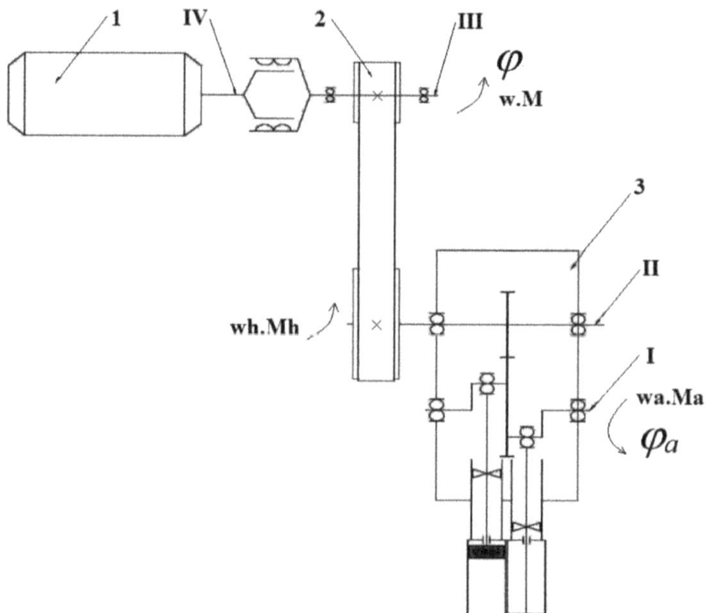

Figure 1. MPU Flow Chart: 1, pumping power; 2, power transmission; and 3, pump.

To study the moments of forces acting on the transmission shaft, it is important to calculate the power and speed properties of the drive elements. The first stage of the calculation is to refine the output parameters of the electric drive section IV (from the motor to the coupling). It is important to note that the scheme of the drive elements sequence significantly influences the accuracy of the calculation of the power and kinematic properties of the pump unit [12,14]. The drilling rig under study has the following

MPU line-up flow chart: motor→open gear→closed gear→coupling→working machine (M→OG→CG→C→WM). Let's determine the torque of the electric motor as:

$$P_m, n, T_m = \frac{P_m}{\omega_r} \tag{1}$$

where: P_m is motor power (P_m = 30 kW); ω_r, rated angular velocity; n, number of revolutions (n = 1500 rpm); and $M_{dr} = T_{II}$. Gear module 2, α = 20, β = 6 from 20 to 50 increases torque to 238.7 N·m.

The driving gear of the transmission shaft of the pump performs the work that drives the pump (drilling rig). Therefore, to reduce energy consumption and improve the reliability of the pump it is necessary to investigate the principle of forces and stresses distribution in the kinematic relations of the gear shaft and gear wheel of the eccentric shaft of the pump [15–20]. The power of the gear and wheel was studied when determining the energy components of the pump in the interaction of the transmission and eccentric shafts.

As a result, the dependence of the pump energy costs during drilling was upgraded taking into account the studied moments of forces and accelerations for a given angular rotation velocity ω of the transmission shaft [17–21]:

$$N_1 = \sum_{i=1}^{n} \Phi_{\tau i} \sum R_i + F_{t1} r_1 (n_1 - n_K) K_a \tag{2}$$

According to formula (2) N_1 is mesh gear power; $\Phi_{\tau i}$ is the tangent force of inertia; R_i is the shoulder of the force; F_{t1} is the force applied to the gear; r_1 is the radius of the pitch circle of the gear; n_1 is the frequency of the gear rotation; n_k is the frequency of the wheel shaft rotation; and K_a is the load factor.

Studies of the principle of the dynamic load effect on the contact surface of the transmission shaft drive gear were carried out step by step, examining how the circumferential, radial, and axial forces in the F_{t1} gear change (Table 2).

Table 2. Change in the forces in the transmission shaft mesh gear.

Type of Forces	Calculation Terms	Designations	Values
Circumferential force	$F_{t1} = \frac{2T_{II}10^3}{d_1}$; $T_{II} = T_{III}\eta_{ot}\eta_b U_{ot}$	F_{t1} is circumferential force, d_1 is the dividing diameter of the gear, mm; T_{II} is the torque of the transmission shaft H ·, η_{ot} is the efficiency of the open transmission (belt), η_b is Bearing efficiency factor, U_{ot} is gear ratio of the open transmission (belt).	F_{t1} = 10.602 kN. T_{II} = 639.59 N·m
Radial force	$F_{r1} = F_{t1}\frac{tg\alpha}{cos\beta}$	F_{r1} is radial force, α is the gearing angle α = 20° β—is the angle of cylindrical wheel teeth inclination, $\beta = 6°$	F_{r1} = 3880.151 N
Axial force	$F_{a1} = F_{t1}tg\beta$	F_{a1} is axial force, F_{t1} is circumferential force, β—is the angle of cylindrical wheel teeth inclination.	F_{a1} = 1114.315 N

Analyzing the results obtained, it was found that the change in the value of the action of the circumferential force F_{t1} in the mesh gear teeth engagement of the transmission shaft largely depends on the open gear drive. Namely, from the power of the electric drive P_m and its angular velocity ω_r, which form the torque T_{II} and T_{III}. According to the established dependencies, it is possible to study the operating and operational properties of the drilling rig pump and to estimate the energy intensity of each operation of the core drilling process. The dependence of the torque M_{K1} on the change in the electric drive operation mode may be expressed through the angular velocity open transmission and described by the equation $M_{K1} = 1.32\omega^{1.006}$. All of these factors will vary depending on the start-up time t_1, braking time t_3 and angle α of the tooth mesh corresponding to Δt_i.

Hence, the circumferential force acting in the mesh gear of the transmission shaft depends on the input properties of the electric motor, which characterize its efficiency. Taking into account that during operation the electric motor operates at relatively stable modes, and the kinematics of shock loads from the drive of the "transmission shaft—gear" system is limited by sliding bearings, the circumferential force in the mesh can be conditionally assumed as F_{t1} = const. Studying the influence of the radial F_{r1} and axial F_{a1} forces on the quantitative values of the mesh force F_{m1}, dependence on the structural properties (tgα and cosβ) of the kinematic "gear-wheel" system is observed. This dependence can become apparent during the transmission shaft operation.

4. Investigation of the Gear Wear Process with Uneven Distribution of Moments of Forces and Displacement of the Contact Area Relative to the Axis of the Tooth Symmetry

With the natural wear of the gear tooth, the degradation of its contact surface during engagement occurs. With minor wear of the material, the tooth width changes from 1 μm to 1 mm, which leads to a change in the engagement angle tgα, changing the area of the contact surface of the tooth [20–22]. As a result of mechanical wear, the deviation of the gearing angle tgα by 1′ significantly changes the action of the radial force F_{r1}, and changes the vector of its application per unit of the tooth contact area [23–27]. For example, at tgα = 20°, the radial force is F_{r1} = 3880.151 N, which ensures effective gearing of the kinematic gear-wheel pair of the pump transmission shaft. With the same mechanical wear, the deviation of the gearing angle is tgα = 19°40′ the value of the radial force F_{r1} = 3838.04 N is reduced by 47 N. Wear of the contact surface leads to a change in the structural gaps Δi between the teeth, and under the influence of dynamic moments of forces and angular velocities ω, a dynamic impact occurs in the contact gearing surface. The maximum values of the impact load can be observed at the beginning of the movement and the stop of the gear. This is a short time interval when the element gains acceleration and deceleration corresponding to the times t_1 and t_2. Local fatigue stress zones are formed by the action of dynamic loads causing microcracks, cracks, and fractures of the teeth. Even the process of partial tooth destruction leads to a distortion of the gearing angle by at least 1°. For example, subject to a partially painted tooth, the gearing angle changes to tgα = 19°, and the radial force changes to F_{r1} = 3670.67 N, which is 210 N less than the effective force F_{r1} = 3880.151 N. Analyzing the calculations, it can be observed that reducing the forces in the gearing reduces the torque, which ensures effective gearing and operation of the pump. However, operating at a lower torque of the tooth gearing in practice actually shows a progressive degradation of the contact surface, resulting in chipping and breaking off the teeth. This phenomenon can be explained by the fact that the amplitude of the destructive stresses of the material acting on the contact area of the worn surface increases under the action of small moments in the gearing, but in short time intervals t_1, t_2 = 0.3 s.

When calculating the gear shaft, it is necessary to consider the actual distribution of forces and moments over the entire length of the shaft during operation, taking into account all of its structural elements. In particular, it is necessary to take into account the role of bearings in the effective load distribution in different sections and planes of the studied shaft-gear. To do this, when calculating the forces in the engagement, it is necessary to take into account the degree of influence of the reaction of the forces and moments transmitted from the bearings. This will describe the process of redistributing forces and torques throughout the entire "Bearing-Shaft-Gear-Gear" system under study. From the calculation, it can be seen that if the reaction forces in the bearings are affected, then there is a significant increase in the transverse forces in the vertical and horizontal planes of the gear engagement. If we consider the actual operating conditions of the pump with structural elements of bearings, the values of the same forces are reduced several times from 8100 N to 100 N, which is more consistent with the real load and confirms the effectiveness of the bearings. This means that their action and role in the redistribution of loads is of great importance for the methodology of calculating the loads acting in the gearing and calculating the durability of the gear. If the reaction forces of the bearings are neglected, overestimated numerical values of the load are formed, which will lead to a significant error in the design calculations.

Thus, a scientific problem arises, namely why (with the existing system of distribution of forces and moments in the vertical and horizontal planes) wear is observed in the gear engagement at small values of forces and moments. A theoretical assumption is made that the change in the durability of the gear and the gear wheel is largely influenced not by the magnitude of the acting force and moment, but by its direction vector, which is proposed to be characterized as the coefficient of displacement of the contact spot relative to the designed axis of symmetry.

Therefore, the actual task is to justify the load distribution factors on the gear teeth, taking into account the natural and progressive wear and the acting moments of the reaction forces. In order to study the principles of dynamic load distribution along the transmission shaft during operation, it is necessary to study the formation of stresses on the limited contact surface of the tooth.

Thus, a theoretical assumption that the wear of the contact surfaces of the gear shaft tooth results in the irregular increase in the structural gaps of the gear Δ_{i1}, and the shift of the contact spot relative to the symmetry axis of the tooth leads to the increase of stresses acting on a small cross-sectional area, thereby reducing the service life factor.

The design scheme of the shaft is shown in Figure 2.

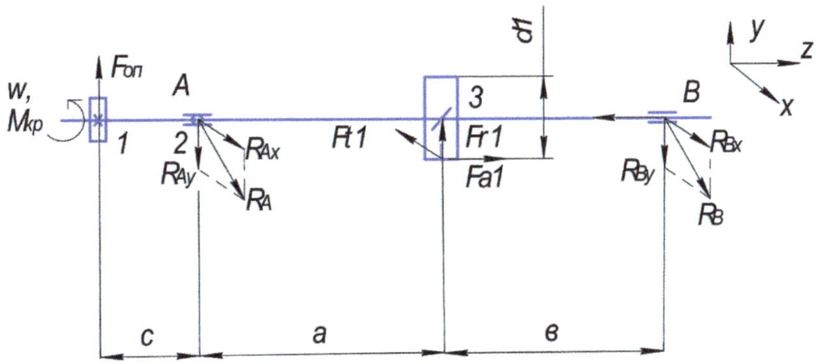

Figure 2. Design Scheme of the Transmission Shaft: *x*, *y*, *z* are coordinate systems of axes; *a*, *b*, *c* are distances between bearings; d1is the gear diameter; *A*, *B* are designations of shaft bearings; w, Mt is the torque; *Fct* is the cantilever power from belt drive; F_{t1}, F_{r2}, F_{a1} are the direction of forces on the gear and on the wheel; R_A, R_{Ax}, R_{ay}, R_B, R_{Bx}, R_{By} are the directions of radial reactions in bearings A, B, respectively.

Let's determine the reactions in the bearings acting in the vertical and horizontal planes (Figure 2). The study of the principle of force reaction distribution along the pump transmission shaft elements shall allow the identification of the most stressed parts of the shaft.

In the vertical plane (3) and (4):

$$-F_{OG}\cdot c+F_{r1}\cdot a+M_a-R_{By}(a+b)=0 \tag{3}$$

$$R_{By}=\frac{-F_{og}\cdot c+F_{r1}\cdot a+M_a}{(a+b)} \tag{4}$$

$$-F_{og}(c+a+b)+R_{Ay}(a+b)+M_a-F_{r1}b=0 \tag{5}$$

$$R_{Ay}=\frac{F_{og}(c+a+b)-M_a+F_{r1}b}{a+b} \tag{6}$$

In the horizontal plane (7) and (8):

$$F_{t1}a-R_{Bx}(a+b)=0 \tag{7}$$

$$R_{Bx} = \frac{F_{t1}a}{2a} = \frac{F_{t1}}{2} \quad (8)$$

$$R_{Ax} = \frac{F_{t1}}{2} \quad (9)$$

The results of the calculation show that in the vertical plane, the reaction action of the forces R_{By} (2) at point (B) of the force application is 83.54 N, and at point (A) R_{Ay} = 8006.93 H (3). Such a spread of values can be explained by the action of the torque M_t from the forces of the clear transmission F_{og}. Consequently, the sections of the shaft in zones 2 and 3 (Figure 3) are subject to the greatest action of dynamic forces, which conditionally tend to shift it relative to the design axis of rotation. This calculation explains the reason for the greatest amount of wear in the gearing, zone 3, the area of application of the radial force F_{r1}.

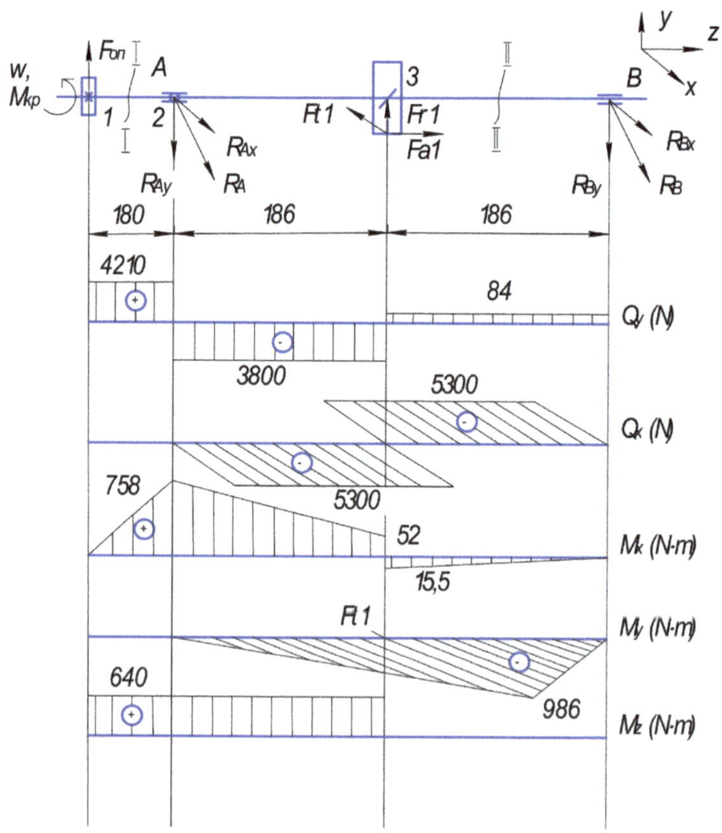

Figure 3. Design scheme of the shaft with plotting bending and torsional torques: *x*, *y*, *z* are coordinate systems of axes; *I*, *II* are shaft sections; *A*, *B* are designations of bearings on the shaft; *w*, *Mt* is the torque; *1*, *2*, *3* are shaft sections; *Fog* is the cantilever force from the belt drive; *Ft1*, *Fr2*, *Fa1* are the directions of forces on the gear and on the wheel; *RA*, *RAx*, *Ray*, *RB*, *RBx*, *RBy* are the directions of radial reactions in bearings *A*, *B*, respectively; *Qy*, *Qx* are transverse forces; *Mx*, *My*, *Mz* are bending moments.

In the horizontal plane, the reaction of the forces R_{Ax} and R_{Bx} equal to 5301 N (8,9) is balanced. This means that in the horizontal plane, in critical zones 2 and 3 and points A and B (Figure 3), the system is in an equilibrium state (Table 3).

Table 3. Calculated equations of transverse forces and bending moments acting in the transmission shaft.

Plotting Transverse Forces in 1–4 Sections, N			Plot the Bending Moment, N·m	
Plane under study	Equality of forces in the *i*-th section	Force values	Equality of forces in the *i*-th section	Values of bending moments
Vertical plane	$Q_{y1} = F_{og}$	4210.32	$M_{x1} = 0$	0
	$Q_{y2} = F_{og} - R_{Ay}$	−3796.61	$M_{x2} = F_{on}c$	−757.86
	$Q_{y3} = F_{og} - R_{Ay} + F_{r1}$	83.541	$M_{x3} = F_{og}(c+a) - R_{ay}a$	51.69
	$Q_{y4} = R_{By}$	83.541	$M_{x4} = 0$	0
Horizontal plane	$Q_{x1} = 0$	0	$M_{y1} = 0$	0
	$Q_{x2} = -R_{Ax}$	−5301	$M_{y2} = 0$	0
	$Q_{x3} = -R_{Ax} + F_{t1}$	5301	$M_{y3} = -R_{Ax}a$	−985.986
	$Q_{x4} = R_{Bx}$	5301	$M_{y4} = 0$	0

As a result of tooth wear, the contact spot shifts relative to the axis of its symmetry. The calculation of the reaction of the forces acting in the bearings in the vertical and horizontal planes confirmed the assumption that the displacement of the contact spot leads to a displacement of the center of application of the radial force F_{r1} to $(\Delta h/2)$-i. This means that the action of the M_t moment shall shift by the value $(\Delta h/2)$-i, which will lead to a sharp surge of stresses on the worn part of the tooth.

Plotting the transverse forces in 1–4 cross-sections.

In the vertical plane (10) and (11):

$$Q_{y1} = F_{og};\ Q_{y2} = F_{og} - R_{Ay} \tag{10}$$

$$Q_{y3} = F_{og} - R_{Ay} + F_{r1} \tag{11}$$

$$Q_{y4} = R_{By} \tag{12}$$

In the horizontal plane (14) and (15):

$$Q_{x1} = 0;\ Q_{x2} = -R_{Ax} \tag{13}$$

$$Q_{x3} = -R_{Ax} + F_{t1} \tag{14}$$

$$Q_{x4} = R_{Bx} \tag{15}$$

Plotting bending moments taking into account wear.

Is the vertical plane (relative to the x axis) (16) and (17):

$$M_{x1} = 0;\ M_{x2} = F_{og}c \tag{16}$$

$$M_{x3} = F_{og}(c+a) - R_{ay}a \tag{17}$$

In the horizontal plane (relative to the y axis) (18):

$$M_{y1} = 0;\ M_{y2} = 0;\ M_{y3} = -R_{Ax}aM_{y4} = 0 \tag{18}$$

Next, let's plot the torques taking into account the deviation of the worn area (19):

$$M_t = M_z = F_{t1}\frac{d_1}{2} \tag{19}$$

Let's determine the total radial reactions in the wear planes (20) and (21):

$$R_A = \sqrt{R_{Ax}^2 + R_{Ay}^2} \tag{20}$$

$$R_B = \sqrt{R_{Bx}^2 + R_{By}^2} \tag{21}$$

Let's determine the total bending moments in the *max* loaded cross sections (22) and (23):

$$M_3 = \sqrt{Mx_3^2 + My_3^2} \quad (22)$$

$$M_2 = Mx_2 = 757.86(\mathrm{N \cdot m}) \quad (23)$$

5. Investigation of the Dependences of the Internal Stress Formation in the Contact Surface of the Wear of the Drilling Pump Transmission Shaft

Localization of concentration zones of internal fatigue stresses during the operation of drilling pumps is a problem in increasing the service life [22–29]. Understanding the principle of fatigue phenomena redistribution in the degraded metal structure during operation will allow effective development of technical solutions for the design and manufacturing technology.

5.1. Determination of Permissible Contact Voltages

The procedure for calculating the permissible contact voltages according to the full algorithm of actions, let's start with this equality.

$$[\sigma]_H = 0.45([\sigma]_{H1} + [\sigma]_{H2} \quad (24)$$

Material-40X, improvement; 269-302 HB; σ_B = 900 MPa sigma time permissible; σ_T = 750 MPa yield strength of the material; σ_{-1} = 410 MPa contact stress-destruction.

$$[\sigma]^*_{H1,2} = \sigma_{\mathrm{Hlim1,2}} = 2\mathrm{HB_{av}} + 70 \quad (25)$$

$$[\sigma]_{H1,2} = \frac{\sigma_{Hlim1,2} \cdot K_H}{S_H} \quad (26)$$

where: K_H, life factor (K_H = 1); S_H, safety factor; and S_H = 1, 1 (for normalization and thermal improvement).

Let's check the contact stresses on the wear area.

$$\sigma_\mathrm{H} = \mathrm{K}\sqrt{\frac{\mathrm{F_t}(\mathrm{U}_\phi + 1)\mathrm{K_{H\alpha}K_{H\beta}K_{HV}}}{\mathrm{d_2 b_2}}} \leq [\sigma]_H \quad (27)$$

where: f is the auxiliary factor for bevel gears, f = 345 for chevron gears.

$$\mathrm{Z_2 = Z_1 U_\Pi;\ d_2 = \frac{Z_2 m}{\cos\beta}}$$

where Z_2 is the number of teeth of the desired gear, U_{gt} is the gear ratio. K_{H2} is the factor considering the load distribution between the teeth. ϑ is the gear rim speed, m/c.

$$\vartheta = \frac{\omega_2 d_2}{2 \cdot 10^3};\ \omega_2 = \frac{\omega_\Pi}{U_{ZP}}$$

(9 degree of accuracy)

where: $K_{H\alpha}$ = 1.13; factor of load distribution between the teeth; ω is the angular velocity of the wheel, ϑ is the circumferential velocity of the wheel; d_2 is the dividing diameter; K_{HV} = 1.06 is the dynamic load factor depending on the circumferential speed of the wheels and the degree of transmission accuracy; and $K_{H\beta}$ is the factor of the load unevenness along the tooth length.

5.2. Determination of the Permissible Bending Stresses in the Case of a Worn Surface during Operation

At the second stage of calculations, we will determine the permissible bending stresses for the worn surface during the operation of gears in drilling pumps.

$$[\sigma_F]_{1,2} = \frac{\sigma_{Flim1,2}}{S_{F1,2}} \gamma_A \gamma_{N1,2} \tag{28}$$

where σ_{Flim} is the limit of the teeth endurance along the bending stresses; σ_{Flim} = 1.8 HB, σ_{Flim} = 1.8σ·285.5 = 513.9 MPa; S_F is the safety factor, S_F = 1.75; γ_A *is* the factor considering the impact of two-way load application; γ_A = 1 (one-way); γ_Hm service life factor, γ_H = 1.

Checking the bending stresses of the gear teeth and the wheel.

$$\sigma_{F1} = \sigma_{F2} \frac{\gamma_{F1}}{\gamma_{F2}} \leq [\sigma_F]_1;\ \sigma_{F2} = \gamma_{F2} \gamma_\beta \frac{F_t}{b_2 m} K_{F\alpha} K_{F\beta} K_{FV} \leq [\sigma]_{F2} \tag{29}$$

where: $K_{F\alpha}$ = 1.35, $K_{F\beta}$ = *1*, K_{FV} =1.12; γ_F is the factor of the tooth form.

5.3. The Nominal Value of the Maximum Bending Stress in the Worn Part of the Transmission Shaft

The calculation method continues by determining the nominal value of the maximum bending stress in the worn part of the transmission shaft (22):

$$\sigma_{\max} = \frac{M_{\max}}{W} \tag{30}$$

where *W* is the axial moment of the resistance of the shaft section, m^2 considering the wear of the contact diameter of the tooth cavity.

$$W = \frac{\pi d_{f1}^3}{32} \tag{31}$$

where is d_{f1} = tooth cavity diameter, mm.

$$d_{f1} = d_1 - 2,4m \tag{32}$$

5.4. Nominal Value of the Maximum Tangential Stress (Torsion)

The next step of the strength calculation is the determination of the nominal value of the maximum tangential stress (torsion) (33):

$$\tau_{\max} = \frac{M_{kp}}{W_\rho} \tag{33}$$

$$W_\rho = \frac{\pi d f_1^3}{16} \tag{34}$$

5.5. Let's Determine the Concentration Factor of Normal and Tangential Stresses for the Calculated Shaft Cross-Sections (37,38)

3 cross-section

$$(K_\sigma)_D = \left(\frac{K_\sigma}{K_d} + K_F - 1 \right) \frac{1}{K_y} \tag{35}$$

$$(K_\tau)_D = \left(\frac{K_\tau}{K_d} + K_F - 1 \right) \frac{1}{K_y} \tag{36}$$

where K_σ, K_τ are effective stress concentration coefficients; σ_B = 900 MPa, σ_{-1} = 410 MPa; r = 3.5 mm is a bearing chamfer; t = 3.5 mm is a fender height.

$$\frac{t}{r} = 1 => \frac{r}{d} = \frac{3,5}{88} = 0.04 => K_\sigma = 1.8, K_\tau = 1.5 \tag{37}$$

where K_d is the coefficient of influence of the cross-section absolute dimensions, K_d = 0.635 (bending for alloy steels + torsion); K_F *is* fatigue strength surface condition factor, K_F = 1.50 (R_a 10 ⇒ R_z 40); Ky- fatigue surface hardening factor, K_y = 2.4 quenching with heating of HDPE [σ_B core, 600/800].

5.6. Let's Define the Criteria and Limits of Endurance in the Calculated Shaft Cross-Sections

To determine the possibilities of the strength resource, we will determine the criteria and limits of endurance in the calculated sections of the shaft, as in Figures 4 and 5 (38):

$$(\sigma_{-1})_D = \frac{\sigma_{-1}}{(K_\sigma)_D}; (\tau_{-1})_D = \frac{\tau_{-1}}{(K_\tau)_D} \tag{38}$$

where σ_{-1}, τ_{-1} is limits of endurance of smooth samples under a symmetric cycle of bending and torsion (39);

$$\tau_{-1} = 0.58; \sigma_{-1} \approx 0.58 \cdot 410 = 237.8MPa \tag{39}$$

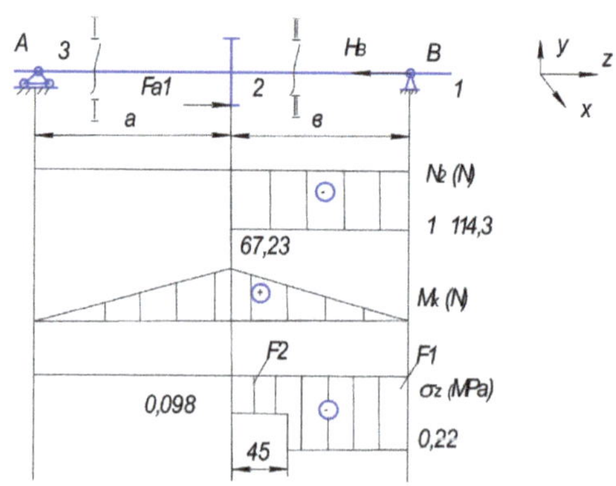

Figure 4. Design scheme of the shaft with plotting under the action of longitudinal forces: *I, II* are shaft sections; *x, y, z* are coordinate systems of axes; *a, b* are distances between bearings; *A, B* are designations of bearings on the shaft; *1, 2, 3* are shaft sections; *Fog* is the cantilever force from the belt drive; *Fa1* is the direction of forces on the gear and on the wheel; *FB* is the direction of force; *F1, F2* is cross-sectional area; *N2* is longitudinal force; *Mx* is bending moment; σ_z is the stress in the corresponding shaft section.

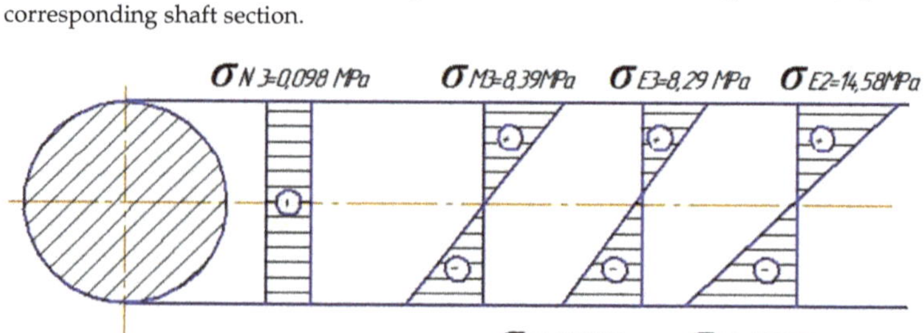

Figure 5. Design scheme with plotting the highest voltages in the cross section of the shaft: *σN3, σM3, σE3, σE2*, are stresses in the corresponding shaft cross-sections.

5.7. Plotting under the Action of Longitudinal Forces

$$\sum F_2 = 0\ F_{a1} - H_B = 0 H_B = F_{a1} = 1114.3\ (\mathrm{H})\mathrm{I}-\mathrm{I}:\ N_1 = 0 \quad (40)$$

$$\mathrm{II}-\mathrm{II}:\ F_{a1} + N_2 = 0 N_2 = -F_{a1} = -1114.3 M_{x\mathrm{I}} = 0 \quad (41)$$

$$M_{x2} = F_{a1} \cdot \frac{d_1}{2} M_{x3} = 0 \quad (42)$$

$$\sigma_Z = \frac{N}{F} \quad (43)$$

where: N is longitudinal force, F is the cross-sectional area; $\sigma_{z1} = 0$

$$\sigma_{Z2} = \frac{-N_2}{F_2} \quad (44)$$

$$F_1 = \frac{\pi d_1^{-2}}{4} \quad (45)$$

5.8. The Greatest Stress in the Cross-Section of the Shaft

Let's determine the greatest stress in the cross section of the shaft and build a plot. Cross-section 3, Cross-section 2.

$$\sigma_{\mathrm{Nmax}} = -0.098 MPa;\ \sigma_{\mathrm{Nmax}} = -0.22 MPa \quad (46)$$

$$\sigma_{\mathrm{Mmax}} = 8.39 MPa;\ \sigma_{\mathrm{Mmax}} = 14.8 MPa \quad (47)$$

$$S_{\sigma 3} = \frac{(\sigma_{-1})_D}{\sigma_\varepsilon} \quad (48)$$

When studying the premature failure of gears due to fatigue chipping, let's take the maximum stress of crumpling according to Hertz at the elastic touch site (49) as an indicator of the load on the tooth surface:

$$\sigma_H = z_g \sqrt{\frac{q}{2\rho_{\mathrm{dr}}}} \leq [\sigma_H] \quad (49)$$

where z_g is the number of gear teeth; ρ_{rr} is the reduced radius of curvature; q is the specific contact load; and $[\sigma_H]$ is the maximum permissible crumple stress in Hertz.

The maximum Herzian bearing stress at the elastic touch site is represented as (50):

$$\sigma_H = Z_H Z_M Z_\varepsilon \sqrt{\frac{\omega_{Ht} u \pm 1}{d_{\omega 1} u}} \leq [\sigma_H] \quad (50)$$

where Z_H is the factor considering the longitudinal distribution factor; Z_M is the contact strength factor; Z_ε is the factor considering the influence of the value ε_α on the load capacity of spur gear; ω_{Ht} is the angular velocity of gear wheels; d_{w1} is the initial diameter of the gear wheel; and u is the gear ratio.

Formulas (2) and (3) indicate that the conditions for the occurrence of critical crushing stresses σ_H depend on the area of the contact spot formed on the working width b_w of the gear ring taking into account the radius ρ_{dr} of curvature in the gearing area. It is the area of the tooth contact spot that will limit the circumferential force P_{Ht} from the specific contact load acting relative to the diameter d_ω of the initial circle:

$$q_{\mathrm{k}} = \frac{P_n}{l_{\mathrm{k}}} K_{H\beta} K_{Hv} K_{H\alpha} \quad (51)$$

where P_n is the force acting to the normal cross-section; l_c is the length of the contact spot, m; $K_{H\beta}$ is the load distribution factor across the width of the ring gear; K_{Hv} is the dynamic load factor; and $K_{H\alpha}$ is the factor of load distribution between the teeth.

The long-term operation introduces light surface wear, contributing to the loose fit of the contact surfaces and, as a consequence, the deviation of the design axes of the contact gear (changes in the structure of the mechanism), which causes the redistribution of the load on the teeth surface [15,22,23].

The hardness of the working surfaces was taken as the basis for determining the contact endurance limit and its base number of cycles before destruction. The experience of operating pumps, the results of metallographic examination, and the analysis of the acts of repair and restoration of gears that have failed allowed us to conclude that the surface hardness is not (in most cases) the main cause of failure. When studying the cut of the tooth and its contact surface, the integrity of the coatings is observed, and the grain size and color scheme of pearlite inclusions in the phase structure change from the focus of the load application to the base of the tooth. Rate and moment of fatigue stress formation largely depends on the metal base, its phase composition, and the dispersion of inclusions of martensitic and austenitic grains. Based on the conducted studies, it is proved that the cyclic durability depends more on the value of the applied specific load acting for 1 s.

The action of large moments in a short period of time, for example, when starting up (t_1) or braking (t_3) the pump transmission shaft (for 0.2–0.35 s), leads to a violation of the critical bending stresses equality $\sigma_H \leq [\sigma_H]$. Violation of this condition causes the teeth of the pivot wheel to break, regardless of the amount of wear. This means that the λ factor of the contact spot displacement from the axis of the design symmetry of the tooth (32) plays a significant role in the formation of moments and affects the assessment of the service life of the mechanism:

$$\lambda = \frac{l_k Z_\varepsilon^2 \Delta i}{b_w \left(d_w - d_f\right)} \tag{52}$$

where Δi is the deviation of the contact spot from the design axis; and d_w and d_f are the diameters of the initial circle and the circle of the toothed wheel depressions.

It is also essential to evaluate the role of the factor $Z^2{}_\varepsilon$, which considers the influence of the value ε_α on the load capacity of straight gears. More precise factors are set by the obtained polynomial Equation (53):

$$\lambda = -7.0434\left(Z_\varepsilon^2\right)^2 + 20.616\left(Z_\varepsilon^2\right) - 13.218 \tag{53}$$

The most adequate values of the moment considering the contact strength can be determined using the λ factor of the contact spot deviation from the design axis (54):

$$M_{H1} \leq [M_{H1}] = \frac{b_w d_{w1}^2 [k_0] \lambda \theta}{2(u \pm 1)} \tag{54}$$

where $[M_{H1}]$ is the permissible moment; $[k_0]$ is the service life factor; and θ is the distribution factor across the tooth.

As established by research, a significant contribution to the change in the value of the effective bending moments is made by the λ factor of deviation from the design axis. More precisely, this property can be described by the dependence (55):

$$M_{H1} = 5.161\lambda^2 - 1.605\lambda + 7.693 \tag{55}$$

Therefore, it is possible to estimate how the service life factor ${K_{HL}}^2$ changes, taking into account the influence of the equivalent number N_{HE} of stress cycles, depending on the λ factor for a given surface hardness:

$$K_{HL}^2 = \sqrt[3]{\frac{N_{H0}}{N_{HE}}} \tag{56}$$

Taking into account the approximation of the data, the number N_{HE} of stress cycles and the λ factor, the equation shall be as follows (57):

$$N_{HE} = -\,63.56ln(\lambda) + 43.365;\ R^2 = 0.935 \tag{57}$$

Based on the results of the research, a nomogram was developed that determines the dependence of the change in the service life factor and the load distribution factor on the factor of the contact spot deviation from the design axis λ at a given surface hardness (Figure 6).

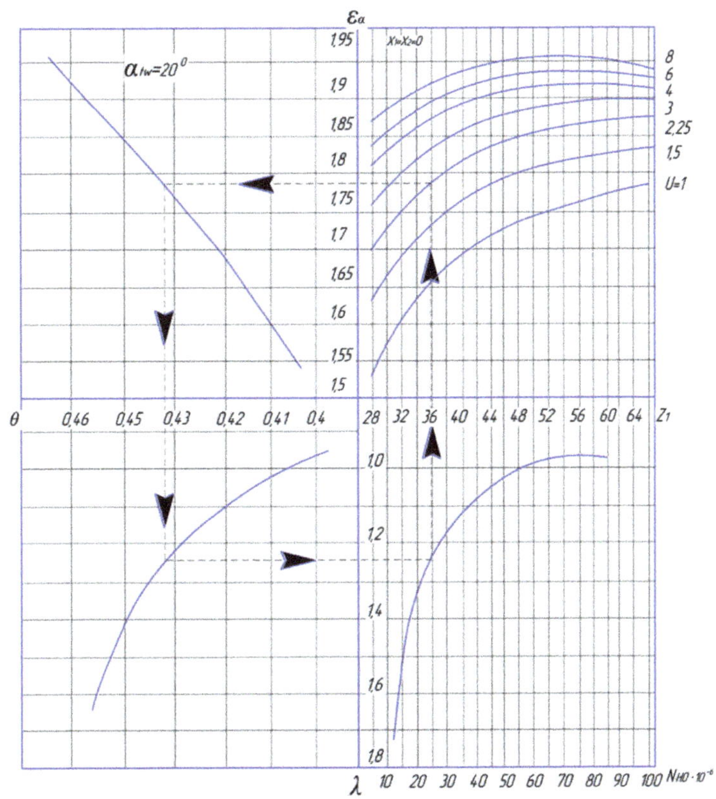

Figure 6. Nomogram for determining the dependence of the basic cycles of contact stresses from the change in the load distribution factor and the factor of the contact spot deviation from the design axis λ. θ is the coefficient of uneven load distribution along the length of the tooth, and λ is the coefficient of deviation of the contact spot from the design axis of symmetry.

The developed nomogram (Figure 6) allows you to predict the life of the pump without errors, taking into account dynamic loads and operational characteristics of natural wear of the teeth. Analyzing the curves of the graph (Figure 6), it can be seen that with an increase in the value of λ = 1.3 of the coefficient of deviation λ of the contact spot from the design axis of symmetry, the service life of the pump gear engagement decreases sharply to NHO = 25×10^{-4}. However, it is observed that changes in the coefficient of deviation of the contact spot $\lambda = 0.9 \div 1.1$ do not significantly affect the service life of NHO = $60 \times 10^{-4} \div 80 \times 10^{-4}$ cycles, which is the optimal range of strength resource. A sharp degradation of the tooth surface and a decrease in the resource is observed from the values of $\lambda = 1.25$. This is explained by the fact that the displacement of the contact spot to $\lambda = 1.25 \div 1.3$ causes an uneven load distribution along the length of the tooth, as evidenced by the coefficient θ. At the same time, having drawn imaginary perpendiculars according to the graphs (Figure 6), we see how the load distribution coefficient at angular accelerations in different operating modes affects the technical and dynamic parameters of the gear wheel. Thus, the decrease in the wear resistance of the wheel tooth depends not only on the applied force to the area, but also on the degree of displacement of the contact spot at the load distribution coefficient along its length. Moreover, the phase structure of the martensite + perlite base plays an important role in the reliability of the tooth. This structure allows you to withstand high dynamic loads, even with an uneven distribution of forces. The proposed nomogram, in addition to forecasting the resource, allows us to reasonably choose the optimal technological and dynamic parameters of the pump during its operation.

The proposed system approach, which takes into account the formation of stresses depending on the phase structure of the metal and the deviation of the contact spot during the redistribution of a cyclically changing load, will allow more accurately predicting failure.

6. Simulation, Experimental 3D Loading of Gears for Verification and Confirmation of Theoretical Research Results

In order to obtain adequate research data using the SOLIDWORKS software, a loading scheme was clearly developed in which torque was applied through the shaft to the drive gear and the meshing ring. To simulate the loading of kinematic pairs of engagement, the Static II Pro submodule of the SOLIDWORKS software with the GearTrax application was used. When developing the loading scheme, to identify the input parameters, the main characteristics were set: the type of gear support (cylindrical support) and the possible degrees of freedom that determine the directions of the applied engagement forces, accelerations, etc.

Based on the calculations of the geometric characteristics, power and strength calculations, a model of interaction of the kinematic pair "the drive gear of the transmission shaft—driven of the eccentric shaft" in the form of assembly units was obtained, which will qualitatively and quantitatively characterize the picture of the load distribution over the contact surface of the kinematic pair of gearing of the rotary platform reducer (Figure 7).

Figure 7. Element-wise modeling of a pair of gearing.

In addition to quantitative characteristics, we also determined qualitative indicators. Let us set the applied load in the contact patch at the base of the tooth cavity up to its pitch diameter. The places subject to the greatest stress are indicated by the color spectrum.

The origin of contact and bending stresses on the gear rim of the eccentric shaft is small, but the depth of propagation and the color spectrum of the spectral diagram allow us to conclude that the transmission shaft with the gear works under conditions of large alternating loads occurring in a short period of time (Figure 8). The concentration of critical stresses has a local character and is short in time. The color scale characterizes the contact area of the gear operating with stresses σF = 68,347 ÷ 123,025 MPa. This load mode proceeds within the permissible limits under the condition of strength σF ≤ [σF].

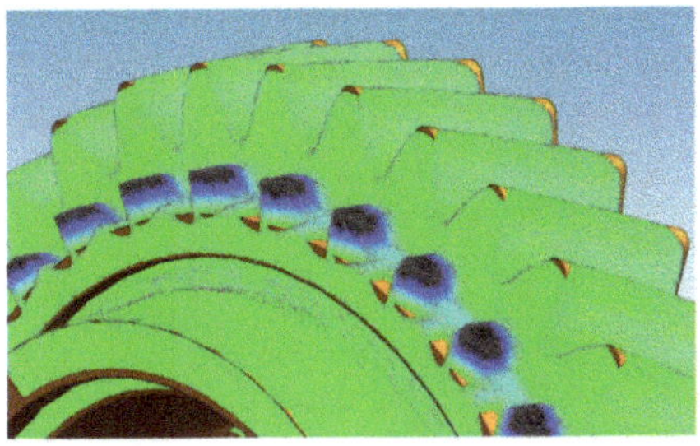

Figure 8. Load distribution on the contact surfaces of the teeth.

Spectral analysis of the drive gear of the transmission shaft of the pump showed that with equidistant engagement, the load and forces are distributed evenly over the contact surface of the tooth. Normal stresses are also distributed at the base of the tooth without going beyond the critical values of the durability coefficient. However, at the same effective load, but with the deviation of the contact spot in the engagement from the design axis of symmetry, the stresses increase both at the crown of the tooth and at the base. Critical stress distribution zones are shown in yellow and red (Figure 8). This process confirms the satisfactory convergence of the effect of the proposed contact spot displacement coefficient on the service life of the gear when the design axis of the wear spot symmetry is shifted.

When the gear is turned, the torque from the transmission shaft is transmitted through the meshing of the teeth to the driven gear. Accordingly, the action of the forces of resistance and the forces of inertia of its own masses will primarily affect the drive gear. Pump accelerations and angular speeds reach peak values during acceleration and deceleration. In the period t_1 (acceleration time during the operation of the mud pump), the initial stage of interaction of the engagement pair occurs and the internal stresses σ_F and σ_H sharply increase. This period corresponds to the contact of two rolling toothed surfaces at a given force of 20 kN with stresses σ_F = 68,347 − 123,025 MPa. The load will reach its maximum value during the final engagement period t_3. During braking, forces and a moment of inertia are added to the torque to exert pressure on the contact surface area. The magnitude of pressure is progressively increasing. Maximum stress values arise, tending to disrupt the power balance and deform the geometry of the tooth (Figure 9).

Figure 9. Zones of maximum stress concentration when rotating the turntable.

Next, we investigate the distribution of stresses in the teeth of the drive gear. The first power disturbances are transmitted to the shaft at the base of the gear. This is also evidenced by the red color of the spectrum (Figure 10).

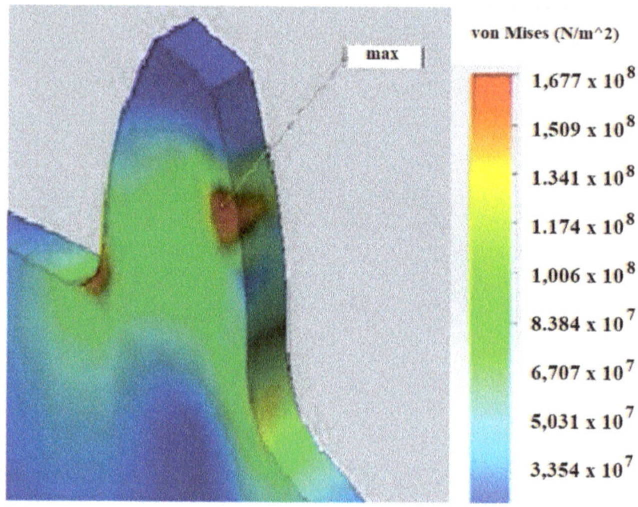

Figure 10. Localization of internal stresses at a given load.

Further, the stresses are transmitted to the working surface of the tooth. When the maximum values of stresses are reached, structural breakdown of the metal occurs and, as a result, fatigue deformation and fracturing occurs. With an uneven distribution of the load over the displaced contact patch caused by fatigue stresses that increase in a short period of time, the fracture zone of the material surface increases along the length of the deceleration path (Figure 10). The traversed path of the drive gear at the moment of braking is accompanied by a slow rolling of the teeth (i.e., a decrease in the contact area with increasing voltage (Figure 10)).

It is this moment that is critical, since the condition of equilibrium of the balance of forces and acting stresses of contact and bending is violated. The investigated processes explain the nature of the fracture of the teeth on the gear rim of the eccentric shaft. Fractures, as a rule, occur in the same zone, corresponding to the maximum operating range of the gears at the moment of braking the drill and the pump.

The presented results of modeling the process of engagement of the kinematic pair "drive gear of the transmission shaft—driven eccentric shaft" indicate the high energy consumption of the transmission shaft of the mud pump. The force and strength analysis of the interaction of the contact surfaces of the engagement with the classical arrangement of the mechanism showed the zones of a dangerous section, on which the concentration of bending and contact stresses is localized. A high level of stress values concentrated in the contact spot of the engagement reaches critical values at some time intervals Δt_i (i.e., the conditions of strength $\sigma > [\sigma]$ are violated), which leads to the destruction of the metal under cyclic loads and, as a consequence, the flaking of the teeth. Critical values of stresses arise at the moment of deceleration of the pump and drill, which occurs at the maximum value of the zenith angle of drilling $\alpha = (35 \div 60)^\circ$. The obtained values of forces and stresses in a pair of engagement during modeling are in satisfactory agreement with theoretical calculations and reveal the essence and nature of the appearance of tooth breakage at a certain mode of the time interval Δt_3.

Analyzing the results of the research carried out, we can draw the following conclusions:

- The distribution of the load over the contact surface of the kinematic pair of engagement is uneven and inconsistent in time;

- Due to the prevailing torques, the gear rim of the transmission shaft is a more loaded part;
- The hardness of the tooth surface does not play a decisive role in the durability of the mechanism. The most significant factor is the factor that determines the number of working cycles (resource), namely the degree of uniformity of the distribution of forces in the critically extreme periods of acceleration t_1 and deceleration t_3 with an increasing value of the moment of inertia.

The results of theoretical and laboratory studies of simulation models confirm that one of the more effective methods for increasing the durability and energy efficiency of the rotary mechanism is the timely redistribution of forces in the engagement.

7. Discussion of the Displaced Spot of the Contact Surface Wear

The influence of the misalignment of the axes on the work of the kinematic gearing pair was studied along with the process of teeth load distribution. To calculate the gearing pair, the superposition method was used considering the presented factors separately.

Thus, the proposed mathematical model of load distribution along the tooth surfaces of the kinematic pair of "transmission shaft driving gear—eccentric shaft rim gear" can more accurately assess the impact of the factor of the λ contact spot deviation from the design axis on the service life of the mechanism. As established by research, the λ factor significantly affects the value of the moment that occurs in the gearing, and therefore shall characterize the power spent by the gearbox of the rotary platform to overcome the resistance forces.

With an increase in the wear asymmetry in the contact spot, the energy intensity of the drilling process also increases, which means that the equation describing the gear power in the gearing and the distribution of the bending moment will be as follows (58):

$$\begin{cases} N_{\mathrm{g}} = \sum\limits_{i=1}^{n} \Phi_{\tau i} \sum R_i + \left(F_g r_g + M_{H1}\right)\left(n_g - n_k\right) K_a; \\ \sum\limits_{i=1}^{n} \Phi_{\tau i} \sum R_i = \frac{\varphi m r^2}{t_1^2}; \\ M_{H1} = \frac{b_w d_{w1}^2 [k_0] \lambda \theta}{2(u \pm 1)}. \end{cases} \tag{58}$$

where N_g, gear engagement power; $\varphi_{\tau i}$, tangential force of inertia; R_i, force application shoulder; F_g, the force applied to the gears; r_g, the radius of the dividing circle of the gear; n_g, gear rotation speed; n_k, speed of rotation of the gear wheel; K_a, a coefficient that takes into account the impact of the load, $[M_{H1}]$, acceptable moment; $[K_o]$, a coefficient that takes into account the durability; θ, coefficient of uneven load distribution along the length of the tooth, b_ω, working width of the tooth; and d_ω, diameter of the initial circle of the gear wheel.

The proposed method for calculating the service life of the mechanism crown teeth, as well as the established relationship between the value θ (60) and the values z_g and λ, allows for considering the impact of external loads and the axial displacement of the contact spot when calculating the bending moments. As can be seen from the Equation (33), the greater the λ and θ factors are, the greater the M_{H1} values is. It was determined that the load distribution on the gear teeth during the displacement of the gearing axes occurs unevenly and at different speeds. Increasing the deviation of the contact spot from the design axis increases the energy intensity of the drilling process by 17–42%.

The proposed method allows the determination of the dependence of the basic operation cycles with variable loads N_{HO} on λ and θ. For a more accurate determination of fatigue stresses, along with dynamic loads, it is necessary to take into account the physical and mechanical properties of the material (HB or HRC hardness). The calculations showed the greater the λ and M_1, the greater the wear *i* and the higher the probability of tooth breakage.

When determining the dependence of the service life factor $K^2{}_{HL}$ on the bending stress, the following methods were used: A. S. Pronikov's theory of the transition from one type of interaction to another, $h/r \geq K_o\ (c \cdot \sigma_T/E)^2$; transition from elastic deformation to fracture, σ_x = 0.33 HB/1 − ψ; P.F. Dunayev's theory; $[\sigma]_H$ = (1.8 HB + 67)/1.1 or $[\sigma]_H$ = (14 HRC + 170)/1.1, here $[\sigma]_H$ describes the permissible contact-surface stress, and the 1.1 factor describes the margin of safety [22–32].

The dependence of λ on the contact bending stresses σ and—equation (32) [32] was obtained taking into account the above-mentioned theories of contact surfaces interaction and the method developed by the author.

8. Conclusions

The power N_g of the gearing of the drive transmission shaft gear and the eccentric shaft gear, which characterizes the energy consumption of the drill bit depth stroke, was determined.

The established ratios of the studied indicators shall help to develop a set of measures to reduce the influence of external inertia forces and to optimally redistribute the load in the gearing during drilling at the angle φ without loss of power in the gearing, especially in the critical time interval t_i.

The proposed method of calculating the load of the kinematic pair of the drilling unit pump gearing allowed taking a broader look at the problem of the energy intensity of the drilling operation. For example, GOST 16162-78 regulates determining the limit of contact endurance by the hardness of the working surfaces. The contact stress base is also taken as a function of the average hardness.

This condition does not fully determine the cycle base that corresponds to the achievement of critical values of contact stresses at the maximum service life. This discrepancy can be confirmed by the fact that the transmission shaft with external gearing is made of 40X steel (GOST 1050-74) after heat treatment, provided by the manufacturing technology with the hardness of the active teeth surfaces of 30–35 HRC. The drive gear is made of 40X steel (GOST 4543-71), and the surface hardness is 260 HB (GOST 8479-70). The specified hardness is high enough to withstand significant loads. According to the manufacturing technology, only the upper layers of the metal are quenched or cemented, since it is the surface that must have high physical and mechanical properties, and the metal base must be viscous (load damper). However, with such high physical and mechanical properties of the surfaces in the practice of operating pumps, the breakage of the teeth of the transmission shaft is quite common.

As shown by theoretical studies conducted according to the proposed method, a significant role in the service life of the mechanism is played not by the hardness of the surfaces, but by the concentration of forces applied to the contact area per unit of time. It is the magnitude of the applied force per second of time per contact area that forms the stress concentration zones. The localization of fatigue stresses over time exceeds the limit of possible bending stresses and the fluidity of the metal, which leads to the destruction of the tooth.

Author Contributions: Conceptualization, formulation and research of scientific hypotheses and project administration V.V.S.; methodological component, formal analysis and writing—original draft preparation S.N.K.; data curation and reviewing A.V.S.; software study of strength calculation O.V.I.; validation and reviewing P.V.; analytics, editing, graphic drawing Z.Z.Z. All authors have read and agreed to the published version of the manuscript.

Funding: This work was supported by a publications grant of the Gheorghe Asachi Technical University of Iasi (TUIASI), project number GI/P6/2021.

Institutional Review Board Statement: Not applicable.

Informed Consent Statement: Not applicable.

Data Availability Statement: All required data are provided in the article.

Conflicts of Interest: The authors declare no conflict of interest.

References

1. Pachurin, G.V.; Filippov, A.A.; Goncharova, D.A.; Kuzmin, A.N.; Gevorgyan, G.A. Fatigue of structural materials in corrosive environments. *Sci. Tech. Prod. J. Vestn. Mashinostroeniya* **2020**, *117*, 78–84.
2. Lin, H.; Bergadà, J.M.; Zeng, Y.; Zhang, Y.; Wang, J. Rotor spinning transfer channel design optimization via computational fluid dynamics. *Text. Res. J.* **2018**, *88*, 1244–1262. [CrossRef]
3. Tikhonov, V.S.; Baldenko, F.D.; Bukashkina, O.S.; Liapidevskii, V.Y. Effect of hydrodynamics on axial and torsional oscillations of a drillstring with using a positive displacement motor. *J. Pet. Sci. Eng.* **2019**, *183*, 257–265. [CrossRef]
4. Feng, Z.-M.; Guo, C.; Zhang, Y. Variable speed drive optimization model and analysis of comprehensive performance of beam pumping unit. *J. Pet. Sci. Eng.* **2020**, *49*, 152–163. [CrossRef]
5. Brijesh, R.N.; Sagar, M.P. The Effect of Venturi Design on Jet Pump Performance. *J. Res.* **2016**, *2*, 23–28.
6. Savinkin, V.V.; Vizureanu, P.; Sandu, A.V.; Ratushnaya, T.Y.; Ivanischev, A.A.; Surleva, A. Improvement of the turbine blade surface phase structure recovered by plasma spraying. *Coatings* **2020**, *10*, 62. [CrossRef]
7. Khaing, H.; Lwin, Y.M.; Lwin, Y. Design and Calculation of Centrifugal Pump (Impeller) For Water Pumping. *Int. J. Sci. Eng. Technol. Res.* **2019**, *28*, 321–324.
8. Vetter, S.; Leidich, E.; Hasse, A. Probability of mine survival. In Proceedings of the Nodal Connections, the 18th International Conference on New Trends in Fatigue and Fractures of Any Scale, (ICIE 2018), IDMEC Instituto de Engenharia Mecanica, Lisbon, Portugal, 20 July 2018; pp. 225–228.
9. Neikes, K.; Schlecht, B.; Vetter, S.; Leidich, E. *Untersuchungen zum Einfluss von Mittelspannungen auf die Ermüdungsfestigkeit von Wellen und Achsen (Investigations into the Influence of Medium Stresses on the Fatigue Strength of Shafts and Axles)*; Final report no. FVA-. 321 VI, 1321 .th ed; Frankfurt am Main, Germany, 2019; Volume 16, p. 157.
10. Baldenko, D.F.; Baldenko, F.D. Wheel Tooth Profiles of Hydraulic Machines and Mechanical Gears: Traditions and Innovations. *Glob. J. Res. Eng. A Mech. Mech. Eng.* **2020**, *20*, 211–219.
11. Lyagov, I.A.; Baldenko, F.D.; Lyagov, A.V.; Yamaliev, V.U.; Lyagova, A.A. Methodology for calculating technical efficiency of power sections in small-sized screw downhole motors for the «perfobur» system. *J. Min. Inst.* **2019**, *240*, 694–700. [CrossRef]
12. Sheha, A.A.; Nasr, M.; Hosien, M.A.; Wahba, E.M. Computational and experimental study on the water-jet pump performance. *J. Appl. Fluid Mech.* **2018**, *117*, 1013–1020. [CrossRef]
13. Kyznetsova, V.N.; Savinkin, V.V.; Ratushnaya, T.Y.; Sandu, A.V.; Vizureanu, P. Study of the spatial distribution of forces and stresses on wear surfaces at optimization of the excavating part of an earthmoving machine transverse profile. *Coatings* **2021**, *182*, 1–16.
14. Barsoum, I.; Khan, F.; Barsoum, Z. Analysis of the torsional strength of hardened splined shafts. *Mater. Des.* **2014**, *54*, 130–136. [CrossRef]
15. Kuznetsova, V.N.; Savinkin, V.V. More Efficient Rotation of Excavator Platforms. *Russ. Eng. Res.* **2017**, *37*, 667–671. [CrossRef]
16. Gorlenko, A.O.; Davydov, S.V.; Shevtsov, M.Y.; Boldyrev, D.A. Wear-Resistance Increase of Friction Surfaces of Steel Machine Parts by Electro-Mechanical Hardening. *Mech. Eng. Metallurg. Steel Transl.* **2019**, *49*, 800–805. [CrossRef]
17. Jeong, H.S.A. Comparative Study on the Equation of Motion about Models of Variable Displacement Piston Pump through MATLAB & DAFUL. *Mach. Complexes* **2016**, *12*, 75–76.
18. Burov, A.; Bazbetov, A.; Ponomarev, N.; Sapozhnikov, M. Recalculation of the loading mode of drilling pumps of single and double action is made: Oil-producing equipment and complexes. In Proceedings of the Volgograd State Technical University, Volgograd, Russia, 22 February 2019; pp. 74–77.
19. Koleda, E.V.; Kireev, S.O.; Korchagina, M.V.; Efimov, A.V. Determination of reactions in the crank shaft supports of the drive part of the high-pressure pump. In Proceedings of the 13th International Scientific and Practical Conference (State and Prospects of Development of the Agro Industrial Complex) In the Framework of the XXIII Agroindustrial Forum of the South of Russia and the Exhibition Interagromsh, Don State Technical University, Rostov-on-Don, Russia, 25 June 2020; pp. 605–609.
20. Kadhimm Zarzoor, A.; Almuramady, N.; Hussein, E.S. Stress analysis for spur gears using solid works simulation. *Int. J. Mech. Eng. Technol.* **2018**, *259*, 927–936.
21. Savinkin, V.V.; Kuznetsova, V.N.; Ratushnaya, T.Y.; Kiselev, L.A. Method of integrated assessment of fatigue stresses in the structure of the restored blades of CHP and HPS. *Bull. Tomsk. Polytech. Univ. Geo Assets Eng.* **2019**, *330*, 65–77.
22. Savinkin, V.V. *Development of the Theory of Energy Efficiency of Single-Bucket Excavators*; Siberian State Automobile and Road University: Omsk, Russia, 2016.
23. Savinkin, V.V.; Kuznetsova, V.N. Investigation of the influence of the displacement of the contact spot of engagement on the durability of the crown of the rotary wheel of the excavator. In *Scientific, Technical and Industrial Journal: Scientific Peer-Reviewed Journal "Construction and Road Machines"*; Publishing House of Technical Literature SDM-Press: Moscow, Russia, 2017; Volume 1, pp. 12–15.
24. Xiaofeng, Q.; Jie, L.; Xingguo, Z.; Li, F.; Ruiqiang, P. Fracture failure analysis of transmission gear shaft in a bidirectional gear pump. *Eng. Fail. Anal. Mach. Reliab.* **2020**, *118*, 138–141. [CrossRef]
25. Rossetti, A.; Macor, A.; Scamperle, M. Optimization of components and layouts of hydromechanical transmissions. *Int. J. Fluid Power* **2017**, *318*, 123–134. [CrossRef]
26. Das, S.; Nayak, B.; Sarangi, S.K.; Biswal, D.K. Condition Monitoring of Robust Damage of Cantilever Shaft Using Experimental and Adaptive Neuro-fuzzy Inference System (ANFIS). *Technol. Eng.* **2016**, *144*, 328–335. [CrossRef]

27. Xiang-Yu, Y.; Shi-Jeng, T. Computerized method for analyzing the contact of loaded teeth of spur gears with a high contact ratio with or without modification of the side surface, taking into account the angular contact of the tip and shaft misalignment. *Theory Mech. Mach.* **2016**, *97*, 190–214.
28. Tan, J.-J.; Hu, Z.-H.; Wu, L.-J.; Feng, Z.H.; Song, C.S. Influence of Geometric Design Parameters on Static Strength and Dynamics of Spiral Bevel Gears. *Int. J. Rotating Mach.* **2017**, *148*, 211–223.
29. Pronikov, A.S. *Parametric Reliability of Machines M.*; Publishing House of Bauman Moscow State Technical University: Moscow, Russia, 2002; p. 560.
30. Danube, A.V. Non-Traditional Tribology. In *The Modification of the Surfaces of Friction*; Lamdert Academic Publishing: Sunnyvale, CA, USA, 2013; pp. 270–281.
31. Ren, Y.; Chang, S.; Liu, G.; Wu, L.; Lim, T.S. Impedance Synthesis Analysis of Gear Train Vibration. *Shock. Vib.* **2017**, *214*, 425–438.
32. Van der Linden, F.L.J. Modeling Gear Positioning Systems: An Object Oriented Model of Gear Contact with Validation. *J. Mech. Eng. Sci.* **2016**, *230*, 1084–1100. [CrossRef]

applied sciences

MDPI

Article

Parametric Study of the Corrosion of API-5L-X65 QT Steel Using Potentiostat Based Measurements in a Flow Loop

Arjun Ravikumar [1], Paul Rostron [2], Nader Vahdati [1,*] and Oleg Shiryayev [3]

1 Department of Mechanical Engineering, Khalifa University of Science and Technology, Abu Dhabi 127788, UAE; arjun.ravikumar@ku.ac.ae
2 Centre for Defense Chemistry, Cranfield University, Defense Academy of the United Kingdom, Shrivenham SN6 8LA, UK; paul.rostron@cranfield.ac.uk
3 Department of Mechanical Engineering, University of Alaska, Anchorage, 3310 UAA Drive, ECB 301, Anchorage, AK 99508, USA; oshiryayev@alaska.edu
* Correspondence: nader.vahdati@ku.ac.ae

Abstract: Low-carbon steel is widely used in industrial pipelines, and corrosion studies are focused mostly on erosion-corrosion, its prediction and control. In this paper, the corrosion rate in pipelines is modeled using a flow loop and measured by linear polarization resistance method (LPR) using a 3-electrode corrosion setup for API-5L-X65 QT steel. Optical microscopy and SEM studies are conducted to examine the surface of the sample and the corrosion products. The effect of NaCl concentration on the corrosion rate is studied at different pH, temperature range, and flow velocities with dissolved oxygen content in the solution maintained at 6 mg/L (6ppm). The corrosion rate is found to be varying from 1 mil per year (0.0254 $mmyr^{-1}$) to 10 mils per year (0.254 $mmyr^{-1}$), and the corrosion rate increases with the flow velocity and reaches a maximum at Reynolds Number above 10,000. Further increase in fluid velocity shows corrosion is flow insensitive, and uniform corrosion is predominant in the region.

Keywords: pipeline corrosion; flow loop; linear polarization resistance; predictive equation

Citation: Ravikumar, A.; Rostron, P.; Vahdati, N.; Shiryayev, O. Parametric Study of the Corrosion of API-5L-X65 QT Steel Using Potentiostat Based Measurements in a Flow Loop. *Appl. Sci.* **2021**, *11*, 444. https://doi.org/10.3390/app11010444

Received: 4 December 2020
Accepted: 30 December 2020
Published: 5 January 2021

1. Introduction

Corrosion is a natural phenomenon associated with metals that leads to material destruction. Corrosion is an engineering problem, as well as an economic problem as the financial losses associated with corrosion, is enormous. The cost of corrosion worldwide in 2017 [1] was shown to be approximately 3.4% of the global GDP ($2.5 trillion). The same published study showed the cost of corrosion in the US to be 3.1% of GDP ($276 billion). In India, the cost of corrosion was estimated as 2.4% of the GDP in 2011-12 [2]. Pipeline accidents, due to corrosion, are frequent in the oil and gas industries. A study conducted by Saudi Aramco in 2013 concluded that the cost of corrosion for their operations was around $900 million per year [3]. Corrosion is not completely avoidable, so the industry needs to find ways to monitor, control, mitigate, and reduce corrosion. Implementing a proper corrosion management approach in the industry requires an in-depth understanding of corrosion mechanisms and implementing reliable methods to assess the corrosion rate [4].

Corrosion assisted by flow causes severe damages to oil and gas pipelines, heat exchanger systems in process industries. Maintenance and replacement of these corroded components are very expensive. The corrosion rates depend upon many factors, such as dissolved oxygen concentration, temperature, total dissolved salts, pH, fluid dynamics, and presence of scale on internal surfaces [5]. H.R Copson [6], in his studies summarizes that the corrosion assisted by flow varies according to the material under investigation and with the exposure conditions. He also states that flow generally increases the corrosion rates, but in some special cases, the effect can be the opposite. Rotating disk electrode method and impingement jet systems are two conventional methods used to measure the corrosion rates in a flow system. In the work of Namboodhiri et al. [7], a correlation between mass transfer and flow is developed using the rotating cylinder method to evaluate the corrosion of High Strength Low Alloy (HSLA) steel in 3.5% NaCl solution. However, it is not always possible to reproduce the exact flow patterns occurring in the pipeline system in the rotating cylinder device. A flow loop system provides a more realistic environment to evaluate the actual corrosion rates in a pipeline system [8].

The corrosion rate of steel in a flow loop system can be estimated using weight loss methods, electrochemical methods, and other methods, such as corrosion characterization, and acoustic emission measurements. Y. Utanohara et al. [9] investigated the corrosion rate at a pipe elbow using the electric resistance method to study the effect of the thermal flow field on corrosion rates. The corrosion rates were measured at a temperature range of 50–150 °C and flow velocities of 0–6 m/s, and the corrosion rate was found to have a nonlinear relationship with velocity [9]. Temperature and fluid velocity were the only two parameters that varied. Weight loss methods give an average corrosion rate over a period of time, but the measuring process takes a long time. The great difficulty with the weight loss method is the ability to hold all process parameters constant for a long period of time to see the effect of one parameter on the corrosion rate.

For a pipeline, there are so many parameters that can affect its corrosion rate, each of which tends to have an exponential effect on the corrosion rate that most methods give poor and irreproducible results. Electrochemical techniques are ideal for the study of corrosion in pipelines—particularly the linear polarization resistance (LPR) method, since the measurement is rapid (less than 5 min), does not affect the sample, and can be automated to perform continuous measurement of corrosion rate. For this method to work, the material should be polarized typically on the order of ± 10 mV compared to the Open Circuit potential, when there is no net current is flowing. By taking the slope of the potential versus current curve, as the current flow is induced between the working and counter electrodes [10] and using the Stern-Geary equation, the resistance can then be used to find the corrosion rate of the material. Corrosion rate measurements in the flow loop have been done by various researchers to evaluate the flow accelerated corrosion and erosion. Zafar et al. [11] used a flow loop to investigate the corrosion resistance of SA-543 and X65 steels in an oil-water emulsion containing H_2S and CO_2 using polarization curves. LPR was measured using a potentiostat and two-electrode system for 24hrs at an interval of 30 min. The effect of salt concentration and pH on corrosion rate was not considered in this work. Ajmal et al. [12] investigated the flow accelerated corrosion of oil field water in the loop system at pipe elbows at turbulent conditions using the LPR method and correlated the corrosion rates with fluid velocities and shear stresses, and the electrochemical results were validated using CFD simulation results. Huang et al. [13] used the polarization method to evaluate the corrosion behavior of X52 steel at an elbow of a loop system and its correlation with flow velocity, shear stress, and the volume fraction of particles. The work by Sun et al. [14] did a parametric study to evaluate the effects of CO_2 on corrosion rates of carbon steels (C1010, C1018, and X65) using a flow loop. Localized corrosion rates were determined at two temperatures (40 °C and 90 °C), pH, salt concentrations, flow regimes, and CO_2 partial pressure using the LPR method. In [14], the salt concentration was varied from 0 to 1%. The effect of flow rate on corrosion rate was not considered in this work, and the superficial gas and liquid velocities were assumed as 10 m/s and 0.1 m/s, respectively.

One of the earliest corrosion prediction models was by de Waard and Milliams [15] in 1975, where the authors studied the relationship between the corrosion rate on grit blasted steel specimens (X52) and partial pressure of CO_2. De Waard-Milliams provided a model to determine the corrosion rates incorporating the effects of pH and temperature. The relation was modified in his later works and provided a modified equation [16], which accounts for the effect of dissolved iron at lower and higher temperatures. Corrosion studies of 316 Stainless Steel were performed by Jepson et al. [17] using oil-water in a flow loop, and the rates were determined using electrical resistance probes and weight loss methods. The findings were used to develop a predictive equation for corrosion rates at different temperatures, flow velocities, and carbon dioxide partial pressure for the oil-water mixture. In recent works, the CO_2-multi-phase flow corrosion model is presented by Nesic et al. [18], which considers the kinetics of electrochemical reactions, diffusion effects, and the effect of steel type. The model also considers the kinetics of scale growth and precipitations and accounts for the effect of steel types, multi-phase flow, and H_2S. Nothing is mentioned of the steel types in the paper except they are mild steels. Pots [19] developed a corrosion prediction tool HYDROCOR, a computer-based engineering spreadsheet for localized corrosion in carbon pipelines considering various scenarios in the field. The model accounts for various environments with corroding agents like CO_2, H_2S, (organic) acids, bacteria, and oxygen, and the spreadsheet model is coupled with models for water chemistry, multi-phase flow, oil protectiveness, heat transfer, and thermodynamics using FORTAN to predict different scenarios. Khajotia et al. [20] used "a Case-based Reasoning-Taylor Series (CBR-TS) model" to predict the corrosion behavior in field pipelines considering operation parameters. Case-based reasoning model and Taylor expansion method are employed to predict the corrosion rate as a function of pipeline material, pH, CO_2, and H_2S content and temperature. The CBR-TS model was tested using a field database and a hypothetical database. NORSOK M-506 [21] developed a semi-empirical corrosion rate model using laboratory data that investigates the corrosion rates in a flow loop by varying parameters, such as temperature, solution pH, and CO_2 partial pressure, but it has limitations to predict the interaction between CO_2 and H_2S.

The aforementioned corrosion rate models do not include the interaction of chlorine ions and the corrosion effects. Whilst predictive software for pipeline corrosion is commercially available (11 commercial software packages), the results of each predictive package do not agree with each other. Even different versions of the same software have been known to give different corrosion rate results. There are many reasons for this, but it is clear that there is still a need for self-consistent predictive software for pipeline corrosion. Rolf Nyborg [22] has reviewed the models available to predict Corrosion rate in different CO_2 partial pressures, temperatures, and other flow parameters, and the work concludes that all the models have limitations depending upon the philosophy used in developing these models. A general prediction of the corrosion rates is difficult as the corrodents and the environmental factors are varying in different field data. The work concludes that the prediction of corrosion rates by these models vary considerably from case to case—most models are successful in predicting their respective data, but very unsuccessful when predicting other field data. This is concluded in Figure 1 below, where six different models are analyzed by our research team, which is similar to the comparison done by Nyborg in his work, and the results show a large deviation in corrosion prediction in all these models.

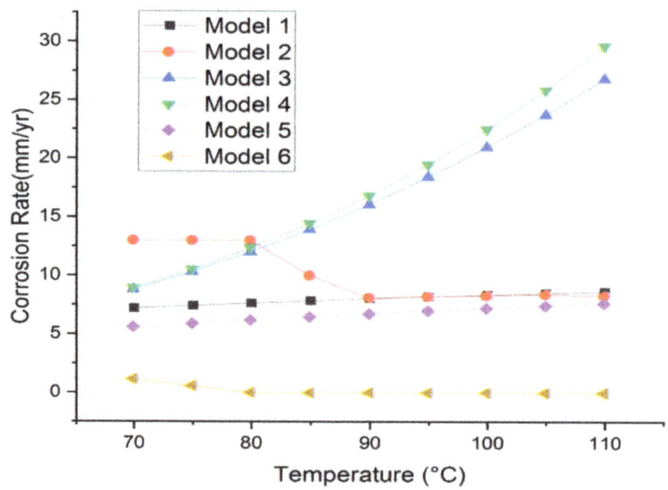

Figure 1. Different Corrosion models and predicted corrosion rates (Field data from [22] for the oil well 5).

In the oil industries, the corrosion rates of pipelines are intensified when the hydrocarbons phase is transported with corrosive brines, and the composition and its effects should be analyzed to predict the rates [23]. The presence of NaCl complicates the kinetics and the mechanism of pipeline corrosion depending on whether it is deaerated or aerated systems. The effect of chloride ions on corrosion rates varies depending upon the environment as it affects anodic kinetics, scale formation, CO_2 solubility, and pH [24]. However, in the literature, there are limited studies that provide corrosion rate models with detailed parametric studies in NaCl solutions. To improve the prediction of corrosion rates a parametric study incorporating the different parameters which affect corrosion is desirable, especially in pipelines in the industries where the flow conditions and rates are not stable. In his later work, Nesic et al. updated his previous work [18] to develop an H_2S Corrosion model [25] which includes the kinetics of iron sulfide growth with experimental results at very low temperatures (5 °C to 20 °C) and high salinity brines (up to 25 wt% NaCl). In CO_2-saturated solutions, the effect of NaCl concentration on corrosion rates in carbon steel is studied by [26] exposing the sample for 100 h by changing NaCl concentrations from 0.001 wt% to 10 wt% at room temperature. The corrosion rates were determined using Electrochemical Impedance Spectroscopy (EIS) and linear polarization method under the freely corroding condition and authors also studied anodic and cathodic kinetics using microelectrode technique. According to this study, the corrosion rate tends to decrease with NaCl concentration, which is opposite to the naturally aerated seawater. The effect of NaCl concentration on mild steel in CO_2-saturated brines is studied by [24,27] using a mechanistic model that considers mass transfer and electrochemical kinetics. The corrosion rates predicted by the model is compared with corrosion rates obtained using experiments conducted in a deaerated, CO_2-saturated brine environment. The study was conducted using a three-electrode electrochemical glass cell, and the corrosion rates were measured using LPR and EIS techniques. The flow rates are changed by changing the stirrer speed from 100 to 800 rpm. The authors studied the effect of pCO_2, NaCl concentrations on corrosion rates of the mild steel, and the experimental results show corrosion rates are independent of flowrate at high salinity. The proposed model could not predict this flow insensitive behavior at high flow rates.

In this paper, corrosion studies of the pipeline carbon steel (API-5L-X65 QT steel) are done in a flow loop using the LPR method to investigate the parameters which will affect the corrosion rates. Parametric studies have been conducted by changing flow rate, NaCl concentration, temperature, and pH, while bubbling oxygen into the solution. Key to this work, however, was the rigorous control of all parameters save one. This parameter was varied, and the corrosion rate was determined as a function of flow velocity, which was continuously varied from 0 to 0.8 m/s. Upstream pipelines in oil industries exhibit flow fluctuations, and the material degradation will be maximum because of corrosion and erosion. This work focuses on identifying the flow regions where the uniform corrosion is dominating and to predict a range of flow velocities where the corrosion rates can be kept constant. In that flow regions, parametric studies are conducted to develop a correlation with flow conditions and other parameters aiming to develop a predictive equation to obtain corrosion rates. In this research, we intend to measure the corrosion rate in a pipeline system using tight control of parameters. This is necessary, since corrosion rates are affected by so many different parameters, some of which have exponential effects.

2. Materials and Methods

2.1. Experimental Test Set-Up

The test solutions are prepared from deionized water (1×10^4 Ohm-cm^{-1}) produced from a Metrohm Ion exchange still, with pH 5.8 supplied from a 20L reservoir. Solutions are then made according to the required parameters. The concentration of NaCl in the test solution is varied from 1% to 3.5%. The pH of the test solution is maintained to the desired values using acetic acid sodium acetate buffer solutions. The pH of the test solution is measured using an Orion pH meter and monitored at infrequent intervals. The Orion pH meter is manually calibrated before each use. The dissolved oxygen content in the solution, which is a vital factor in the corrosion rate, is maintained by bubbling air into the reservoir at 6 mg/L (6ppm) and measured using a Milwaulkie D.O meter every hour. The corrosion electrode sensor is a 3-electrode sensor where the material sample is the working electrode, a graphite electrode is the counter electrode, and Ag/AgCl is the standard reference electrode. The three-electrode sensor is designed to fit into the flow and arranged such that the working electrode faces the oncoming flow. Microbial corrosion is prevented by the periodic addition (each day) of disinfectant (Dettol) to the solution.

A circulating loop system, as shown in Figure 2, was used for measuring the corrosion rate. The schematic diagram of the flow loop and the actual flow loop are shown in Figures 3 and 4. The flow loop consists of a plastic reservoir, a 0.50 horsepower/0.37 kW submersible pump (made of plastic) with a float switch, a pipe (made of PVC plastic) of internal diameter 28.9 mm and an outer diameter of 33.55 mm, PVC pipe fittings, PVC valves, a plastic calibrated flow meter, and the electrode test section. A 20 L test solution is supplied from the reservoir using the submersible pump, and the flow velocity is changed and continuously monitored using the flow meter. The entire flow loop is purposely made from plastic to make sure the only corroding component in the flow loop is the test specimen. The operating conditions are maintained in desired values, and the temperature of the test solution is varied from 25 °C to 40 °C (±0.5 °C) in the study and the operating conditions are maintained at desired values. The temperature is maintained using a heating and cooling system and continuously measured using a thermocouple and a mercury thermometer at infrequent intervals. The corrosion electrode system is connected to Gamry 600 potentiostat and a computer with Gamry instruments software. After mounting the electrode system in the flow loop, the required flow rate is obtained by adjusting the manifold valves in the loop and varying the amount of bypass, and monitoring the LPM values in the digital gauge. Once the flow is stable, the potentiostat is turned on, the readings can be recorded, and further analysis can be performed using the respective Gamry software. After recording one reading, the sample is removed and cleaned before continuing the experiments. There is a 2.02 m run of straight unaffected pipe, equivalent to 69 diameters, to allow for a fully developed flow to occur and for pump noise to dissipate. The test specimen is at the bottom of a vertical flow loop, as shown in Figure 4. This is to ensure that only single-phase flow impinges on the test specimen. On the top (return) side of the loop, there is an additional deaerator that catches bubbles and ensures that, after a few minutes of flushing, the loop has no trapped bubbles in it and is a single phase.

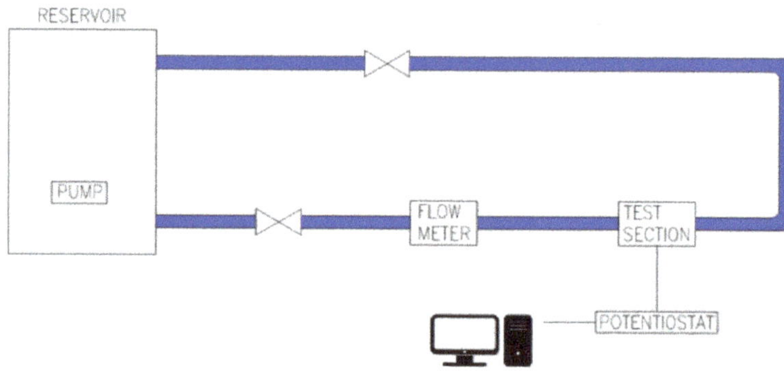

Figure 2. Flow loop for investigating corrosion using LPR method.

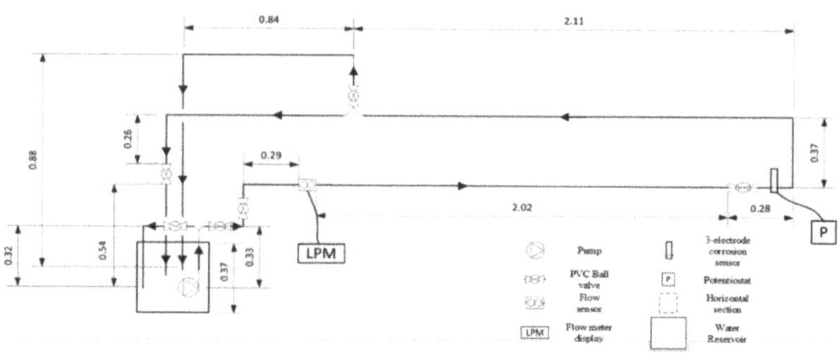

Figure 3. A layout of the experimental setup.

Figure 4. The actual flow loop.

2.2. Pipeline Steel

The working electrode was a 0.5 cm diameter API-5L-X65 QT (Quenched and Tempered) material. The microstructure of the steel is shown in Figure 5a,b from the optical micrograph of the samples consisting of ferrite and pearlite. The volume fractions of ferrite and pearlite have been measured to be ~77% and ~23%, respectively. The average ferrite to pearlite ratio is around 80% to 20%, as evident from the figure. The average grain size has been found to be ~15 μm. Figure 5c shows the SEM photograph of the mild steel sample after polishing. Like the optical micrograph, the grain distribution of the samples is clearly visible in the SEM images. The microstructure shows the carbon content is very low.

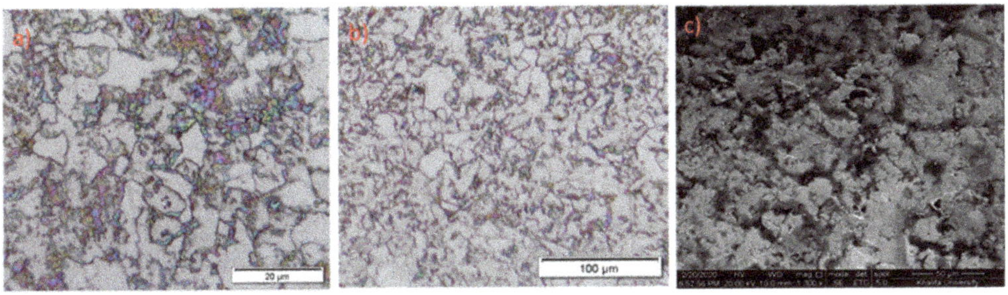

Figure 5. The microstructure of the steel is studied using optical microscopic images of the metal sample and shown (**a**) 20-micrometer (**b**) 100-micrometer (**c**) SEM images of the surface.

Three electrode (Specimen, Graphite, Ag/AgCl) setup was constructed using a nylon support material (shown in white color) and the electrodes fixed in place using marine grade epoxy (PC 10—marine) to prevent movement and crevice corrosion, as shown in Figure 6. The electrode, covered by an orange cap, is the Ag/AgCl electrode, the dark gray electrode is the graphite, and the light gray electrode, is the API-5L-X65 QT specimen. The electrodes were designed to sit in the full flow of the system.

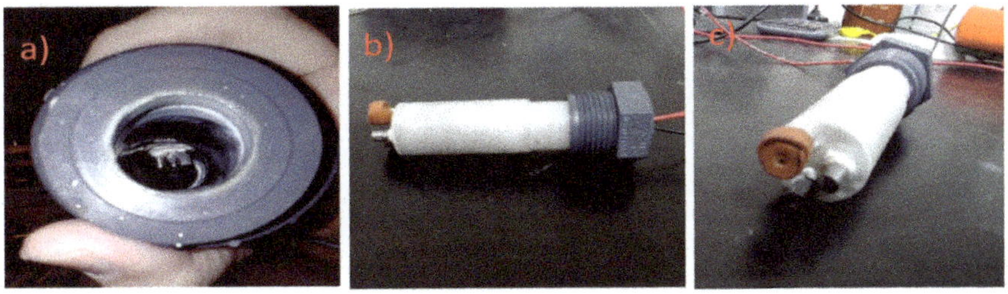

Figure 6. (**a**) Projection of sensor electrodes into the flow (**b**) isometric view of the sensor (**c**) side view. The orange rubber cap keeps the reference sensor (Ag/AgCl) from drying out.

2.3. Linear Polarization Resistance

In this model, the corrosion process is assumed to be controlled by the kinetics of electron transfer reaction at the metal surface, and the electrochemical reaction can be expressed using the Tafel equation,

$$I = I_o e^{\frac{2.303(E-E_O)}{\beta}} \quad (1)$$

where I is the current resulting from the reaction, I_o is a reaction-dependent constant called the exchange current, E is the electrode potential, E_o is the equilibrium potential, and β is the Tafel constant (volts/decade).

The Tafel equation is generally used to express the behavior of one reaction, and it is modified to Butler-Volmer equation, which can express both anodic and cathodic reactions, and it is given by,

$$I = I_{corr} e^{\frac{-2.303(E-E_{corr})}{\beta_a}} e^{\frac{-2.303(E-E_{corr})}{\beta_c}} \quad (2)$$

where I is the current (amperes), I_{corr} is the corrosion current, E is the electrode potential, E_{corr} is the corrosion potential, and β_a and β_c are the anodic and cathodic Tafel constants in volts/decade.

From the Tafel analysis shown, the current-voltage curve at E_{corr} can be considered as linear, and Equation (2) can be modified to obtain the Stern-Geary equation.

$$I_{corr} = \frac{1}{R_P} \times \frac{\beta_a \beta_c}{2.303(\beta_a + \beta_c)} \quad (3)$$

The corrosion current can be related to the corrosion rate using Faraday's law,

$$Q = n \times F \times M \quad (4)$$

where Q is the charge in coulombs, n is the no of electrons transferred, F is Faraday's constant (96,485 coulombs/mole), and M is the number of moles equal to $M = m/AW$, where AW is the atomic weight. Substituting the equivalent weight in terms of M in Equation (4), the mass of the species m can be written as,

$$m = \frac{(EW)Q}{F} \quad (5)$$

where EW is the equivalent weight. Modifying Equation (5) and substituting the value of Faraday's constant, the corrosion rate (CR) is found to be:

$$CR = \frac{I_{corr} \times K \times EW}{d(A)} \quad (6)$$

where d is the density (g/cm^3), A is the sample area in cm^2, and K is a constant.

Electrochemical measurements are performed using Gamry Reference 600 potentiostat. The experiments were conducted till the system reached the open circuit potential (OCP) before linear polarization resistance was measured. Figure 7 shows the OCP measurement where potential is plotted against time for 1%, 2%, and 3.5% NaCl concentrations at 25 °C when the flow velocity is kept at zero. OCP, also called corrosion potential, is the Potential difference between the working and reference electrode when no external current is flowing in the cell. OCP measurement allows us to determine the E_{corr} and gives a broad view of the stability of the system. The plot shows the OCP value becomes stable after 3600 s for three concentrations, and the test time for the experiments are kept at 1 h throughout the study. The area of the cylindrical electrode is $1.6cm^2$ with a density of 7.87 g/cm^3, the potential was swept between −0.02 Volts to 0.02 volts with a scan rate of 0.2 mV/s. The sample period was kept as 2 s, and Tafel analysis is conducted to determine the Tafel constants by running a potentiodynamic scan from −250 mV to 250 mV.

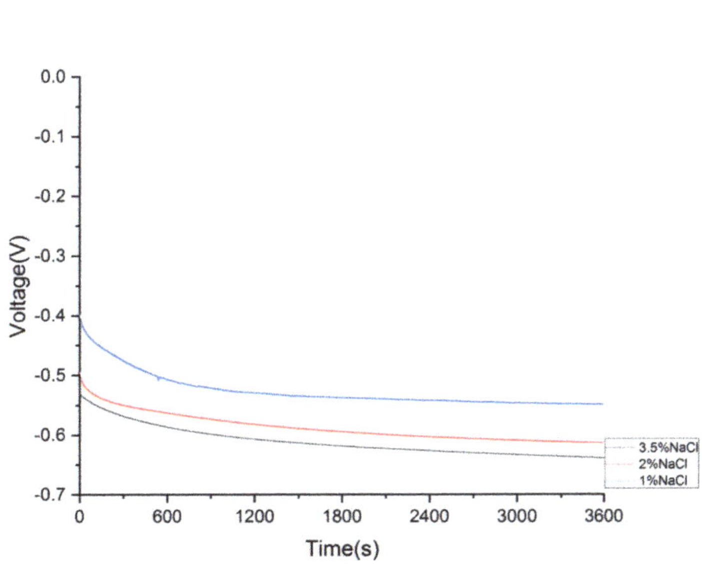

Figure 7. Potential is plotted against time for 1%, 2%, and 3.5% NaCl concentrations at 25 °C when the flow velocity is kept at zero.

3. Discussion of Results

3.1. Experimental Observations and Analysis

Prediction of corrosion rate in the flow loop is a challenging task as various parameters affect the rates depending upon the material properties. The dissolved oxygen in the system is an important factor that controls the corrosion rate, since the corrosion mechanism is different whether the system is aerated or deaerated. The temperature of the fluid in contact, the flow rate of the fluid, the concentration of the chloride ions, dissolved salts, and pH are parameters that can enhance and reduce the corrosion rate of the mild steel in a flow system.

In this study, we investigated the effects of flow rate, temperature, pH, and NaCl concentration on corrosion rate by changing one of the parameters and keeping the other parameters constant. The study is conducted in two different pH levels, namely, pH-7 and pH-5. The pH is maintained by adding sodium acetate buffer solution and is monitored using a well-calibrated pH meter. The temperature of the feed solution is changed from 25–40 °C every 5 °C degrees, the flow rate of the solution in the loop is varied from 0–0.8 m/s (Reynolds number up to 22,000), and the salt concentration of the solution is varied from 1% to 3.5% by adding NaCl. The effect of velocity on corrosion rate has been studied previously, and the application of these findings in practical flow systems is very difficult. The corrosion rate and its relationship with flow velocity varies differently for different metals and alloys; thus, it is very challenging to form a specific conclusion. The changes in velocity alter the oxygen supply to the material and remove corrosion products from the specimen surface, thus resulting in different corrosion rates. Generally, at low flow rates, the corrosion rate will be uniform at the surface, and at higher flow rates, corrosion will be more localized, resulting from the formation of oxygen concentration cells. At higher flow rates, corrosion rates can sometimes increase because of increased oxygen supply and can decrease due to the destruction of thermal and concentration gradients. Here are the corrosion rate results, obtained from the flow loop of Figure 4.

Figure 8 presents the corrosion rate of API-5L-X65 steel in millimeter per year (1 mm = 0.0393 inches = 39.3 mils) versus flow velocity (between 0 and 0.8 m/s), as NaCl concentration varies from 1% to 3.5%, and temperature varies from 25 °C, 30 °C, 35 °C to 40 °C with the feed solution maintained at pH-7. The change in corrosion rate with the parameters, such as flow velocity, temperature, and NaCl concentrations is discussed in the figure. For flow velocities up to 0.2 m/s, the corrosion rate increases as the oxygen supply to the material increases, and at higher flow velocities (0.2 to 0.8 m/s), the corrosion rate appears to reach a steady value. It is assumed a stable oxide film is formed on the material surface, which provides a constant corrosion rate. At zero velocity, the corrosion rate is significant and shows an increasing trend with an increase in NaCl concentration and temperature. The formation of the black iron oxide layer is also observed on the metal surface when there is no flow. It can be observed from Figure 8, that at pH-7, the corrosion rate varies from 0.0254 mm/yr to about 0.2032 mm/yr when the NaCl concentrations vary from 1% to 3.5%. At a temperature of 25 °C, the corrosion rate is less than 0.0254 mm/yr for 1% NaCl concentration and the corrosion rates reach 0.0508 mm/yr and 0.1143 mm/yr at 2% and 3.5% NaCl, respectively. A corrosion rate of 0.0558 mm/yr is observed for 2% NaCl solution at 30 °C, and 0.0635 mm/yr at 35 °C. At 40 °C, the corrosion rates increase from 0.0508 mm/yr to 0.08636 mm/yr when NaCl concentration changes from 1% to 2%. Corrosion rate further increases and reaches a value of 0.1727 mm/yr is at 3.5% NaCl concentration.

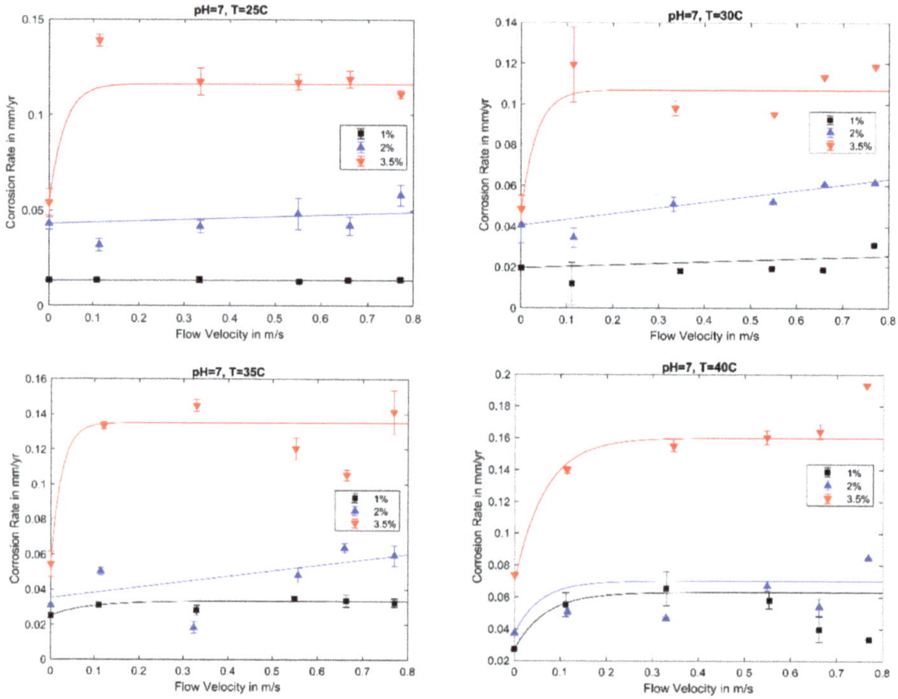

Figure 8. Corrosion rates of API-5L-X65 QT are shown at different flow velocities for feed solution maintained at pH-7, varying NaCl concentrations and temperatures (25 °C, 30 °C, 35 °C, and 40 °C).

The effect of flow velocities on the corrosion rates of API-5L-X65 QT is shown at different temperatures in Figure 9. When the feed solution is maintained at 1% NaCl concentration shown with black color lines (see Figure 9), the corrosion rate reaches 0.0782 mm/yr at 30 °C, 0.1028 mm/yr at 35 °C, and 0.1424 mm/yr at 40 °C. An increase in NaCl concentration to 2%, shown with blue color lines, causes an increase in the corrosion rates in the range of 0.1143 mm/yr to 0.2032 mm/yr when flow velocity and the temperature is varied. The red line shows corrosion rats at 3.5% NaCl concentration, and the corrosion rates are increased up to 15–28% than the 2% NaCl concentration conditions. At 40 °C, the corrosion rate reaches a maximum value of about 0.254 mm/yr (10mpy) at a flow velocity of 0.4 m/s, and a further increase in flow velocity is not affecting the corrosion rates. The increase in temperature shows a 5% to 10% increase in corrosion rates in all concentrations except for some data points.

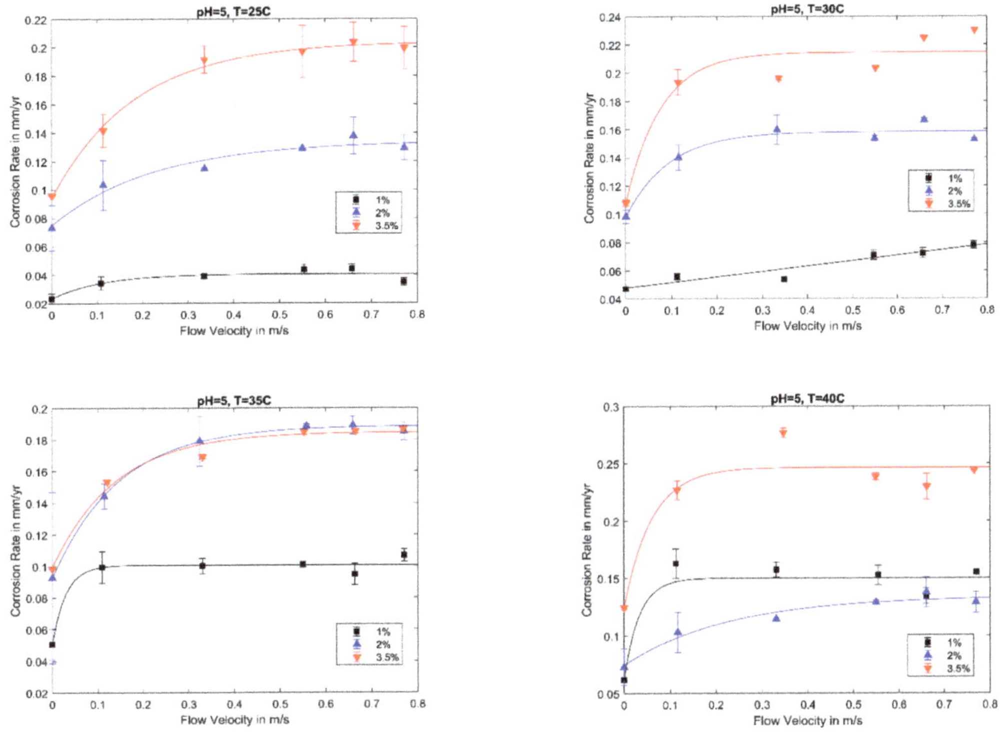

Figure 9. Corrosion rates of API-5L-X65 QT are shown at different flow velocities for feed solution maintained at pH-5 varying NaCl concentrations and temperatures (25 °C, 30 °C, 35 °C, and 40 °C).

In Figure 10, a study is conducted to determine the influence of pH on the corrosion rate. The results are shown at a temperature of 25 °C at different flow rates and NaCl concentration. In the literature, significant studies have been done that analyze the relation between pH and the corrosion rates, but very limited studies to model the effects in a flow loop, as discussed earlier. The physical and chemical properties of the corrosion products and the corrosion layer formed on the surface lead to variation in corrosion rates in different pH and flow velocities [28]. The studies show conflicting results at different pH as the high testing period leads to different corrosion mechanisms depending upon the material property [28]. As the LPR technique is short term method, observed trends eliminate the possibilities of different corrosion mechanisms at different pH and show a consistent trend in corrosion rates. At zero velocity for pH 7, the corrosion rate observed is about 0.04321 mm/yr for 2% concentrated solution and 0.05419 mm/yr for 3.5% NaCl solution. When the pH is maintained at 5, at zero velocity, the corrosion rate observed is 0.0732 mm/yr for 2% NaCl solution, and at 3.5% NaCl concentration, the corrosion rate shows an increase of 75% than pH 7 and records a value of 0.0956 mm/yr. As shown in the previous results (Figures 8 and 9) at different pH and temperatures, the majority of the data suggests that the corrosion rate is insensitive to flow at high velocities and maintains a steady rate. The results of Figures 8 and 9 show the corrosion rate tends to increase when the hydrogen ion concentration in the feed solution increases.

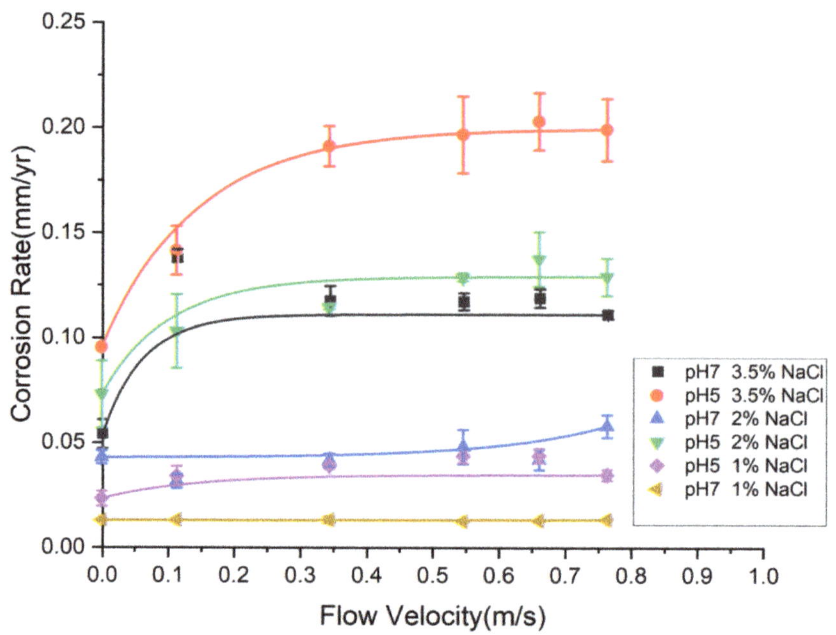

Figure 10. Corrosion rates of API-5L-X65 QT metal are shown at different flow velocities maintaining the feed solution pH of 5 and 7 and varying NaCl concentrations at 25 °C.

3.2. Parametric Study in Flow Insensitive Region

The results discussed in the previous section consistently shows two different regimes. A flow-sensitive region, where small changes to the flow or Reynolds number (Re) results in an exponential increase in corrosion rate and a flow insensitive region where the increase in flow rate does not affect the corrosion rate of the sample. This constant rate corrosion flow region is visible in most graphs, but not all.

Figure 11 shows the flow insensitive region, the corrosion rates are shown at different Reynolds number for 1% NaCl concentration for pH of 7 and 5 at different temperatures. In all the cases shown in Figure 11, the flow insensitive region is visible, corrosion rates are increasing with Reynolds number, and when Re is above 5000, the corrosion rates reach a plateau. It is assumed that a corrosive oxide layer formed on the surface limits the flow to the metal surface, and it minimizes the diffusion of hydroxides from the steel surface to the flowing water, which reduces the ion concentrations in the interfaces. The corrosion rate in the corrosion insensitive flow region is where uniform corrosion is predominant, and that region can be utilized to derive correlations for corrosion behavior at different operating parameters. The identification and analysis of this flow insensitive region in pipelines will be valuable in corrosion prediction and control.

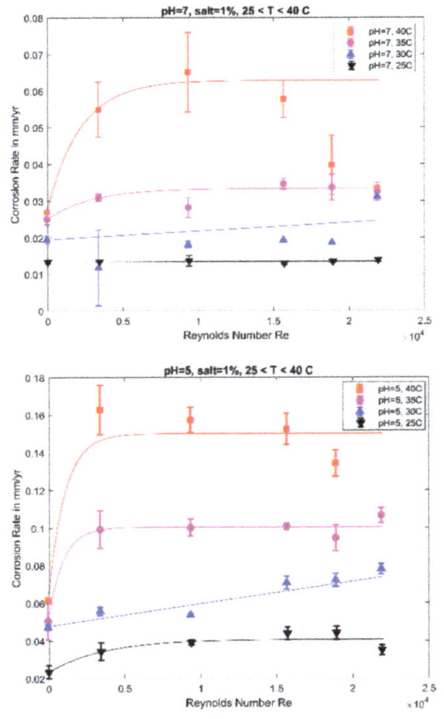

Figure 11. Corrosion rates are shown at different Reynolds number for feed solution maintained at pH–7 on the left, and at pH-5 on the right at 1% NaCl, at following temperatures (25 °C, 30 °C, 35 °C, and 40 °C).

The corrosion rates in the flow-insensitive region are taken as constant when the Reynolds number are greater than 10,000, and the corrosion rates are plotted with temperature, salinity, and pH to find the correlation between the parameters. pH and NaCl concentrations have a significant effect on corrosion rate like temperature, and the relationship is derived using the corrosion rates obtained in the flow insensitive region. The analysis, shown in Figure 12, provides the relation between NaCl concentration and corrosion rates. The corrosion rate, influenced by the presence of chloride ion, can cause the acceleration of both the anodic and cathodic reactions. The plot is analyzed, and the corrosion rate seems to increase as the power of NaCl concentration (X = NaCl concentration) and from the figure, the experimental data shows a good match with the power equation. It can be seen from Figure 12 that there is a good fit between the experimental data and the power equation with the coefficient of determination (R^2) above 0.9. To create one equation that can define the corrosion rate as a function of NaCl concentration (X), an average constant exponent value around 1.4 was chosen. The corrosion rates can be expressed as $Ax^{1.4}$ and a constant, which depends on the other flow parameters. Thus, the following relations CR = A (NaCl Concentration)$^{1.4}$ can be obtained to correlate the corrosion rates with NaCl concentration.

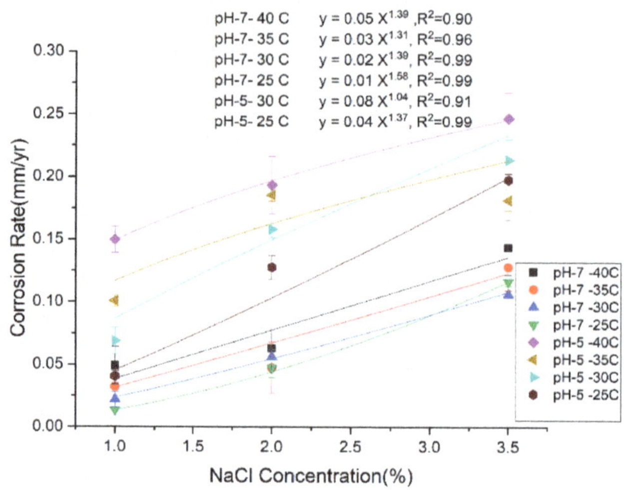

Figure 12. Corrosion rates for flow insensitive region are fitted with varying NaCl Concentration.

The corrosion rates from the flow insensitive region are plotted against the temperature, and the corrosion rates at pH 7 and 5 are plotted for 1%, 2%, and 3.5% NaCl solution. The corrosion rate in the mild steel samples seems to increase with the temperature in the flow insensitive region. At pH 7 for 2% NaCl concentrations, the corrosion rates increase from 0.0484 mm/yr to 0.0660 mm/yr when the temperature increases from 25 ° to 40 °C, whereas for the 3.5% NaCl concentration the corrosion rate changes from 0.1143 mm/yr to 0.1524 mm/yr. The data in the flow insensitive region is analyzed to predict the corrosion change with temperature. The exponential relationship between corrosion rates and temperature given by Arrhenius is widely used in the corrosion prediction. Arrhenius gave the equation in static conditions as $K = A \exp(-E_a/RT)$, where K = rate constant, R = gas constant (8.314 J\mole), T = temperature in degree Kelvin, E_a = activation energy (J\mole.K), and A = Modified frequency factor (pre-exponential factor). An attempt was made to fit the Arrhenius equation to the experimental data, but the fit was not good. The analysis, shown in Figure 13, reveals that the corrosion rate can be expressed as a power equation of temperature, showing a much better fit than the Arrhenius equation. In all the cases studied, the power constant varies when the NaCl concentration changes from low (1%) to high (3.5%). For both pH 5 and 7 at 1%NaCl concentration, the power constant is found to be about 2.7, but when the NaCl concentration is increased to 3.5%, the power constant is found to be about 0.9. The corrosion rates can be expressed as $AT^{0.9}$ for high NaCl concentration. Therefore, the following relation between the corrosion rate and the temperature is proposed: Corrosion Rate = A $(\text{Temperature})^{0.9}$ for high salinity. The value of the coefficient A was found to be varying with both NaCl concentration and pH. The correlation for temperature concludes that more data points need to be obtained in order to be able to generalize an empirical formula.

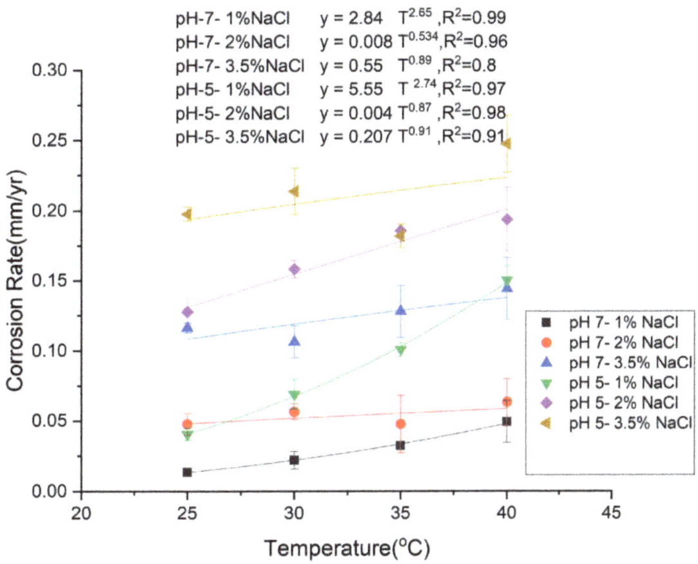

Figure 13. Corrosion rates at different temperatures for flow insensitive regions are fitted using Power-law.

The parametric study was conducted by maintaining constant oxygen concentration in the system. The oxygen concentration affects the corrosion rate of mild steel in NaCl, and the solubility of the oxygen will be an important factor in the predictive equation. In the work presented in this paper, individual equations (as shown in Figures 12 and 13) relating the corrosion rate in low flow velocities are developed. In future studies, the work will look at the synergistic effects of changing two or more parameters at the same time at different oxygen concentrations. This eventually leads towards a predictive model of the corrosion in pipelines and identifying the flow regions where corrosion rates can be minimized.

4. Conclusions

Using the linear polarization resistance (LPR) method and the work presented in this paper, we have been able to make repetitive measurements of the corrosion rate of API-5L-X65 QT steel—taking an average of at least three measurements per parameter. For each test condition, one parameter was varied, while all other parameters were kept constant. This level of detail is essential to assess the true effect of each parameter on the corrosion rate.

Experimental data presented in this work indicates a complex relationship between pH level, temperature, salt concentration, flow velocity, and corrosion rate. For the majority of combinations of experimental parameters, one can observe a range of flow velocities where the corrosion rate reaches a quasi-steady state or exhibits a linear behavior. The complex behavior where the corrosion rate is best described by exponential functions is mostly observed for cases with pH = 5. For pH = 5, the exponential function, that fits the experimental data very well, has the form $c_r = a + be^{-cv}$ where c_r is the corrosion rate, v is the fluid velocity, and a, b, and c are the fitting constants. The majority of the experimental data, but not all, suggests that in the velocity range of 0.2 m/s to 0.8 m/s, the corrosion mechanism appears to be uniform corrosion, and erosion corrosion is not occurring. Near fluid velocity of 0.8 m/s, an increase in corrosion rate is observed in most cases but not all. Turbulence-induced erosion corrosion maybe occurring. The highest corrosion rate occurs at 40 °C and 3.5% salt concentration, and the corrosion rate is about 0.25 to 0.3 mm/yr.

In this paper, we have also produced individual equations relating to the corrosion rate to a given parameter. In the future, it is our intention to look at the synergistic effects of changing two or more parameters at the same time and study their effects on the corrosion rate of oil pipelines. Eventually, this work will lead towards a predictive model of the corrosion rate in mild steel pipelines. Since API-5L-X65 material is widely used in the oil industry, the experimental data collected and published in this paper can be very useful to corrosion engineers working in the oil industry.

Author Contributions: A.R. conducted all the experimental testing, and wrote the first draft of this paper, P.R. was responsible for the design of the flow loop, supervision of this project and experimental data analysis, N.V. was the overall principle investigator of the research project that included this work, is the corresponding author, and was responsible for detailed review of this journal paper, and O.S. was responsible for experimental data analysis. All research members (the authors) met weekly and gave technical inputs/suggestions to the first author throughout the duration of this research project. All authors have read and agreed to the published version of the manuscript.

Funding: This research was funded by Khalifa University (KU) and we are grateful to the Chemistry Department of the Khalifa University for providing laboratory space for this project.

Data Availability Statement: Data sharing not applicable.

Conflicts of Interest: The authors declare no conflict of interest.

References

1. Koch, G. *Cost of Corrosion*; Woodhead Publishing Series in Energy: Cambridge, UK, 2017; pp. 3–30.
2. Shekari, E.; Khan, F.; Ahmed, S. Economic risk analysis of pitting corrosion in process facilities. *Int. J. Press. Vessel. Pip.* **2017**, *157*, 51–62. [CrossRef]
3. El-Sherik, A. *Trends in Oil and Gas Corrosion Research and Technologies: Production and Transmission*; Woodhead Publishing: Cambridge, UK, 2017.
4. Liu, J.; BaKeDaShi, W.; Li, Z.; Xu, Y.; Ji, W.; Zhang, C.; Cui, G.; Zhang, R. Effect of flow velocity on erosion–corrosion of 90-degree horizontal elbow. *Wear* **2017**, *376*, 516–525. [CrossRef]
5. Rostron, P. Critical review of pipeline scale measurement technologies. *Indian J. Sci. Technol.* **2018**, *11*. [CrossRef]
6. Copson, H. Effects of velocity on corrosion. *Corrosion* **1960**, *16*, 86t–92t. [CrossRef]
7. Namboodhiri, T.; Upadhyay, S. Flow-assisted corrosion of API X-52 steel in 3.5% NaCl solution. *Can. J. Chem. Eng.* **2002**, *80*, 456–464.
8. Zhang, G.; Zeng, L.; Huang, H.L.; Guo, X.P. A study of flow accelerated corrosion at elbow of carbon steel pipeline by array electrode and computational fluid dynamics simulation. *Corros. Sci.* **2013**, *77*, 334–341. [CrossRef]
9. Utanohara, Y.; Murase, M. Influence of flow velocity and temperature on flow accelerated corrosion rate at an elbow pipe. *Nucl. Eng. Des.* **2019**, *342*, 20–28. [CrossRef]
10. Compton, R.G.; Sanders, G.H. *Electrode Potentials*; Oxford University Press Oxford: Oxford, UK, 1996.
11. Zafar, M.N.; Rihan, R.; Al-Hadhrami, L. Evaluation of the corrosion resistance of SA-543 and X65 steels in emulsions containing H2S and CO_2 using a novel emulsion flow loop. *Corros. Sci.* **2015**, *94*, 275–287. [CrossRef]
12. Ajmal, T.; Arya, S.B.; Udupa, K.R. Effect of hydrodynamics on the flow accelerated corrosion (FAC) and electrochemical impedance behavior of line pipe steel for petroleum industry. *Int. J. Press. Vessel. Pip.* **2019**, *174*, 42–53. [CrossRef]
13. Huang, H.; Tian, J.; Zhang, G.; Pan, Z. The corrosion of X52 steel at an elbow of loop system based on array electrode technology. *Mater. Chem. Phys.* **2016**, *181*, 312–320. [CrossRef]
14. Sun, Y.; Nesic, S. A parametric study and modeling on localized CO_2 corrosion in horizontal wet gas flow. In *Corrosion 2004*; NACE-04380; Nace International: New Orleans, LA, USA, 2004.
15. De Waard, C.; Milliams, D. Carbonic acid corrosion of steel. *Corrosion* **1975**, *31*, 177–181. [CrossRef]
16. De Waard, C.; Lotz, U.; Milliams, D. Predictive model for CO_2 corrosion engineering in wet natural gas pipelines. *Corrosion* **1991**, *47*, 976–985. [CrossRef]
17. Jepson, W.P.; Bhongale, S.P.; Gopal, M.P. Predictive model for sweet corrosion in horizontal multiphase slug flow. In *Corrosion 1996*; NACE-96019; Nace International: Denver, CO, USA, 1996.
18. Nesic, S.; Cai, J.; Lee, K.-L. A multiphase flow and internal corrosion prediction model for mild steel pipelines. In *Corrosion 2005*; NACE-05556; Nace International: Houston, TX, USA, 2005.
19. Pots, B.F. Prediction of corrosion rates of the main corrosion mechanisms in upstream applications. In *Corrosion 2005*; NACE-05550; NACE International: Houston, TX, USA, 2005.
20. Khajotia, B.; Sormaz, D.; Nesic, S. Case-based reasoning model of CO_2 corrosion based on field data. In *Corrosion 2007*; NACE-07553; Nace International: Nashville, TN, USA, 2007.

21. Olsen, S. CO_2 corrosion prediction by use of the Norsok M-506 model-guidelines and limitations. In *Corrosion 2003*; NACE-03623; Nace International: San Diego, CA, USA, 2003.
22. Nyborg, R. CO_2 Corrosion Models for Oil and Gas Production Systems. In *Corrosion 2010*; NACE-10371; Nace International: San Antonio, TX, USA, 2010.
23. Cabrera-Sierra, R.; Cosmes-López, L.J.; Castaneda-López, H.; Calderón, J.T.; López, J.H. Corrosion Studies of Carbon Steel Immersed in NACE Brine by Weight Loss, EIS and XRD Techniques. *Int. J. Electrochem. Sci.* **2016**, *11*, 10185–10198. [CrossRef]
24. Han, J.; Carey, J.W.; Zhang, J. Effect of sodium chloride on corrosion of mild steel in CO2-saturated brines. *J. Appl. Electrochem.* **2011**, *41*, 741–749. [CrossRef]
25. Nesic, S.; Wang, S.; Fang, H.; Sun, W.; Lee, J.K. A new updated model of CO_2/H_2S corrosion in multiphase flow. In *Corrosion 2008*; NACE-08535; Nace International: New Orleans, LA, USA, 2008.
26. Zeng, Z.; Lillard, R.; Cong, H. Effect of salt concentration on the corrosion behavior of carbon steel in CO_2 environment. *Corrosion* **2016**, *72*, 805–823. [CrossRef]
27. Han, J.; Carey, W.; Zhang, J. A coupled electrochemical–geochemical model of corrosion for mild steel in high-pressure CO_2–saline environments. *Int. J. Greenh. Gas Control* **2011**, *5*, 777–787. [CrossRef]
28. Scheers, P. The effects of flow velocity and pH on the corrosion rate of mild steel in a synthetic minewater. *J. S. Afr. Inst. Min. Metall.* **1992**, *92*, 275–281.

applied sciences

MDPI

Article

Corrosion and Tribocorrosion Behaviors for TA3 in Ringer's Solution after Implantation of Nb Ions

Chenfei Xue [1,2], Pingze Zhang [1,2,*], Dongbo Wei [1,2,3], Hengmei Hu [1,2], Fengkun Li [1,2] and Kai Yang [1,2]

1 College of Materials Science and Technology, Nanjing University of Aeronautics and Astronautics, Nanjing 211106, China; xuechenfei@nuaa.edu.cn (C.X.); weidongbo@nuaa.edu.cn (D.W.); hhm@nuaa.edu.cn (H.H.); lifengkun@nuaa.edu.cn (F.L.); ykcl@nuaa.edu.cn (K.Y.)

2 Materials Preparation and Protection for Harsh Environment Key Laboratory of Ministry of Industry and Information Technology, Nanjing 211106, China

3 Aero-engine Thermal Environment and Structure Key Laboratory of Ministry of Industry and Information Technology, Nanjing 211106, China

* Correspondence: zhangpingze@nuaa.edu.cn; Tel.: +86-139-5188-3686

Received: 1 November 2020; Accepted: 21 November 2020; Published: 24 November 2020

Featured Application: A potential application of this work is in the corrosion and tribocorrosion behaviors of Ti alloys after surface modification.

Abstract: Ti alloys are prone to corrosion and wear due to the hostile environment in bodily fluids, but the Ti-45Nb alloy is considered to be a promising titanium alloy with excellent biocompatibility and resistance to physiological corrosion. In this study, Nb ions were implanted into a TA3 alloy and the effect on the biological corrosion as well as tribocorrosion behavior of TA3 in Ringer's solution was systematically investigated. The surface microstructure and XRD results revealed that the implanted samples showed a smoother surface due to the sputtering and radiation damages, and the Nb ions mainly existed in the alloy as the solid solution element. The electrochemical polarization tests showed that the implantation of Nb ions can increase the corrosion potential of the samples, showing a better thermodynamic stability. The tribocorrosion tests showed that the implanted samples exhibited a better thermodynamic stability in a corrosive environment accompanied by wear behavior, and the worn surface showed fewer pitting pits, indicating a better corrosion resistance. However, the abrasive wear and oxidation wear degree of the sample increased because of partial softening of the surface and brittle passivation film.

Keywords: TA3; ion implantation; corrosion resistance; tribocorrosion resistance; Ringer's solution

1. Introduction

Titanium and its alloys have been considered as some of the most suitable implants for the replacement and repair of human hard tissues because of their good corrosion resistance, mechanical properties and biocompatibility [1–4]. Among the various titanium alloys, Ti–6Al–4V and Ti–6Al–7Nb have been widely used owing to their good properties [3,5,6]. However, they are easy to dissolve and can undergo aseptic loosening in physiological environments due to the complex electrolyte environment in human bodily fluids, as well as the comprehensive action of active cells [2,5]. Furthermore, some studies have revealed that the corrosion and wear behavior of Ti–6Al–4V leads to the diffusion of aluminum and vanadium into the blood, causing local inflammation, allergies, poisoning, and diseases such as Alzheimer's disease [6,7]. In addition, the two-phase ($\alpha+\beta$) Ti–6Al–4V has poorer corrosion resistance than the single-phase due to the galvanic coupling effect [8]. Therefore, there has been a continuing interest in exploring new materials with excellent properties and no toxicity. Surface modification treatment on existing materials is a feasible solution.

Ion implantation is a surface modification technique, in which high-energy ion beams are implanted onto the surface of a material and eventually remain inside to optimize the surface performance [9]. At present, ion implantation has been widely used in the surface modification of Ti and its alloys in the medical field [10–14]. Yan et al. [13] implanted Zr ions into a titanium surface and revealed that a Ti–Zr alloy layer on titanium exhibited good mechanical properties and corrosion resistance. Zhao et al. [15] observed that the generation of Nb_2O_5/TiO_2 and implanting Nb ions into a Ni-Ti alloy significantly improved the corrosion resistance of the composite film in Hank's solution. Zu et al. [16] implanted Nb ions to obtain a wear-resistant modified layer in Ti-Al-Zr, and the wear-resistance improved because of the formation of intermetallic $NbTi_4$. Studies have shown that the presence of Nb can stabilize the passivated film of Ti alloys, improve the corrosion resistance, and improve its wear-resistance [15,17].

Titanium alloys are prone to corrosion and wear phenomena coming from the hostile environment of the body and friction processes, and sometimes the wear and corrosion behaviors of titanium alloys occur simultaneously when used as an implant material. Therefore, it is necessary to study the tribocorrosion problem in bodily fluids while studying corrosion behavior, as it is more meaningful for practical application and provides data references for practical application.

In this paper, Nb ions were implanted into the single-phase TA3 with a dose of 6.85×10^{16} ions/cm^2 at an energy of 125 keV. The surface morphology, phase structure, mechanical properties, corrosion and tribocorrosion behavior of the samples in Ringer's solution were characterized by scanning electron microscopy (SEM), X-ray diffraction (XRD), friction and wear instrument and nanoindentation instrument, combined with electrochemical measurements.

2. Experimental Details

2.1. Test Specimens and Ion Implantation

The pure titanium TA3 was cut into block samples of 15 mm × 15 mm × 4 mm. Prior to Nb ion implantation, the samples were prepared by standard metallographic technique to the extent that the surface roughness was below 0.2 μm. Following this, they were polished to a mirror finish with a silica suspension, washed by ultrasonic in anhydrous ethanol and finally dried in air. Niobium (Nb, 99.99 wt%) ions were injected at an extraction voltage of 125 keV and a vacuum of 10^{-3} Pa with a dose of 6.85 × 1016 ions/cm^2.

2.2. Surface Characterization and Crystal Structure Analysis

The usage of scanning electron microscopy (SEM) and X-ray diffraction (XRD) was to observe the change of surface morphology and the phase composition of samples after they were injected, while the composition distribution on the surface of samples was determined by EDS.

2.3. Characterization of Mechanical Properties

The nanoindentation instrument (DuH-W201) can be used to reflect the mechanical properties of the sample surface. In this paper, the static load control mode was selected with a loading load of 10 mN kept for 10 s, and the Load–Displacement curves were obtained. The average values were calculated from five different test points selected on each sample.

2.4. Electrochemical Test

A classical three-electrode system was utilized for electrochemical measurement, in which the reference electrode is a Saturated Calomel Electrode (SCE), the counter electrode is a platinum electrode, and the working electrode is the specimen. All the samples were coated with AB glue, except for an exposed area of 1 cm^2 soaked in Ringer's solution at 37 °C for electrochemical testing. The constituents of Ringer's solution were NaCl 8.61 g/L, $CaCl_2$ 0.49 g/L and KCl 0.30 g/L. After a stable open circuit potential (OCP), the potentiodynamic polarization was conducted from −0.8 V to 0.8 V with a scan rate of 1 mV/s. All the measurements were carried out by the ZAHNER ZENNIUM.

2.5. Tribocorrosion Tests

The tribocorrosion behavior of samples was studied by using a reciprocating tribometer. Sliding wear conditions against GCr15 steel balls with a diameter of 6 mm were established in Ringer's solution. The performed load was 2 N, and the friction test reciprocating length was 5 mm with a frequency of 1 Hz for 10 min. In this test, the curves of open circuit potential as well as the coefficient of friction (COF) were recorded before, during and after the wear process, where the samples were immersed in Ringer's solution with an exposed area of 1 cm^2. In addition, the SEM and 3D Measuring Laser Microscope were adopted to observe the morphology and measure the size of the worn surface.

3. Results and Discussion

3.1. Surface Structure and Morphology

The surface morphology and composition of TA3 before and after the implantation of Nb ions are shown in Figure 1. Compared to the substrate, the surface morphology of TA3 implanted with Nb ions appears smoother, while some small pits appear on the metal surface. Studies have suggested that some pits are formed on the surface due to the bombardment and cascade collisions caused by high-energy ion implantation [18,19]. Additionally, the etching effect produced during the continuous bombardment of TA3 with Nb ions reduces the relative height of the surface bulge, which explains the flattening of surface morphology [20]. In general, the smoother the surface, the better its corrosion resistance due to the smaller potential difference between the concave and convex [21,22]. Therefore, the effect of ion implantation on a metal's surface can improve the corrosion performance appropriately. Moreover, as shown in Figure 1, the content of Nb ions is approximately 2.24 at%, indicating that the Nb ions were implanted into the sample successfully. However, the content of Nb is lower, and this may be due to the far greater effective penetration depth of EDS than the implanted depth of Nb.

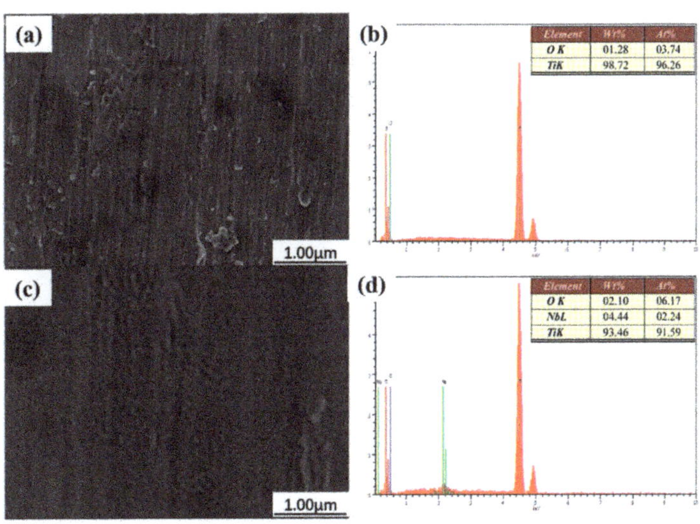

Figure 1. Backscattered SEM images and EDS analysis: (**a**,**b**) substrate, (**c**,**d**) Nb-implanted sample.

Figure 2 shows the XRD pattern of samples before and after Nb ions were implanted. Compared to the substrate, no new diffraction peaks were observed after Nb ions were implanted, indicating that no new phase was formed and Nb ions mainly existed in the solid solution element, which would improve the mechanical properties of the metal surface owing to the solid solution intensification. Figure 2 shows that the diffraction peaks of Ti (101) and Ti (002) shifted towards the higher angle ride after implantation. It is well known that the crystal imperfections resulting from ion implantation may lead to lattice distortion, leading to the shift of diffraction peaks [23]. As the ionic radius of Nb is greater than that of Ti, the implanted Nb ions occupy the position of Ti and may lead to localized compressive strain, explaining the shift direction of the diffraction peak. Different from tensile stress, the presence of residual compressive stress does not induce stress corrosion properties.

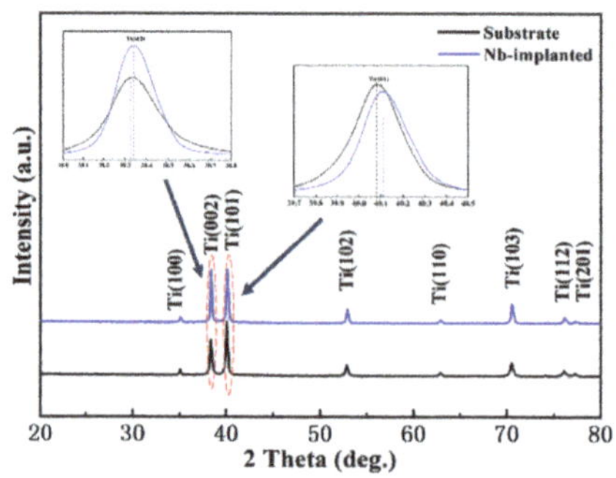

Figure 2. XRD patterns of substrate and the Nb-implanted sample.

3.2. Nanomechanical Properties

Figure 3 shows the Load–Displacement curves of nanoindentation tests for samples under 10 mN normal load. Table 1 shows the mechanical properties obtained by the L–D plots. In general, a higher ratio between hardness and elastic modulus (H/E) serves as a better wear-resistance [24]. In this test, the elastic modulus (E) and hardness (H) of the implanted samples decreased to 78% and 61.7% of the substrate, respectively. In addition, the H/E was approximately 78% of the substrate. Studies have shown that the irradiation effect during the implantation may increase the surface temperature of the samples, causing the substrate to soften [25]. The associated bombardment and cascade collisions will form a large number of defects on the surface, such as vacancy, dislocation and dislocation group, strengthening the substrate to some extent.

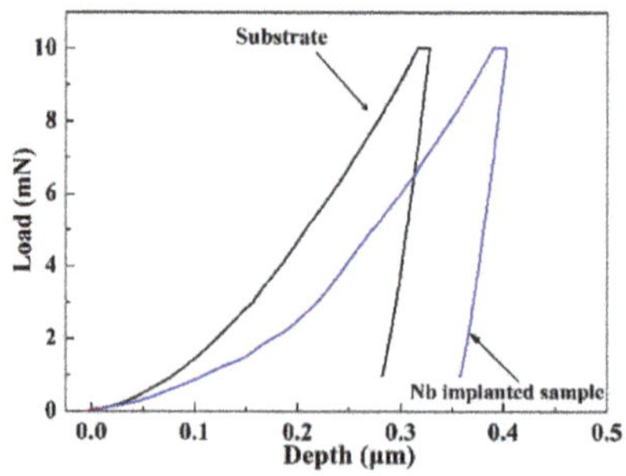

Figure 3. The nanoindentation Load–Displacement curves of the substrate and the Nb-implanted sample.

Table 1. Calculated values of elastic modulus and hardness of samples.

Sample	H(GPa)	E(GPa)	H/E
substrate	4.677	146.645	0.032
Nb implanted	2.888	114.743	0.025

3.3. Corrosion Performance

Figure 4 shows the potentiometric polarization results of the samples immersed for 1 h at 37 °C in Ringer's solution, in which the anode branch represents the dissolution of the Ti under higher potential, whereas the cathode branch represents the hydrogen evolution reaction [26]. It can be seen in Figure 4 that Nb ion implantation shifted the anode part towards positive potential, and areas of passivation can be observed obviously, which indicates that the implanted samples have better thermodynamic stability. In addition, the results present a higher slope in the anode part than in the cathode part, indicating that the corrosion process is controlled by the anodic process (active dissolution of the metal) and is conducive to inhibiting the dissolution of Ti alloys [27].

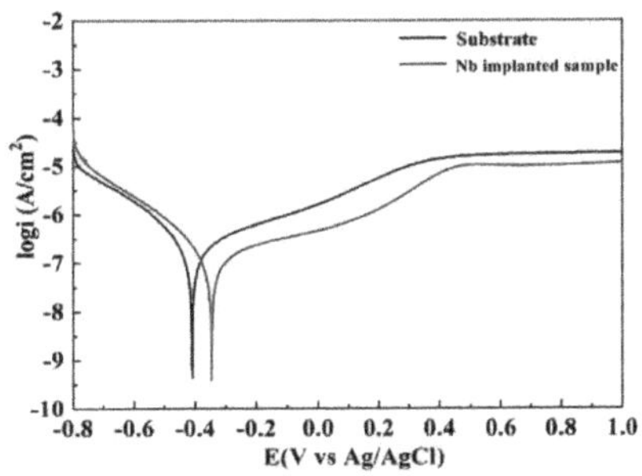

Figure 4. The potentiodynamic polarization curves of the untreated and treated samples in Ringer's solution.

Table 2 shows the electrochemical parameters obtained, including corrosion potential (E_{corr}), corrosion current density (I_{corr}) and polarization resistance (R_p) from the Tafel plots. The results clearly show that, compared with the substrate, the E_{corr} of implanted samples increased by 0.15 V, and the I_{corr} also decreased by an order of magnitude. In addition, the R_p of the implanted samples is about three times as much as that of the substrate. Additionally, it can be observed that the implanted sample has a lower passivation current, which indicates that the passivation layer of the implanted sample has a better corrosion resistance. The above results show that Nb ion modification can effectively improve the thermodynamic and kinetic stability of the passivation layer, exhibiting a better corrosion protection effect in the short–term immersion.

Table 2. Corrosion parameters calculated from the potentiodynamic polarization curves.

Sample	E_{corr} (V)	I_{corr} ($\mu A/cm^2$)	R_p ($k\Omega/cm^2$)
Substrate	−0.349	0.194	340
Nb implanted	−0.199	0.051	1110

3.4. Tribocorrosion Tests

The tribocorrosion tests (combined corrosion and wear tests) were performed according to the experimental procedure described in Section 2.5. The variation in the coefficient of friction (COF) and open circuit potential (OCP) of substrates and Nb-implanted samples is shown in Figure 5. In general, the friction coefficient will vary with the degree of surface damage and wear. Starting load at after 5 min, the COF of the substrate increases rapidly and enters the stable wear stage with an average friction coefficient of 0.395, while that of the Nb-implanted sample remains around 0.15 at the initial stage of wear due to the reduced surface roughness of the sample caused by ion implantation. However, it increases sharply and becomes similar to that of the substrate after about 2 min, which may be due to wear of the implanted layer and the exposure of the substrate as wear deepens.

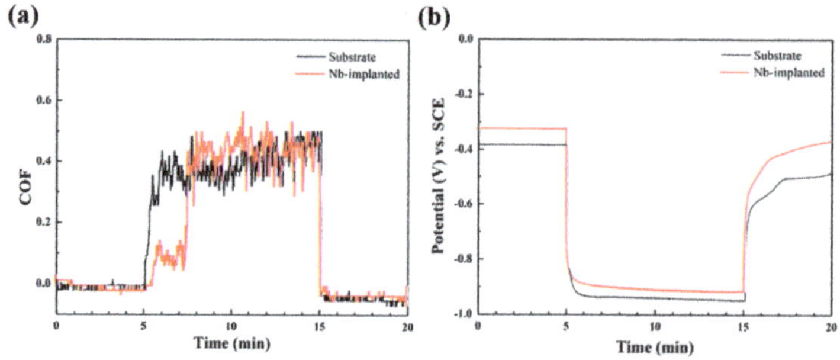

Figure 5. The variation of coefficient of friction and open circuit potential of samples in Ringer's: (**a**) coefficient of friction; (**b**) open circuit potential.

In general, the variation in the OCP is affected by friction and dielectric corrosion [28]. It can be observed in Figure 5b that the OCP of both samples reaches a steady state after 5 min of immersion in Ringer's and before the tribological process, indicating the formation of a stable passivation film. The OCP of samples mutates and moves towards a lower potential due to local damage to the passivation film of the worn surface after a tribological load. The exposed inner metal is more prone to corrosion due to the wear behavior. In addition, the agitation in the friction process promotes the corrosion reaction, which explains the mutation of the OCP. After unloading the load, the OCP of both samples increases rapidly and gradually becomes stable because the metal at the abrasion mark being exposed to the solution promotes the formation of a new passive layer. The presence of defects at the wear marks decreases the OCP to a lower level than it was before the load. It can be observed that the potential of the implanted sample becomes stable faster and is higher than that of the substrate, which indicates that Nb can still improve the corrosion behavior of the substrate after the wear behavior. The OCP of the implanted samples is higher than that of the substrate before, during, and after the wear process, indicating that the implantation of Nb ions can reduce the corrosion sensitivity and improve corrosion resistance.

The morphology and element composition of the wear scar were tested and shown in Figure 6. A certain number of grooves were seen on the wear scar because the wear debris are repeatedly rolled and embedded into the substrate surface. This is a typical feature of abrasive wear. Many cracks and pitting pits were observed on the surface of the wear scar, indicating that corrosion wear occurred. Where cracks and pitting occur, there is a high corrosion tendency [29]. The elemental composition in Figure 6c shows that the surface of the wear scar is mainly composed of elements Ti, O and Cl. When the oxide film ruptures during tribocorrosion, Cl ions diffuse easily along the defect and thus form soluble chlorine salts. In addition, the corrosion effect of Ringer's solution accelerates the initiation and propagation of cracks and finally leads to the shedding of wear debris, thus forming pitting pits, following Equation (1):

$$nMe(O^{2-},2H^{+}) + xCl- \rightarrow MeCl_x + nH_2O \quad (1)$$

where $Me(O^{2-}, 2H^{+})$ represents the metal oxide film in solution and $MeCl_x$ is the soluble chlorine salt. In addition, the heat generated in the wear process will promote the oxidation reaction. Brittle oxide film is more likely to break under the action of wear, and its repeated breaking and repair accelerate the wear of the substrate.

Compared to the substrate, Figure 6d–f show that the cracks and pitting pits in the wear scar are obviously less than that of the substrate, indicating a lower pitting tendency; however, the depth of the wear scar is greater than that of the substrate. One possible reason is that local softening of the Nb-implanted sample reduced the elastic and plastic deformation resistance, thus the wear debris has a more obvious ploughing effect on samples. Compared to the substrate, the Nb-implanted sample has a decreased content of O in the wear scar, and this may be because the faster rate of oxidation film formation increases the wear in tribocorrosion. In addition, a small amount of the Nb element was observed in the wear scar, and its protective effect for the substrate is not completely lost as it explains the higher OCP of the implanted samples than that of the substrate after tribocorrosion. Figure 7 shows the three-dimensional contours and grinding crack profile of the sample. In addition, the calculation of wear volume shown in Table 3 reveals that the wear rate of the implanted sample is about 1.35 times that of the substrate, because the local softening effect of the implanted sample increases the wear debris during the wear process. Moreover, the Nb element accelerates the formation rate of the brittle passivation film, which is easy to break under the load, increasing the wear rate.

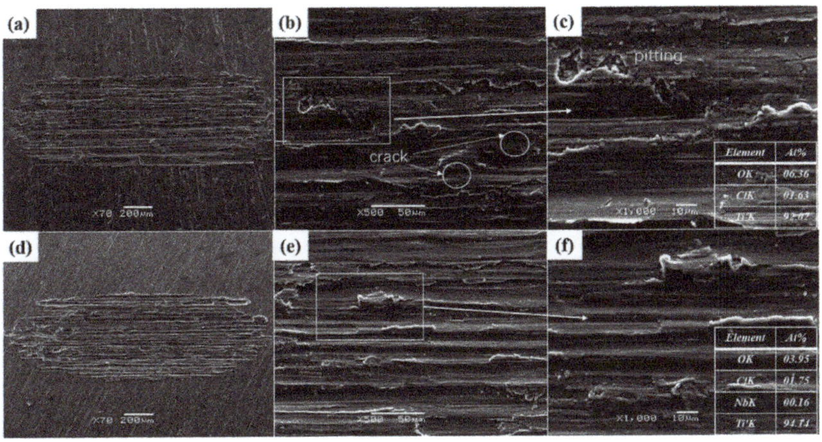

Figure 6. The worn surface and EDS analysis of samples: (**a–c**) substrate; (**d–f**) Nb-implanted sample.

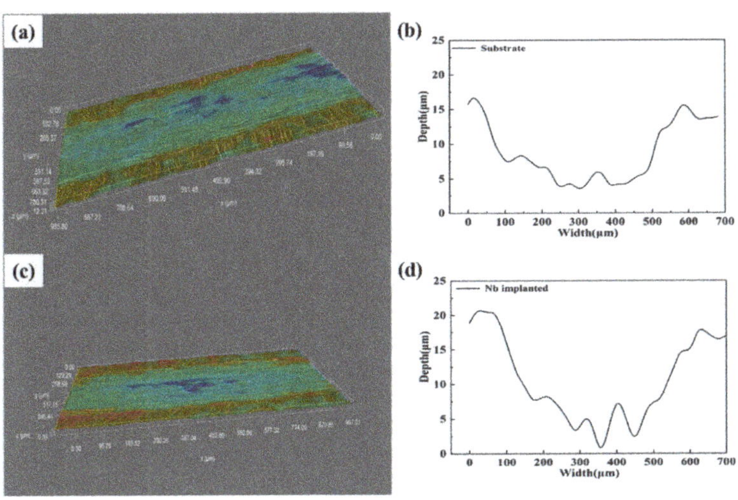

Figure 7. Three-dimensional contours and grinding crack profile of samples: (**a**,**b**) substrate; (**c**,**d**) Nb-implanted sample.

Table 3. Calculation results of wear scar of samples.

Sample	Width (μm)	Depth (μm)	Wear Volume ($\times 10^{-3}$ mm^3)	Wear Rate ($\times 10^{-4}$ $mm^3N^{-1}m^{-1}$)
Substrate	551.2	11.9	13.1	18.3
Nb implanted	539.2	16.5	17.8	24.8

4. Conclusions

In this research, the corrosion resistance and tribocorrosion resistance in Ringer's physiological solution of TA3 implanted by Nb ions were systematically investigated. The observations of this study are as follows:

The implantation led to pits on the surface caused by irradiation damage and cascade collision, but it also made the surface smoother. The XRD patterns showed that Nb ions mainly exist in the form of a solid solution. The nanoindentation tests showed that elastic modulus and abrasion decrease because of the irradiation effect during implantation.

The potentiodynamic polarization curves revealed that the implanted samples indicated a higher corrosion potential and a lower current density, indicating that Nb ions can effectively improve the thermodynamic and kinetic stability of the passivation layer and promote the formation rate of passivation film.

The tribocorrosion tests showed that the ion implanted samples show better thermodynamic stability and can still display better corrosion behavior after the wear behavior. The morphology of the wear scar showed that the Nb-implanted sample has a lower pitting tendency, but due to the surface's partial softening and brittle passivation film, the abrasive wear and oxidation wear degree of the sample increased.

Author Contributions: Conceptualization, P.Z., C.X. and H.H.; methodology, H.H., C.X. and F.L.; validation, P.Z., D.W. and H.H.; formal analysis, P.Z., D.W., H.H. and C.X.; investigation, H.H., C.X., F.L. and K.Y.; resources, P.Z. and D.W.; data curation, P.Z., C.X. and H.H.; writing—original draft preparation, P.Z., C.X. and H.H.; writing—review and editing, P.Z., C.X. and H.H.; visualization, C.X. and H.H.; supervision, P.Z.; project administration, P.Z.; All authors have read and agreed to the published version of the manuscript.

Funding: This project was supported by the Natural Science Foundation for Excellent Young Scientists of Jiangsu Province, China (Grant No. BK20180068), the China Postdoctoral Science Foundation (Grant No. 2018M630555), Opening Project of Materials Preparation and Protection for Harsh Environment Key Laboratory of Ministry of Industry and Information Technology (Grant No. XCA20013-1).

Conflicts of Interest: The authors declare no conflict of interest.

References

1. Yang, K.; Zhou, C.; Fan, H.; Fan, Y.; Jiang, Q.; Song, P.; Fan, H.; Chen, Y.; Zhang, X. Bio-Functional Design, Application and Trends in Metallic Biomaterials. *Int. J. Mol. Sci.* **2017**, *19*, 24. [CrossRef] [PubMed]
2. Turcio-Ortega, D.; Rodil, S.E.; Muhl, S. Corrosion behavior of amorphous carbon deposit in 0.89% NaCl by electrochemical impedance spectroscopy. *Diam. Relat. Mater.* **2009**, *18*, 1360–1368. [CrossRef]
3. Alves, A.C.; Sendão, I.; Ariza, E.; Toptan, F.; Ponthiaux, P.; Pinto, A.M.P. Corrosion behaviour of porous Ti intended for biomedical applications. *J. Porous Mater.* **2016**, *23*, 1261–1268. [CrossRef]
4. Revie, R.W. *Uhlig's Corrosion Handbook*, 2nd ed.; Anti Corrosion Methods & Materials; Emerald Group Publishing Limited: Bingley, UK, 2000; Volume 47.
5. Tamilselvi, S.; Raman, V.; Rajendran, N. Corrosion behaviour of Ti–6Al–7Nb and Ti–6Al–4V ELI alloys in the simulated body fluid solution by electrochemical impedance spectroscopy. *Electrochim. Acta* **2006**, *52*, 839–846. [CrossRef]
6. Fojt, J. Ti–6Al–4V alloy surface modification for medical applications. *Appl. Surf. Sci.* **2012**, *262*, 163–167. [CrossRef]

7. Choe, H.-C.; Saji, V.S.; Ko, Y.-M. Mechanical properties and corrosion resistance of low rigidity quaternary titanium alloy for biomedical applications. *Trans. Nonferrous Met. Soc. China* **2009**, *19*, 862–865. [CrossRef]
8. Atapour, M.; Pilchak, A.L.; Frankel, G.S.; Williams, J. Corrosion behavior of β titanium alloys for biomedical applications. *Mater. Sci. Eng. C* **2011**, *31*, 885–891. [CrossRef]
9. Wu, G.; Li, P.; Feng, H.; Zhang, X.; Chu, P.K. Engineering and functionalization of biomaterials via surface modification. *J. Mater. Chem. B* **2015**, *3*, 2024–2042. [CrossRef]
10. Liu, Y.; Zu, X.; Tang, R.; Xiang, X.; Wang, L.; Ma, W. Role of tantalum and nickel ion implantation on corrosion resistance properties of a Ti–Al–Zr alloy. *Surf. Coat. Technol.* **2007**, *201*, 7538–7543. [CrossRef]
11. Thair, L.; Mudali, U.; Bhuvaneswaran, N.; Nair, K.; Asokamani, R.; Raj, B. Nitrogen ion implantation and in vitro corrosion behavior of as-cast Ti–6Al–7Nb alloy. *Corros. Sci.* **2002**, *44*, 2439–2457. [CrossRef]
12. Shypylenko, A.; Pshyk, A.V.; Grześkowiak, B.; Medjanik, K.; Peplinska, B.; Oyoshi, K.; Pogrebnjak, A.; Jurga, S.; Coy, E. Effect of ion implantation on the physical and mechanical properties of Ti-Si-N multifunctional coatings for biomedical applications. *Mater. Des.* **2016**, *110*, 821–829. [CrossRef]
13. Yan, B.; Tan, J.; Wang, D.; Qiu, J.; Liu, X. Surface alloyed Ti–Zr layer constructed on titanium by Zr ion implantation for improving physicochemical and osteogenic properties. *Prog. Nat. Sci.* **2020**. [CrossRef]
14. Cisternas, M.; Bhuyan, H.; Retamal, M.; Casanova-Morales, N.; Favre, M.; Volkmann, U.G.; Saikia, P.; Diaz-Droguett, D.; Mändl, S.; Manova, D.; et al. Study of nitrogen implantation in Ti surface using plasma immersion ion implantation & deposition technique as biocompatible substrate for artificial membranes. *Mater. Sci. Eng. C* **2020**, *113*, 111002. [CrossRef]
15. Zhao, T.; Li, Y.; Xiang, Y.; Zhao, X.; Zhang, T. Surface characteristics, nano-indentation and corrosion behavior of Nb implanted NiTi alloy. *Surf. Coatings Technol.* **2011**, *205*, 4404–4410. [CrossRef]
16. Zu, X.; Liu, Y.; Lian, J.; Liu, H.; Wang, Y.; Wang, L.; Ewing, R. Surface modification of a Ti–Al–Zr alloy by niobium ion implantation. *Surf. Coatings Technol.* **2006**, *201*, 3756–3760. [CrossRef]
17. Ji, P.; Li, B.; Chen, B.; Wang, F.; Ma, W.; Zhang, X.; Ma, M.; Liu, R. Effect of Nb addition on the stability and biological corrosion resistance of Ti-Zr alloy passivation films. *Corros. Sci.* **2020**, *170*, 108696. [CrossRef]
18. Wan, H.; Si, N.; Chen, K.; Wang, Q. Strain and structure order variation of pure aluminum due to helium irradiation. *RSC Adv.* **2015**, *5*, 75390–75394. [CrossRef]
19. Sumie, K.; Toyoda, N.; Yamada, I. Surface morphology and sputtering yield of SiO2 with oblique-incidence gas cluster ion beam. *Nucl. Instrum. Methods Phys. Res. Sect. B Beam Interact. Mater. Atoms* **2013**, *307*, 290–293. [CrossRef]
20. Wei, D.; Li, F.; Li, S.; Wang, S.; Ding, F.; Tian, T.; Chen, X.; Yao, Z. Effect of Cr ion implantation on surface morphology, lattice deformation, nanomechanical and fatigue behavior of TC18 alloy. *Appl. Surf. Sci.* **2020**, *506*. [CrossRef]
21. Li, W.; Li, D. Influence of surface morphology on corrosion and electronic behavior. *Acta Mater.* **2006**, *54*, 445–452. [CrossRef]
22. Hong, T.; Nagumo, M. Effect of surface roughness on early stages of pitting corrosion of Type 301 stainless steel. *Corros. Sci.* **1997**, *39*, 1665–1672. [CrossRef]
23. Chen, X.-H.; Zhang, P.; Wei, D.-B.; Huang, X.; Ding, F.; Li, F.-K.; Dai, X.-J.; Wang, Z.-Z. Correlation between crystal structure and mechanical performance of Cr-implanted 300M high-strength steel using X-ray diffraction method. *J. Iron Steel Res. Int.* **2019**, *26*, 1106–1116. [CrossRef]
24. Leyland, A.; Matthews, A. On the significance of the H/E ratio in wear control: A nanocomposite coating approach to optimised tribological behaviour. *Wear* **2000**, *246*, 1–11. [CrossRef]
25. Tähtinen, S.; Moilanen, P.; Singh, B.; Edwards, D. Tensile and fracture toughness properties of unirradiated and neutron irradiated titanium alloys. *J. Nucl. Mater.* **2002**, *307*, 416–420. [CrossRef]
26. Wu, G.; Zhang, X.; Zhao, Y.; Ibrahim, J.M.; Yuan, G.; Chu, P.K. Plasma modified Mg–Nd–Zn–Zr alloy with enhanced surface corrosion resistance. *Corros. Sci.* **2014**, *78*, 121–129. [CrossRef]
27. Liu, W.; Li, M.-C.; Luo, Q.; Fan, H.-Q.; Zhang, J.-Y.; Lu, H.-S.; Chou, K.-C.; Wang, X.-L.; Li, Q. Influence of alloyed magnesium on the microstructure and long-term corrosion behavior of hot-dip Al–Zn–Si coating in NaCl solution. *Corros. Sci.* **2016**, *104*, 217–226. [CrossRef]

28. Bayón, R.; Igartua, A.; González, J.; De Gopegui, U.R. Influence of the carbon content on the corrosion and tribocorrosion performance of Ti-DLC coatings for biomedical alloys. *Tribol. Int.* **2015**, *88*, 115–125. [CrossRef]
29. Dearnley, P.A.; Dahm, K.L.; Çimenoğlu, H. The corrosion–wear behaviour of thermally oxidised CP-Ti and Ti–6Al–4V. *Wear* **2004**, *256*, 469–479. [CrossRef]

applied sciences

MDPI

Article

Design of Experiment in the Milling Process of Aluminum Alloys in the Aerospace Industry

Aurel Mihail Țîțu [1,2,*], Andrei Victor Sandu [3,4,*], Alina Bianca Pop [5], Ștefan Țîțu [6], Dragos Nicolae Frățilă [7,*], Costel Ceocea [8] and Alexandru Boroiu [9]

1 Industrial Engineering and Management Department, Faculty of Engineering, "Lucian Blaga" University of Sibiu, 10 Victoriei Street, 550024 Sibiu, Romania
2 The Academy of Romanian Scientists, 54 Splaiul Independenței, Sector 5, 050085 Bucharest, Romania
3 Faculty of Materials Science and Engineering, Gheorghe Asachi Technical University, Blvd. D. Mangeron 71, 700050 Iasi, Romania
4 Romanian Inventors Forum, Str. Sf. P. Movila 3, 700089 Iasi, Romania
5 SC TechnoCAD SA, 72 Vasile AlecsandriStreet, 430351 Baia Mare, Romania; bianca.bontiu@gmail.com
6 The Oncology Institute "Prof. Dr. Ion Chiricuță" Cluj-Napoca, 34-36 Republicii Street, 400015 Cluj Napoca, Romania; stefan.titu@ymail.com
7 Faculty of Dental Medicine, Grigore T. Popa University of Medicine and Pharmacy, 16 University Str., 700115 Iasi, Romania
8 Department of Marketing and Management, The Faculty of Economic Sciences, "Vasile Alecsandri" University of Bacău, 157 Mărăști Street, 600115 Bacău, Romania; costelceocea@gmail.com
9 Automotive and Transport Department, Faculty of Mechanics and Technology, University of Piteşti, 1 Târgul din Vale Street, 110040 Pitești, Romania; alexandru.boroiu@upit.ro
* Correspondence: mihail.titu@ulbsibiu.ro (A.M.T.); sav@tuiasi.ro (A.V.S.); dragos_fratila2002@yahoo.com (D.N.F.)

Received: 26 August 2020; Accepted: 29 September 2020; Published: 4 October 2020

Abstract: For many years, surface has quality received a serious attention due to its influence on various mechanical properties. The main contribution made in this scientific paper is the performance of actual experiments, as well as the experimental processing obtained in order to develop a model for predicting the surface roughness based on the optimization of cutting parameters. The novelty of this paper is brought by the method of obtaining the regression equation of the surface roughness, resulted from a standard end-milling process (standard milling tools, standard milling parameters, recommended by the tool manufacturer, 3 axis CNC machine and standard vice), on aluminum alloy 7136 in temper T76511, through two statistical methods of data analysis. This material is used for the production of extruded parts and is poorly understood for the proposed line of research. This study's aim is to determine the surface roughness equation obtained by the milling of aluminum alloy 7136 in two ways: using Taguchi's experimental design once and the other, using the central composite design. The Taguchi method and the central composite design are used to develop an efficient mathematical model to predict the optimal level of certain processing parameters. Using ANOVA analysis, a comparative study of calculated and experimental surface roughness values is carried out. The initial characteristics (surface roughness) and the controlled factors (cutting speed, depth of cut and feed) are analyzed with the Minitab program. Finally, an analysis of the advantages and disadvantages of the two methods used is presented. This study has a great industrial application, since the main task of every manufacturer is to achieve a better quality of the final product with minimal processing time.

Keywords: aluminum alloy; end-milling process; Taguchi design; central composite design; design of experiments

1. State of the Art

Over the last decade, the use of the aluminum alloys in the manufacturing industry has grown due to the fact that these alloys have an extraordinary ability to combine their two properties: low weight and strength in a single material. Given this, it is imperative to assume knowledge of the machinability properties of these materials when it is desired to provide information for use in industry and by researchers. The aim is to offer the possibility to process these materials by making the best decisions in this regard.

In the automotive industry, aluminum is already well known due to its wide range of applicability [1]. In the aerospace industry, in the 1930s, the most frequently approached aluminum classes were mainly 2xxx, 7xxx and 6xxx [2,3].

The preference felt in this regard in the automotive and aerospace industries is justified by the high strength of these materials against to their low weight but also the fact that they often bring a degree of profitability in the sense that they can replace steel and cast iron in parts manufacturing. Due to the low weight of aluminum, the impact on the environment is also reduced. The supporting argument of this claim is the low energy consumption [3].

Other areas in which aluminum alloys find their applicability, concern the construction area; of course, the electrical, electromechanical, electronic and packaging industries. Aluminum alloys are also used with great success in the manufacture of nanostructures where their property of high mechanical strength and thermal stability is required. A good example of this is the 6061-T6 aluminum alloy [1].

Compared to steels, aluminum alloys have a third of their density and modulus of elasticity, but also a much higher thermal and electrical conductivity, as well as high corrosion resistance, high coefficient of friction, etc. [1].

According to Demir [4] and Singh [5], aluminum is widely used in the manufacture of motor vehicle parts during turns, because it is lightweight. It is desirable to produce high quality surface products as soon as possible.

The surface roughness of mechanical components plays a major role in the relationship between industrial production and cost. Surface roughness usually depends on the cutting parameters, such as: cutting speed, feed rate and cutting depth [6,7]. Proper selection of the control factors is particularly important for the manufacture of surface components and high strength in a short time. With the processing in recent years, much has been done to improve the quality and efficiency of the product. Many aspects of research have yet to be explored.

Ravindra Thamma tested different models to get the best machining parameters for aluminum 6061 [8]. Studies have shown that shaft speed, strength and nose radii have very good surface quality impact.

Somashekara [9] uses control factors such as cutting speed, feed rate and cutting depth to optimize surface roughness when machining Al 6351-T6 alloy with uncoated carbide tools. The Taguchi technique is used to optimize the process parameters and the reinforcement test has been identified to determine the main factors affecting the surface roughness. This test shows that cutting speed has a greater impact on surface roughness.

The effect of cutting parameters (spindle speed, feed rate, and cutting depth) on surface roughness and tool wear is investigated by Devkumar when turning 6061 aluminum alloys [10]. Many regression models were developed for responses and the adequacy of the developed models was tested at 95% confidence interval using variance analysis (ANOVA) method.

Ranganath [11] studied the parameters that influence the roughness during the machining of Al 6061 on CNC lathes. The Taguchi and ANOVA methods were used to study the experimental results. According to his study, the result is that the feed and the speed levels are the most important parameters.

Ali Abdullah [12] developed the cutting parameters: feed, spindle speed and depth of cut, using the Response Surface Methodology method to reduce the surface roughness and increase the material removal rate of Al6061 material during the turning process.

With the Taguchi and RSM technique, Alagarsamyet [13] has conducted the experiment by turning the 7075 aluminum alloy. The author presents an effective approach in order to optimize the cutting parameters (cutting speed, feed rate and feed per tooth) in order to minimize surface roughness and maximize material removal rate. For processing, he used TNMG 115 100 tungsten carbide tool. The author analyzed the performance characteristics in the turning process based on the orthogonal array, the signal/noise ratio and the ANOVA analysis. Finally, he was able to determine a mathematical model of the response surface [13].

Ezuwanizam [14] found that the optimal cutting factors for speed, feed and depth were the optimal life of the TiN coated tool on aluminum 6061. The results show that as the cutting parameter values increase, the tool wear decreases during the final milling. However, the axial cutting depth does not affect the response in the same way as the cutting speed and feed ratio do.

The research in this article begins with the first original contribution made by the authors for this work, namely some comparative graphs about the state of aluminum alloys in the world. The first graphs have been drawn and describe the aluminum alloys used as a percentage in the aerospace industry in the 2005–2019 period (Figure 1).

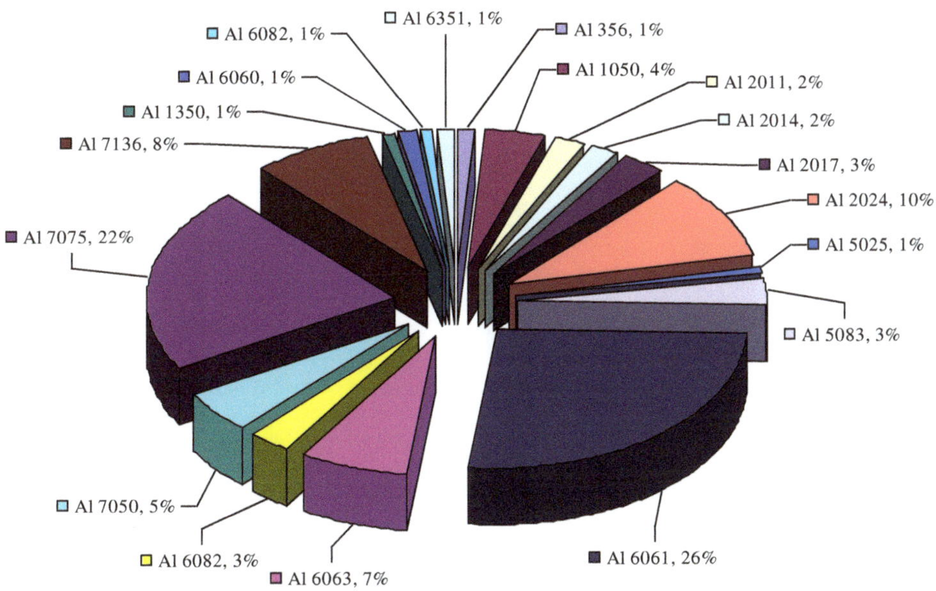

Figure 1. The use of aluminum alloys per year and depending on the number of researches in which they were studied.

As our own finding following the analysis of the graph made, it can be seen that the majority of Al6061 reaches 26%, followed by Al7075 at 22%. The aluminum alloy study is also used in the aviation industry. These alloys have been developed by a number of dedicated companies with the help of principal aircraft manufacturers to improve the production process of various aluminum areas.

This study will focus on Al7136, developed and manufactured by Universal Alloy Corporation. This type of material is used in the aircraft industry. This aluminum alloy, relatively newly developed, has superior properties to other aluminum alloys, among which include its high resistance to traction, also to wear, corrosion, its low value of thermal expansion, high durability, ductility and conductivity. All these properties make Al7136 a versatile material [15].

Another original comparative graph made by the authors is focused on the proportionality of the cutting factors impact exerted on the surface quality.

In Figure 2, this situation is presented. Concerning this case, it can be seen that the most studied parameter is the feed per tooth with 38%, closely followed by the cutting speed with 34% and the cutting depth with 28%.

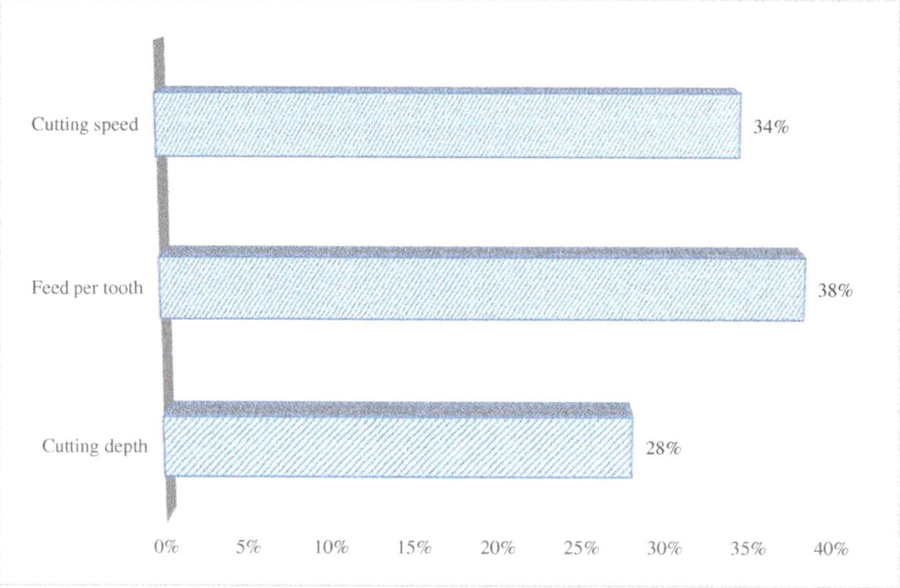

Figure 2. The most studied cutting parameters in the aluminum alloys machining in 2005–2019 period.

Other comparative original graphs are made by the authors on the most studied research directions on aluminum alloy machining (Figure 3), also on the study of the cutting operations frequency of the aluminum alloys (Figure 4), and finally the mathematical methods used to optimize the aluminum alloy cutting processes (Figure 5).

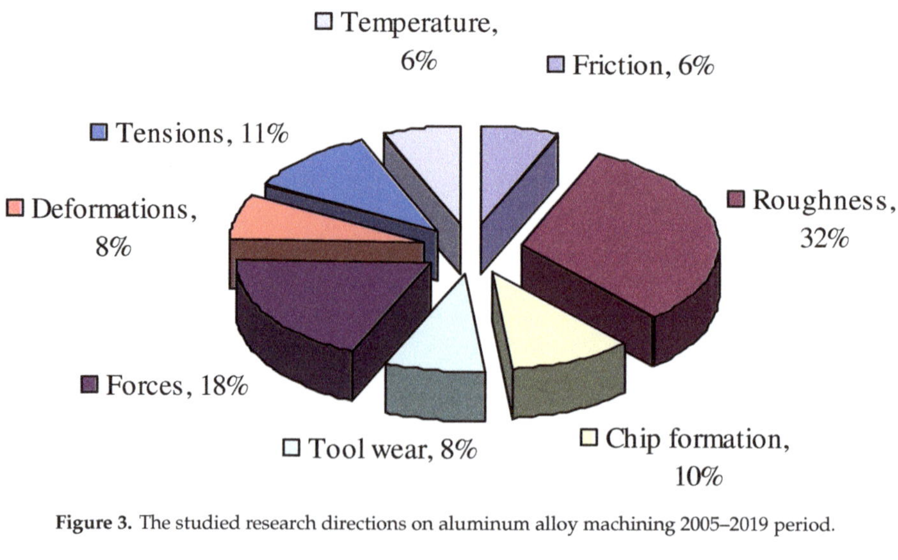

Figure 3. The studied research directions on aluminum alloy machining 2005–2019 period.

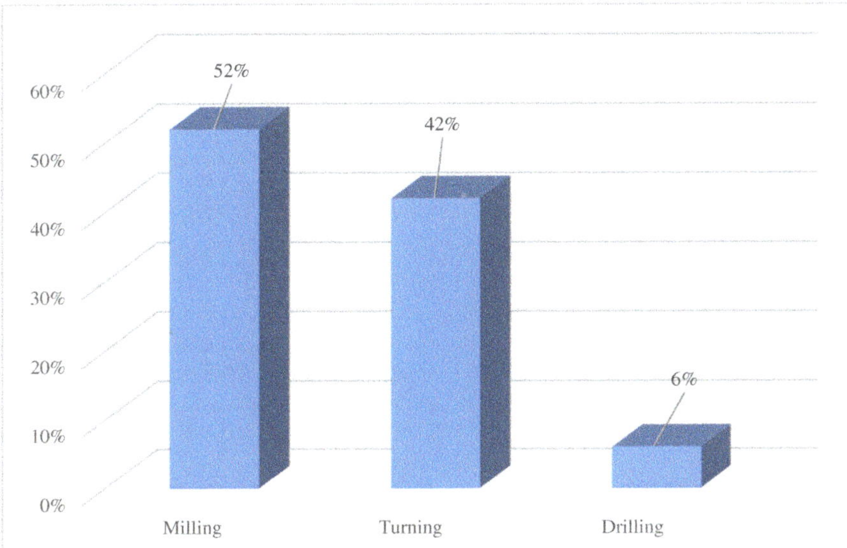

Figure 4. The frequency studying of the cutting operations of the aluminum alloys 2005–2019 period.

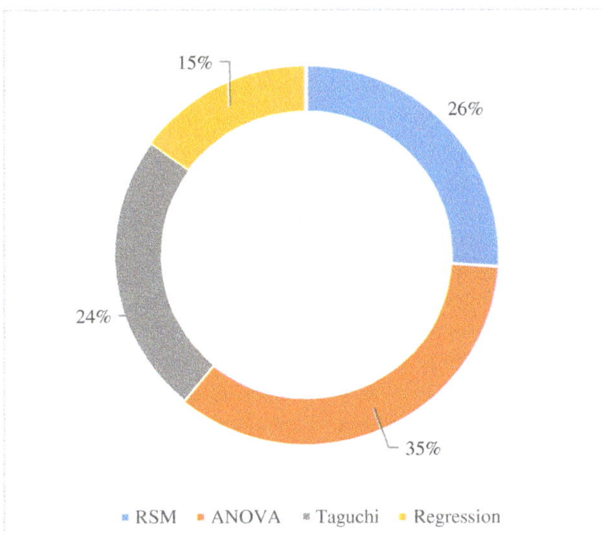

Figure 5. The mathematical methods used to optimize the aluminum alloy cutting processes 2005–2019 period.

As Figure 3 shows, the most studied research direction on aluminum alloy machining is represented by the surface roughness with 32%, followed by the cutting forces with 18%, then by the tensions with 11% and by the chip formation with 10%. The friction, temperature, deformation, and the tool wear were studied less than the above-mentioned parameters.

In the cutting operation case, the most studied machining operation on aluminum alloy is the milling operation with 52%, followed closely by turning with 42%.

Finally, the mathematical methods used to optimize the aluminum alloy cutting processes were analyzed, and the results show that the most studied methods are ANOVA with 35%, RSM with 26% and Taguchi with 24%.

We consider Figures 1–5 important because these graphs are created based on a serious topical documentary.

In this scientific paper, we use the Taguchi, the central composite design and ANOVA methods to obtain the optimal conditions for end-milling process of 7136 aluminum alloy used in the aerospace field. The aim is to obtain the contribution percentage of each parameter in order to confirm the optimal conditions obtained by using the Taguchi and the central composite design methods. Table 1 gives a brief presentation of the research papers in which this 7136 aluminum alloy was also studied.

Table 1. Identified research on Al7136 machining.

Machining Process	Aluminum Alloy	Process Parameters	Objectives	Optimization/Prediction Technique	Ref.
End-Milling	7136	cutting speed, cutting depth	Comparison between the evolution of the machined surfaces quality according to process parameters	Practical experiment	[15]
End-Milling	7136	cutting speed, feed per tooth, cutting depth	Comparison of the surface quality evolutions in different situations related to the cutting regimes resulting from the combination of the process parameters	Practical experiment	[16]
End-Milling	7136	cutting speed, feed per tooth, cutting depth	Optimization of the regression equation of the surface roughness	Taguchi	[17]
Milling	7136	cutting speed, feed per tooth, cutting depth	Determination of the cutting parameters influence, on the surface quality	Taguchi, full factorial design	[18]
Milling	7136	cutting speed, feed per tooth, cutting depth	Determination of the proper configuration of the optimum values of machining parameter and their interactions to obtain the better cutting process performance and to reduce the surface roughness sensitivity to uncontrollable factors	Taguchi	[19]
End-Milling	7136	cutting speed, feed per tooth, cutting depth	To reduce manufacturing costs and processing time. To identify the quantitative relationships between cutting parameters and the surface roughness.	Regression analysis, ANOVA	[20]

This research presents novelty elements in this field due to the experimental programs behind this applied research being permanently improved. Moreover, we consider that the approached subject is very topical, still being associated with the material sampling processes in various industries such as the aerospace industry. We consider that the study we propose represents a topical reference in our field of competence and is argued by topical experimental results that take into account a current stage of knowledge in the field that was the basis of the proposed research. Considering the purpose of this article, which is to determine the surface quality equation produced by Al7136 end-milling, based on the above comparisons, the full test method will be presented.

2. Experimental Procedure

Experiments were conducted to study the cutting effect of the measurement method: cutting speed, cutting depth and feed, applied to the product response: surface roughness.

2.1. Work Material

Tests are performed based on machining of Al7136-T76511 [21]. This alloy is used in aircraft industries due to its properties, such as:

- High strength to weight ratio;
- High wear resistance;
- Low thermal expansion;
- Corrosion resistance;
- Durability;
- Ductility;
- Conductivity, which makes it a versatile material [22].

The Al7136-T76511 is a type of aluminum alloy code 7136. To achieve the T76511 temper, the metal is heat-treated in solution, and stress relieved, after the natural stabilization of over aging. Helping with stress is done by stretching the metal to some degree, depending on the type of product being produced (rack or tube). Aging has been chosen to increase corrosion resistance. The metal is straightened after expansion. This temper stimulus is closely related to T76510, which does not allow such straightening [23].

The chemical composition of the Al7136 conforms to the percentages by weight shown in Table 2, determined in accordance with AMS2355 Standard [24].

According to AMS4415A Standard [23] of the aerospace material specification, the Al7136-T76511 is an aluminum alloy in the form of extruded bars, rods, wire, profiles (shapes) and tubing. In general, these extrusions are found in structural applications that require a combination of high tensile and compressive strength and good corrosion resistance, but the use of these extrusions is not limited to such applications (Table 3).

Table 2. Al7136 composition according to AMS2355 Standard.

Element	min	max
Silicon	-	0.12
Iron	-	0.15
Copper	1.9	2.5
Manganese	-	0.05
Magnesium	1.8	2.5
Chromium	-	0.05
Zinc	8.4	94
Titanium	-	0.10
Zirconium	0.1	0.20
Other Elements, each	-	0.05
Other elements, total	-	0.15
Aluminum	remainder	

Table 3. Minimum tensile properties of Al7136.

Nominal Diameter or Least Thickness (Bars, Rods, Wire, Profiles) or Nominal Wall Thickness (Tubing) Millimeters	Tensile Strength MPa	Yield Strength at 0.2% Offset MPa	Elongation in 50 mm %	Elongation in 5D or $5.65\sqrt{A}$
1.00–6.300	621	593	7	-
6.30–12.50	627	600	7	-
12.50–50.00	634	607	-	6
50.00–80.00	627	607	-	7
80.00–100.00	621	593	-	7

Extrusions must be heat-treated with solution, stress relieved by stretching after treatment of the solution to produce a permanent set nominal value of 1.5 percent, but not less than 1 percent or more than 3 percent and exceeded by T76511 temper, according to AS1990 [25,26]. Extrusions in the T76511 temper can receive a minor recovery, after stretching, of a required quantity meet the tolerance requirements of 3.6.

Extrusions must comply with the following requirements, determined on the mill product according to AMS2355 [24]. To carry out the tests, the parts have the dimensions 500 × 101 × 24.5 mm: the central composite design and 100 mm × 35 mm and 30 mm: the Taguchi method. Each machined sample was performed according to the cutting procedures set based on the combination of the cutting regime in each methodology case.

2.2. Cutting Tool

The test was made with a basic tool for aluminum machining (16 mm cutting tool, 100% engagement, SECO R217.69-1616.0-09-2AN), which has two cutting inserts ISO code XOEX090308FR-E05, H15 [27]. The tool used in the experiment was a brand-new tool and the tool wear was not considered in the study.

2.3. CNC Machine

The machine used for testing is a 3-axis HAAS VF2 CNC. The specifications of the CNC machine are shown in Table 4.

Table 4. HAAS VF-2YT specifications [28].

Travels	Metric
X Axis	762 mm
Y Axis	508 mm
Z Axis	508 mm
Spindle Nose to Table (~max)	610 mm
Spindle Nose to Table (~min)	102 mm
TABLE	METRIC
Length	914 mm
Width	457 mm
T-Slot Width	16 mm
T-Slot Center Distance	125.0 mm
Number of Std T-Slots	3
Max Weight on Table (evenly distributed)	1361 kg
SPINDLE	METRIC
Max Rating	22.4 kW
Max Speed	8100 rpm
Max Torque	122 Nm @ 2000 rpm
Drive System	Inline Direct-Drive
Max Torque w/opt Gearbox	339 Nm @ 450 rpm

Table 4. *Cont.*

Travels	Metric
Taper	CT or BT 40
Bearing Lubrication	Air/Oil Injection
Cooling	Liquid Cooled
FEEDRATES	METRIC
Rapids on X	25.4 m/min
Rapids on Y	25.4 m/min
Rapids on Z	25.4 m/min
Max Cutting	16.5 m/min
AXIS MOTORS	METRIC
Max Thrust X	11343 N
Max Thrust Y	11343 N
Max Thrust Z	18683 N
TOOL CHANGER	METRIC
Type	Carousel (SMTC Optional)
Capacity	20
Max Tool Diameter (full)	89 mm
Max Tool Weight	5.4 kg
GENERAL	METRIC
Air Required	113 L/min, 6.9 bar
Coolant Capacity	208 L

The sample was fixed with one clamp on the CNC table (for the Taguchi methodology) three clamps (for the central composite design) respectively, to obtain the rigidity. In this respect, the samples are parallel with the CNC table and perpendicular to the main shaft.

Throughout the tests, large quantities of Blasocut BC 35 Kombi SW coolant were used in the cutting areas. Among the characteristics of the coolant is the pump pressure, which had a value of 8 bar.

2.4. Response

This paper's goal is to use the Taguchi method and central composite design to analyze the cutting factors impact effects (cutting speed, cutting depth and feed) generated during the milling process on the 7136-aluminum surface. There are several ways to measure the surface quality of a part. Mathematical measurements (R_a) of the surface were obtained and measured in the center of the machine surface using the surface tester Mitutoyo SURFTEST SJ-210 (Figure 6).

The measurement tool resolution was set and tested with a certified gage. The measurement tool available at the time when the experiment was made had a tip diameter of 4 μm. The entire study was based on these data. The measurement tool was set to measure only 5 mm, there were seven measurements conducted in each case. The measurement error was not taken into consideration in order to prove the data reproducible. The sampling length was 2.5 mm with a number of sampling lengths × 7. The type of detector on retractable drive unit type was SJ-210 (4 mN type), the type of filtration used was Gaussian, and standard is ISO 1997, as can be seen in Figure 6. In this figure, it can be seen that the cut-off (λc) is equal to 0.8 mm.

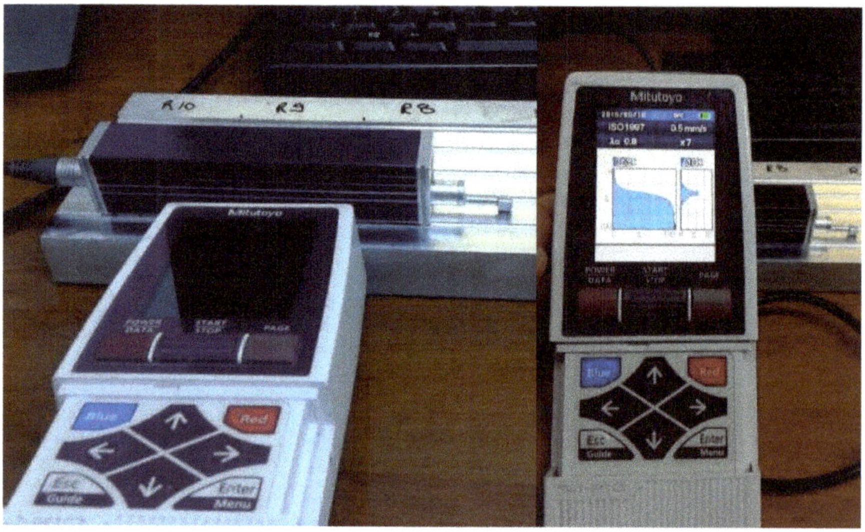

Figure 6. The measurements made with Mitutoyo SURFTEST SJ-210.

2.5. Process Variables and Their Parameters

In this article, cutting experiments were designed taking into account three cutting factors: cutting speed (m/min), feed (mm/tooth) and cutting depth (mm). Selected levels for cutting factors are shown in Table 5 and are in accordance with the SECO Equipment Manufacturer's Guide as well as regarding the machine capabilities and technology, cutting tools and CNC machine. In the Taguchi model, the selected levels reflect the minimum and maximum values of the machining factors, according to the test section. In the central composite design, each factor was assigned to the settled values in the experimental field. The original form of the central composite design was chosen—the circumscribed one. The star points are at an alpha distance from the center, based on the properties desired for the design and the number of factors in the design. The star points establish new extremes for the low and high settings for all factors.

Table 5. Cutting regime parameters and their values for Taguchi's method and central composite design.

Parameter	Taguchi's Method		Central Composite Design
	Value 1	Value 2	Values
Cutting Speed v (m/min)	495	660	495, 530, 570, 610, 660
Cutting Depth a_p (mm)	2	4	2, 2.5, 3, 3.5, 4
Feed per Tooth f_z (mm/tooth)	0.04	0.14	0.04, 0.06, 0.08, 0.11, 0.14

The surface roughness was determined based on the established parameters as shown in Table 5. For each experiment three surface measurements were performed and then the average of the mean square deviation (R_a med) calculation was used, indicating the mathematical mean of the three measurement lines.

3. Taguchi Design versus Central Composite Design

The design of the experiment is a very strong analysis tool with which it can be modeled and analyzed the influence of the control parameters, which is exerted on the followed response. The standard models are difficult to use, especially on a large number of experiments and when the control factors are varied and numerous [29,30].

This research uses Taguchi methodology and the central composite design to analyze the surface roughness by 7136 aluminum alloy machining as an output variable in end-milling process. The Taguchi method is an effective tool that is widely accepted by industrial engineers to approach the highest possible production at the right price and at the right time. The factors influencing the response are arranged in the form of a lattice square model, this pattern is called an orthogonal array. The design and selection of the orthogonal structure and the reasons for the experiments are the main foundations of the Taguchi process [31].

Malvade and Nipanikar [32] noted that the Taguchi method is more focused on shaping the development of production processes to create high-quality output compared to statistical control systems that attempt to control factors affecting quality.

Moshat [33] demonstrated that the Taguchi method is one of the most effective design test tools to create a high-quality product, developed by Genichi Taguchi. The goals of the orthogonal Taguchi (OA) are a step that gives a small number of very balanced and simultaneous tests.

The controls for this test included: cutting speed, cutting depth and feeding of each tooth. A number of eight experiments, as shows the L8 orthogonal array (Table 6), were obtained by applying the design of experiment (DOE) steps. Following the analyzed output data, the optimal machining condition was deducted.

Table 6. The 8 experiments according to L8 (2^7) orthogonal array.

Parameters and Interactions		A -	B	AB	C	AC	BC
Parameter		Cutting Speed [m/min]	Cutting Depth [mm]		Feed per Tooth [mm/tooth]		
				Column			
		1	2	3	4	5	6
Experiment	1	1	1	1	1	1	1
	2	1	1	1	2	2	2
	3	1	2	2	1	1	2
	4	1	2	2	2	2	1
	5	2	1	2	1	2	1
	6	2	1	2	2	1	2
	7	2	2	1	1	2	2
	8	2	2	1	2	1	1

The approach of the problem from the central composite design method perspective consists of the three factors involved, which are taken into account with their corresponding values presented in Table 5, in order to determine their influence level exerted on the response.

The tests were performed respecting the mentioned cutting regimes of the process factors combinations. Minitab 17 design expert software is used to create the tests plan and analyze all responses in a mathematical format.

Overall, eight experiments were carried out for the Taguchi method, each with three R_a measurements and a total of 24 measurements (Table 7).

For central composite design results, 125 experiments and a total of 375 measurements were carried out (Table 8). To determine the surface roughness equation, the contribution of each parameter and their interactions must first be determined.

Table 7. R_a med measurements according to the Taguchi design.

No.	v (m/min)	a_p (mm)	f_z (mm/tooth)	R_a Med (μm)
1	495	2	0.04	0.235
2	495	2	0.14	0.25
3	660	4	0.04	0.565
4	660	4	0.14	0.561
5	495	4	0.04	0.342
6	495	4	0.14	0.27
7	660	2	0.04	0.549
8	660	2	0.14	0.723

Table 8. R_a med measurements according to central composite design [22], pag. 115.

No.	v (m/min)	a_p (mm)	f_z (mm/tooth)	R_a Med (μm)
1	495	2	0.04	0.235
2	495	2	0.06	0.254
3	495	2	0.08	0.266
4	495	2	0.11	0.325
5	495	2	0.14	0.250
6	495	2.5	0.04	0.234
7	495	2.5	0.06	0.247
8	495	2.5	0.08	0.240
9	495	2.5	0.11	0.303
10	495	2.5	0.14	0.322
11	495	3	0.04	0.226
120	660	3.5	0.14	0.529
121	660	4	0.04	0.565
122	660	4	0.06	0.590
123	660	4	0.08	0.526
124	660	4	0.11	0.550
125	660	4	0.14	0.561

3.1. Contribution of Parameters and Their Interactions

A good understanding of the various aspects of different conditions can be seen using the method called ANOVA [5,6]. The purpose of ANOVA is to determine the magnitude associated with each component in the target operation and to reduce the error. ANOVA has also selected the highest quality components from a wide range of options. ANOVA is used to determine which actual metrics affect specified values.

Using the Taguchi method and the central design of the ANOVA analysis, the inclusion and impact of each factor and their interaction with the surface current was determined. Table 9 shows the analysis of variance, using Minitab 17.

The confidence interval in ANOVA was considered to be 95%. In this table (Table 9), the ANOVA analysis revealed that the cutting speed has the highest influence and affects the surface in both cases: Taguchi's method and central composite design.

The reason why the cutting speed has the greatest influence and affects the surface roughness is because if it is not correlated with the rotation speed of the tool, there is a settlement of the processed material that generates tool vibrations and thus leads to higher roughness.

The main contributing factor is the cutting speed and the interaction between the cutting speed and the depth of cut. In this experiment, the effect of the other parameters as well as the interactions between them have a smaller impact on the surface roughness.

Table 9. The variance analysis (Taguchi's method and central composite design).

Source	DF (Degree of Freedom)		Contribution			
	Taguchi	Central Composite Design	Taguchi	p	Central Composite Design	p
Regression	7	7	99.85%	0.078	88.93%	0.000
Cutting Speed (A)	1	1	94.00%	0.270	85.53%	0.283
Cutting Depth (B)	1	1	0.76%	0.455	0.00%	0.017
Feed per Tooth (C)	1	1	0.02%	0.179	1.00%	0.193
A x B	1	1	2.44%	0.345	2.06%	0.003
A x C	1	1	1.43%	0.159	0.00%	0.154
B x C	1	1	0.26%	0.278	0.07%	0.148
A x B x C	1	1	0.95%	0.238	0.27%	0.094
Error	1	118	0.15%	-	11.07%	-
Total	8	125	100.00%	-	100.00%	-

3.2. The Regression Equation of the Surface Roughness

According to the obtained results obtains by Table 8, the R_a regression equation in both situations, the Taguchi method and central composite design, was determined. Using Minitab 17, the surface roughness equation for Taguchi method is (1):

$$R_a = 0.000444A - 0.0955B - 13.51C + 0.000256AB + 0.02708AC + 3.04BC - 0.00634ABC, \quad (1)$$

Using Minitab 17, the surface roughness equation (2) for central composite design (CCD) is:

$$R_a = 0.000393A - 0.340B\text{-}10.17C + 0.00800AB + 0.0199AC + 4.20BC, \quad (2)$$

The normal probability plot in both cases Taguchi and CCD, are presented in Figures 7 and 8. Using the Minitab software, the regression equations related to the R_a surface roughness were determined. The aim is to describe the relationship between the answer (R_a) and the terms in the model (cutting speed, depth of cut, feed per tooth). The regression equation is an algebraic representation of the regression line. In the regression equations, the numerical values represent the estimated coefficients for the terms in the model. These coefficients were calculated based on the statistical methods applied in this scientific research paper.

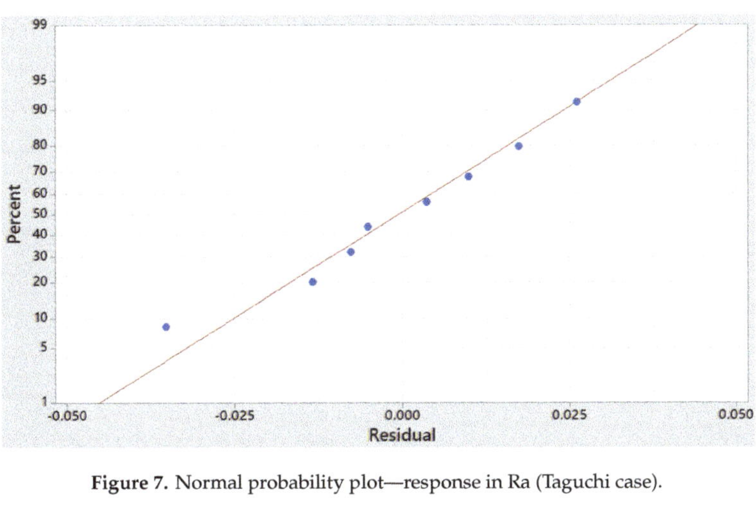

Figure 7. Normal probability plot—response in Ra (Taguchi case).

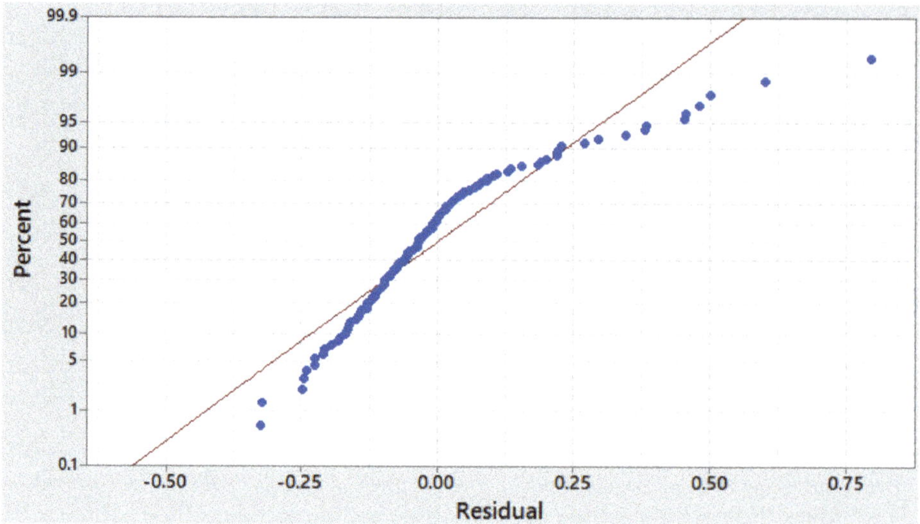

Figure 8. Normal probability plot—response in Ra (central composite design (CCD) case)).

3.3. Comparison Between the Two Regression Equations Determined

To make a comparison between the accuracy of these two determined equations using Taguchi's method and central composite design, Table 10 presents the surface roughness measured values. The surface roughness relative error was calculated between the measure's values and the calculated values of R_a in both situations (Taguchi's method and central composite design).

Table 10. The relative errors between the measured and the calculated surface roughness values [22], p. 178.

No	A	B	C	Ra (measured)	R_a Taguchi (calculated)	R_a Central Composite Design (calculated)	Relative Error εx[%]	
							εx Taguchi	εx Central Composite Design
1	495	2	0.04	0.235	0.270	0.288	15%	23%
2	495	2	0.06	0.254	0.264	0.279	4%	10%
3	495	2	0.08	0.266	0.258	0.269	3%	1%
4	495	2	0.11	0.325	0.249	0.256	23%	21%
5	495	2	0.14	0.250	0.240	0.242	4%	3%
6	495	2.5	0.04	0.234	0.284	0.315	21%	34%
7	495	2.5	0.06	0.247	0.277	0.305	12%	23%
8	495	2.5	0.08	0.240	0.270	0.295	12%	23%
9	495	2.5	0.11	0.303	0.259	0.280	14%	8%
10	495	2.5	0.14	0.322	0.249	0.265	23%	18%
11	495	3	0.04	0.226	0.297	0.341	32%	51%
121	660	4	0.04	0.565	0.578	0.891	2%	58%
122	660	4	0.06	0.590	0.574	0.830	3%	41%
123	660	4	0.08	0.526	0.570	0.770	8%	46%
124	660	4	0.11	0.550	0.563	0.679	2%	24%
125	660	4	0.14	0.561	0.557	0.589	1%	5%

4. Results and Discussions

After the regression equation determining for the R_a, the results were predicted at any area of the experimental domain, based on the determined regression models. Thus, the R_a values of the surface profile were determined using the obtained regression equations. The R_a calculated values results afferent to thc R_a measured values are shown in Table 10.

Starting with these calculations, the comparative graphical representations were made between the experimentally determined measurements and the calculated values using the regression equation determined by the Taguchi method and the central composite design (Figure 9).

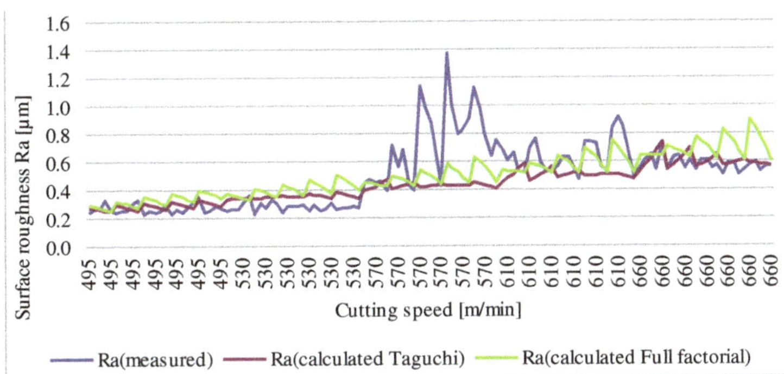

Figure 9. Comparison between the R_a measured versus R_a calculated using Taguchi's method and central composite design.

Figure 10 compares the relative errors obtained in these two situations. In the graphs in Figures 9 and 10, the evolution of the R_a values was presented, depending on the cutting speed, which is the most important factor in the surface roughness.

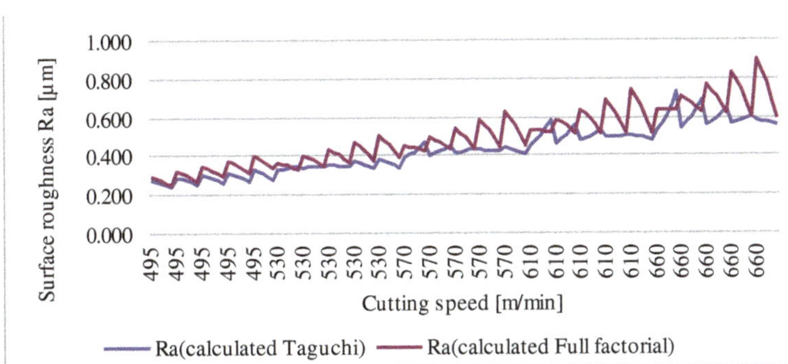

Figure 10. Comparison between the relative errors obtained (Taguchi's method versus central composite design).

Based on the graph analysis, it was found that the calculated arithmetic means deviations of the surface profile based on the regression equations approximate the experimental values with a total average error of 22% for the Taguchi's method and 27% for the central composite design.

Divided by cutting speed intervals, this error accumulates the averages of the following errors presented in Table 11.

Table 11. Average errors of the R_a measurements according to the cutting speed values.

Cutting Speed [m/min]	Medium Error Taguchi [%]	Medium Error Central Composite Design [%]
495	16%	26%
530	28%	48%
570	36%	28%
610	21%	10%
660	8%	23%

The error problem is generated by the values of 570 m/min and 610 m/min. At these cutting speeds, the average roughness of the area, measured by experiments, registers the critical distribution. The roughness value increasing was generated by the vibrations as a result of the machining. It was found that the poorest quality results at 570 m/min and 610 m/min cutting speed. The values of the surface roughness recorded at these speeds are quite high compared to those at lower speeds and upper speeds, respectively.

Vibration events lead to a rough surface. This vibration is the effect of a resonance phenomenon given by tool (piece) materials used. The vibration has not been measured during the experiment, but the statement is based on the fact that all other experiment conditions were respected on all combinations: the same CNC machine, the same setup, the same type of material (certified material by lab tests, mechanical and chemical test). The causes, consequences and production process have to be studied further.

Several papers dealing with this subject have been identified in the literature. For example, [34] in his paper presents the effects of the spindle attributed forced vibrations on the processing characteristics of the vertical milling process. The effects of three spindle levels attributed to forced vibrations together with feed rate and axial cutting depth are evaluated on surface roughness, dimensional accuracy and tool wear under constant conditions of radial cutting depth and cutting speed. He found that the vibration amplitude of machine tools and the axial cutting depth are statistically significant at a 95% confidence level for surface roughness, the vibration amplitude being the most important factor followed by the axial cutting depth. It is found that higher values of vibration amplitude and feed speed result in excessive tool wear, with the vibration amplitude combined with the feed rate and axial depth, resulting in a catastrophic damage to the tool.

Another research belongs to [35], who found that the profile milling processes are very sensitive to vibrations caused by the cutter leakage, especially when it comes to operations in which the diameter of the cut varies on a scale of a few millimeters. The aim of his research was to experimentally analyze the effect of cutter runout on cutter vibration and how it affects chip removal and therefore the topomorphy of the workpiece. He evaluated the effect of vibration phenomena, caused by cutter runout, on the workpiece topomorphy in end-milling.

Overall, Figure 11 shows the significant spatial variation related to cutting speed and feed per tooth. The graph analysis highlights that the speeds of 570 and 610 m/min in addition to a small feed showed an effect on R_a values, which increased in all experiments. We suppose that the vibration may be the reason for this increase. Figure 12 shows the diagrams of the constant level cycles, from which the gradual values increase with increasing cutting speed and feed per tooth value.

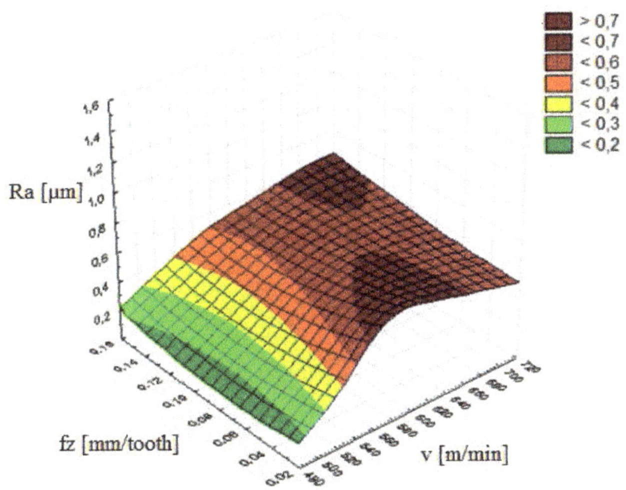

Figure 11. The spatial variation of R_a according to v and f_z.

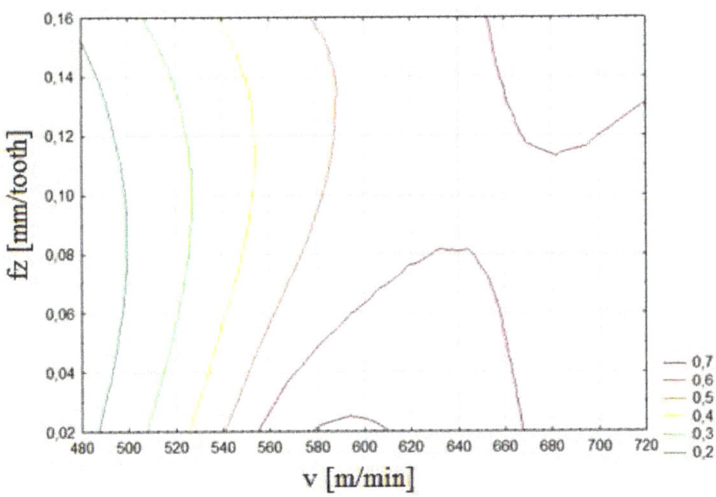

Figure 12. Indication of the spatial variation curves of R_a according to v and f_z.

5. Conclusions

This study's aim was to determine the R_a equations by dimensional process measurement methods in two ways: first using the Taguchi design of experiments and second using the central composite design.

Analysis of variance (ANOVA) is performed to identify key variables and clarify their impact on response characteristics.

The conclusion of both cases is the same: the cutting speed is the factor which has the highest effect on the end-milled surface. It can be seen by the percentage effect of each factor on the surface that there is a very weak correlation between the control factors affecting the surface quality.

The other two parameters, have a meaningless contribution (less than 1%) and can be forgotten.

By comparing relative errors between two measurement equations on the R_a scale it can be seen εx for the Taguchi method, the median value of relative errors is 22% and for the central composite design εx the relative error is 27%.

In general, the values derived from the central composite design are closer to the measured roughness.

The main R_a advantage of the Taguchi method is the reduction in the number of tests, lower costs and shorter time. The disadvantages include the fact that this method is not as good as the central composite design.

On the other hand, the advantage of the central composite design is the accurate comparison according to the Taguchi method. The disadvantages include the large number of tests, the cost of the tests, and the extra time spent on the tests.

It is difficult to say which method is the best. The selection belongs to the researcher and depends on the research experience, time and resources, as well as how accurate the results should be.

As a further research direction, there is the possibility of extending the studies and the research, taking into account other parameters and other indicators in a more complex form.

The vibration leads to the rough surface and is given by the resonance phenomenon of tool (piece) materials used. This can be a direction for further research because the causes, consequences and production process have to be studied further.

Author Contributions: Each author of this paper contributed equally to the research. Conceptualization and Supervision (A.M.T.); Data curation (A.B.) Formal analysis (A.B.P.); Validation (D.N.F.); Validation (C.C.); Visualization (S.T.); Writing—review & editing (A.V.S.), All authors have read and agreed to the published version of the manuscript.

Funding: This research received no external funding.

Conflicts of Interest: The authors declare no conflict of interest.

References

1. Santos, M.C.; Machado, A.R.; Barrozo, M.A.S.; Jackson, M.J.; Ezugwu, E.O. Multi-objective optimization of cutting conditions when turning aluminum alloys (1350-O and 7075-T6 grades) using genetic algorithm. *Int. J. Adv. Manuf. Technol.* **2015**, *76*, 1123. [CrossRef]
2. Tan, E.; Ögel, B. Influence of heat treatment on the mechanical properties of AA6066 alloy. *Turk. J. Eng. Environ. Sci.* **2007**, *31*, 53–60.
3. Troeger, L.P.; Starke, E.A.Jr. Microstructural and mechanical characterization of a superplastic 6xxx aluminum alloy. *Mater. Sci. Eng.* **2000**, *A 277*, 102–113. [CrossRef]
4. Demir, H.; Gündüz, S. The effects of aging on machinability of 6061 aluminium alloy. *J. Mater. Des.* **2009**, *30*, 1480–1483. [CrossRef]
5. Singh, D.P.; Mall, R.N. Optimization of surface roughness of aluminium by ANOVA based Taguchi method using Minitab15 software. *Int. J. Technol. Res. Eng.* **2015**, *2*, 2782–2787.
6. Aamir, M.; Tu, S.; Giasin, K.; Tolouei-Rad, M. Multi-hole simultaneous drilling of aluminium alloy: A preliminary study and evaluation against one-shot drilling process. *J. Mater. Res. Technol.* **2020**, *9*, 3994–4006. [CrossRef]
7. Aamir, M.; Tolouei-Rad, M.; Giasin, K.; Vafadar, A. Machinability of Al2024, Al6061, and Al5083 alloys using multi-hole simultaneous drilling approach. *J. Mater. Res. Technol.* **2020**, *9*, 10991–11002. [CrossRef]
8. Ravindra, T. Comparison between multiple regression models to study effect of turning parameters on the surface roughness. In Proceedings of the 2008 IAJC-IJME International Conference, Nashville, TN, USA, 17–19 November 2008; Volume 133.
9. Somashekara, H.M.; Swamy, N.L. Optimizing surface roughness in turning operation using taguchi technique and ANOVA. *Int. J. Eng. Sci. Technol.* **2012**, *4*, 1967–1973.
10. Devkumar, V.; Sreedhar, E.; Prabakaran, M.P. Optimization of machining parameters on AL 6061 Alloy using response surface methodology. *Int. J. Appl. Res.* **2015**, *1*, 1–4.
11. Ranganath, M.S.; Vipin, R.; Mishra, S. Optimization of process parameters in turning operation of aluminium (6061) with cemented carbide inserts using taguchi method and ANOVA. *Int. J. Adv. Res.* **2013**, *1*, 13–21.

12. Abdallah, A.; Bhuvenesh, R.; Embark, A. Optimization of cutting parameters for surface roughness in CNC turning machining with aluminum alloy 6061 material. *IOSR J. Eng.* **2014**, *4*, 1–10. [CrossRef]
13. Alagarsamy, S.V.; Rajakumar, N. Analysis of influence of turning process parameters on MRR & surface roughness of AA7075 using taguchi's method and RSM. *Int. J. Appl. Res. Stud.* **2014**, *3*, 1–8.
14. Ezuwanizam, M. Optimization of Tool Life in Miling. Ph.D. Thesis, Universiti Malaysia Pahang, Pahang, Malaysia, 2010.
15. Ţîţu, M.; Pop, A.B. A comparative analysis of the machined surfaces quality of an aluminum alloy according to the cutting speed and cutting depth variations. In *New Technoloies, Development and Application II, NT 2019, Lecture Notes in Networks and Systems*; Karabegović, I., Ed.; Springer: Cham, Switzerland, 2020; Volume 76. [CrossRef]
16. Pop, A.B.; Ţîţu, M. A comparative analysis of the machined surfaces quality of an aluminum alloy according to the cutting speed and feed per tooth variations. In *New Technoloies, Development and Application II, NT 2019, Lecture Notes in Networks and Systems*; Karabegović, I., Ed.; Springer: Cham, Switzerland, 2020; Volume 76. [CrossRef]
17. Pop, A.B.; Ţîţu, M.A. Optimization of the surface roughness equation obtained by Al7136 End-Milling. *MATEC Web Conf.* **2017**, *137*, 3011. [CrossRef]
18. Pop, A.B.; Ţîţu, M.A. Experimental research regarding the study of surface quality of aluminum alloys processed through milling. *MATEC Web Conf.* **2017**, *121*, 05005. [CrossRef]
19. Ţîţu, M.; Pop, A.B. Contribution on taguchi's method application on the surface roughness analysis in end milling process on 7136 aluminum alloy. *Mater. Sci. Eng.* **2016**, *1*, 012014. [CrossRef]
20. Pop, A.B.; Ţîţu, M. Regarding to the variance analysis of regression equation of the surface roughness obtained by end milling process of 7136 aluminum alloy. *IOP Conf. Ser. Mater. Sci. Eng.* **2016**, *161*, 012014. [CrossRef]
21. Bonţiu Pop, A.B.; Lobonţiu, M. The influence of feed per tooth variation on surface roughness of 7136 aluminum alloy in end milling. *Appl. Mech. Mater.* **2015**, *808*, 34–39. [CrossRef]
22. Bonţiu Pop, A.B. Surface Quality Analysis of Machined Aluminum Alloys Using End Mill Tool. Ph.D. Thesis, Technical University of Cluj-Napoca, Baia Mare, Romania, 2015.
23. *SAE Aerospace - AMS4451A, Aerospace Material Specification, Aluminum Alloy Extrusions 8.9Zn - 2.2Cu - 2.2Mg - 0.15Zr (7136-T76511, -T76510) Solution Heat Treated, Stress-Relieved, Straightened, and Overaged*; SAE International: Warrendale, PA, USA, 2008.
24. *AMS2355M—Quality Assurance, Sampling and Testing Aluminum Alloys and Magnesium Alloy Wrought Products (Except Forging Stock), and Rolled, Forged, or Flash Welded Rings*; SAE International: Warrendale, PA, USA, 2019.
25. *SAE Aerospace AS1990—Aluminum Alloy Tempers*; SAE International: Warrendale, PA, USA, 1990.
26. *DIN EN515 E, 2017—Aluminium and Aluminium Alloys—Wrought Products—Temper Designations*; German Institute for Standardisation (Deutsches Institut für Normung), Comite Europeen de Normalisation: Brussels, Belgium, 2017.
27. *Catalog and Technical Guide Seco, Milling* 2014, SECO. Available online: https://pdf.directindustry.com/pdf/seco-tools/mn-2014-milling/5699-261905.html (accessed on 8 December 2017).
28. Technical Book HAAS VF-2YT, Haas Technical Publications, Manual_Archive_Cover_Page Rev A June 6. 2013. Available online: https://www.haascnc.com/content/dam/haascnc/en/service/manual/service/english---vf-series-service-manual---2002.pdf (accessed on 8 December 2017).
29. Blasocut BC 35 Kombi SW. Available online: https://www.lastuamisnesteet.fi/wp-content/uploads/2017/02/Blasocut-BC-35-esite.pdf (accessed on 8 December 2017).
30. Hanif, I.; Aamir, M.; Ahmed, N.; Maqsood, S.; Muhammad, R.; Akhtar, R.; Hussain, I. Optimization of facing process by indigenously developed force dynamometer. *Int. J. Adv. Manuf. Technol.* **2019**, *100*, 1893–1905. [CrossRef]
31. Kumar, S. Optimization and process parameters of CNC end milling for aluminum alloy 6082. *Int. J. Innov. Eng. Res. Technol.* **2015**, *2*, 2394–3696.
32. Malvade, N.V.; Nipanikar, S.R. Optimization of cutting parameters of end milling on VMC using taguchi method. *J. Eng. Res. Stud.* **2014**, *5*, 14–17.
33. Moshat, S.; Datta, S.; Bandyopadhyay, A.; Pal, P. Optimization of CNC end milling process parameters using PCA-based Taguchi method. *Int. J. Eng. Sci. Technol.* **2010**, *2*, 95–102. [CrossRef]

34. Zahoor, S.; Mufti, N.A.; Saleem, M.Q.; Mughal, M.P.; Qureshi, M.A.M. Effect of machine tool's spindle forced vibrations on surface roughness, dimensional accuracy, and tool wear in vertical milling of AISI P20. *Int. J. Adv. Manuf. Technol.* **2017**, *89*, 3671–3679. [CrossRef]
35. David, C.; Sagris, D.; Stergianni, E.; Tsiafis, C.; Tsiafis, I. Experimental analysis of the effect of vibration phenomena on workpiece topomorphy due to cutter runout in end-milling process. *Machines* **2018**, *6*, 27. [CrossRef]

applied sciences

MDPI

Article

A Study on the Errors of 2D Circular Trajectories Generated on a 3D Printer

Adriana Munteanu [1], Dragos-Florin Chitariu [1,*], Mihaita Horodinca [1], Catalin-Gabriel Dumitras [1], Florin Negoescu [1], Anatolie Savin [2] and Florin Chifan [1]

1 Digital Manufacturing Systems Department, Gheorghe Asachi Technical University of Iasi, 700050 Iasi, Romania; adriana.munteanu@academic.tuiasi.ro (A.M.); mihaita.horodinca@academic.tuiasi.ro (M.H.); catalin-gabriel.dumitras@academic.tuiasi.ro (C.-G.D.); florin.negoescu@academic.tuiasi.ro (F.N.); florin.chifan@academic.tuiasi.ro (F.C.)

2 Veoneer, 700259 Iasi, Romania; anatolie.savin@yahoo.ro

* Correspondence: dragos.chitariu@tuiasi.ro

Abstract: This paper presents a study on the movement precision and accuracy of an extruder system related to the print bed on a 3D printer evaluated using the features of 2D circular trajectories generated by simultaneous displacement on *x* and *y*-axes. A computer-assisted experimental setup allows the sampling of displacement evolutions, measured with two non-contact optical sensors. Some processing procedures of the displacement signals are proposed in order to evaluate and to describe the circular trajectories errors (e.g., open and closed curves fitting, the detection of recurrent periodical patterns in *x* and *y*-motions, low pass numerical filtering, etc.). The description of these errors is suitable to certify that the 3D printer works correctly (keeping the characteristics declared by the manufacturer) for maintenance purpose sand, especially, for computer-aided correction of accuracy (e.g., by error compensation).

Keywords: 3D printer; circular trajectories; signals processing; errors evaluation

Citation: Munteanu, A.; Chitariu, D.-F.; Horodinca, M.; Dumitras, C.-G.; Negoescu, F.; Savin, A.; Chifan, F. A Study on the Errors of 2D Circular Trajectories Generated on a 3D Printer. *Appl. Sci.* **2021**, *11*, 11695. https://doi.org/10.3390/app112411695

Academic Editor: Gabi Chaushu

Received: 15 November 2021
Accepted: 6 December 2021
Published: 9 December 2021

1. Introduction

The additive fabrication is a common topic in various domains of activity (industry, biology, medicine). The compliance with precision conditions of 3D-printed parts (shape and dimensions tolerances, surface quality, etc.) becomes more and more a critical issue. Their use in some technical applications where precision and accuracy (P&A) are required is severely restricted since, for the present, other manufacturing technologies offer better results. Many studies in engineering and scientific research are focused on ensuring the P&A of 3D printers by error avoidance [1]. Other studies are involved in experimental research (measurements and data processing) in order to verify and to correct the errors generated by the lack of P&A on 3D printers by error compensation [1].

There are many issues involved in the appearance of errors in additive fabrication. Not surprisingly, some of them are not related with the 3D printer features (e.g., structure, P&A of kinematics, dynamics, position control, deposition process, temperature, etc.). For example, in [2] is established that the accuracy of STL (.stl) files (from Standard Tessellation Language or Standard Triangle Language, commonly used by Fused Deposition Modelling on 3D printers) essentially depends on the design of 3D CAD models (six different CAD systems generate STL files with different accuracies). In [3], is established that the conversion in STL format is done with errors by some CAD software products. The software interface used to drive the printer (slicer software, establishing the way the model is built) is often a source of inaccuracy [4] when inappropriate values of setting parameters are suggested by the software and accepted by user [5].

Sometimes the CAD models are generated with errors (and these errors are transferred to the printed object), especially when these models are generated as a virtual copy of a real object, e.g., by imaging, segmentation, and post processing of medical models [6,7] or by reverse engineering using a 3D-scanning and design process [8,9].

Some sources of errors are related to material deposition during additive manufacturing (e.g., flow properties) as susceptible to random variations [5]. The effect of print layer height on the accuracy of orthodontic models is considered in [10]. The effect of orientation (effect of gravity) and the effect of the support are investigated by [1]. A study on the influence of the accuracy related to size/position of printed benchmarks and the position of working planes on 3D printers is revealed in [11].The influence on P&A and mechanical properties (measured by tensile test) of a specimen object related to the print position of a specimen object is presented in [12,13].

The influence of different common printing technologies on the accuracy of mandibular models was considered in the research results shown in [14]. The influence of different types of thermoplastic filament materials used in additive fabrication on surface roughness was evaluated in [15].

The evolution in time of internal material tension, dimension and shape (by aging) as source of errors in 3D printing is a topic studied in [16]. A simple method to appreciate the P&A of 3D printers (and to calibrate it as well) is the measurement of a printed object (the most common being "#3D Benchy" from Creative Tools) or the quality of a printed structure (e.g.,"#All in one test 3D printer"). In [17], an insitu measurement method (by the scanning of layers) during the additive manufacturing process is proposed. The use of a coordinate measurement machine is proposed to describe the accuracy of medical models in [18,19]) and the optical scanning is mentioned in [6]). A metrology feedback procedure is used in [20,21] to improve the geometrical accuracy by errors compensation using a 3D scanning on sacrificial printed objects.

The displacement errors and the errors of the relative position of movable parts are often related by kinematics of a 3D printer. Frequently, the contouring errors due to axis misalignment (also a relevant topic, investigated in our study) are involved in a bad P&A. The measurement of these errors and their compensation in the computer numerical control system of the printer is a major challenge in improving the additive fabrication performances. Thus, in [22] is proposed a simple compensation model for kinematic errors based on measurement results done on the printer using a Renishaw QC10 ball bar device (from Renishaw, UK). For the 3D printers based on parallel robotic systems, in [23,24] some theoretical kinematic error models useful in automatic compensation are proposed. A volumetric experimental compensation technique for kinematic errors is exposed in [25]. In order to minimize the tracking errors of desired trajectories, a feed forward control procedure is proposed in [26].

This brief study of the literature reveals that few reports are focused on non-contact measurement methods of the errors produced by the kinematics of a 3D printer during complex motions of the extruder related to the print bed (especially 2D closed trajectories).

Our approach on this paper was based on this obvious remark: the open-loop computer-aided control of each of three motions of a printer (usually 3D printers don't use feedback control related to real position) is vulnerable to some uncontrollable (constant or variable) phenomena generated by mechanical parts (elastic deformations in toothed belts, mechanical backlashes, hysteresis behaviour, errors in lead screw threads, variation of friction forces in axes carriages, mechanical wear, structural vibrations induced by stepper motors, etc.). Because the 3D printing is achieved by deposition of material layer by layer (in a horizontal plane), an important characterization of P&A for any printer should be done by the P&A displacement, position and, especially, of 2D complex closed trajectories with the printer running in absence of the printing process. On this line of thinking we consider that a circular trajectory (of the extruder system related to the print bed) is one of the best theoretical approaches mainly because two simultaneous, strictly correlated linear motions are involved (on the x and y-axis). With a constant speed on the circular trajectory, these motions should be described by two almost identical harmonic evolutions (except the $\pi/2$ phase shift between), with periodic changes of the position, direction, speed and acceleration. An important argument for our approach based on circular trajectories is this one: the 2D circular trajectories are systematically involved in ISO standards methods for P&A evaluation of CNC manufacturing systems (ISO 230-4, [27]).

The benefits of this approach with 2D circular trajectories in P&A evaluation is confirmed by the work from [22].

In addition, from an experimental point of view, it is appropriate to work with circular trajectories generated by two simultaneous, simple harmonic linear motions (cosine motion of the extruder system on the x-axis, sine motion of the print bed on the y-axis) measured with a computer-assisted experimental setup based on two optical (non-contact) position sensors. We consider that this is a better approach, in contrast with the measurement system proposed in [22], which uses a ball bar measurement device placed between the extruder and the print bed. Because of this device, the relative motion of the extruder related to the print bed is not totally free.

The Section 2 of this paper presents the computer-aided experimental measurement setup, the Section 3 presents the theoretical and experimental considerations and comments related with the results (signal processing and P&A estimation, based mainly on real trajectory circular fitting), and the Section 4 is dedicated to conclusions and future work.

2. Experimental Setup

The experimental research was done on an Anet A8 3D printer [28], previously used in additive manufacturing for 230 operating hours. The 3D printer has the print bed movable on the y-axis, while the extruder system moves along the x and z-axes independently. On the x-axis, motion is numerically controlled (in open loop) via a stepper motor and a toothed belt (similarly on the y-axis); on the z-axis, the motion is controlled (in open loop) with two synchronized stepper motors with screw-nut systems. The absolute displacement measurement on the y-axis is done with a non-contact ILD 2000-20 [29] laser triangulation sensor (from MICRO-EPSILON MESSTECHNIK GmbH & Co, Ortenburg, Germany) with 20 mm measurement range (1 µm resolution and 10,000 s^{-1} sampling rate) with the laser beam placed perpendicularly on the target (the print bed) as Figure 1 indicates. Here the point of incidence is placed in the center of the red rectangle. An identical sensor is firmly placed on the print bed, with the laser beam placed perpendicularly to the extruder system (which moves as a sensor target in x-direction), as Figure 2 indicates.

Figure 1. A partial view of the setup with optical sensor on y-axis. 1-optical sensor; 2-print bed; 3-the extruder system; 4-toothed belt for y-axis displacement; 5-first screw for z-axis displacement.

Figure 2. A partial view of the setup with optical sensor on *x*-axis; 6-optical sensor; 7-toothed belt for *x*-axis displacement; 8-s screw for *z*-axis displacement.

The signals generated by these sensors are two voltages proportional with the displacements (with 1.975 mm/V as proportionality factor for the *x*-axis sensor and 2.0185 mm/V for the *y*-axis sensor) related to the middle of the measurement range.

These signals are simultaneously numerically described (by sampling and data acquisition)with a PicoScope 4824 numerical oscilloscope (from PicoTechnology UK, 8 channels, 12 bits, 80 MS/s maximum sampling rate, 256 MS memory) and delivered in numerical format to a computer for processing and analysis. Figure 3 presents a scheme of the computer-assisted experimental setup with only the movable parts of the 3D printer involved in 2D circular trajectories (the extruder and the print bed), the optical sensors for *x* and *y*-motion non-contact measurement, the power supply for sensors, the numerical oscilloscope and the computer.

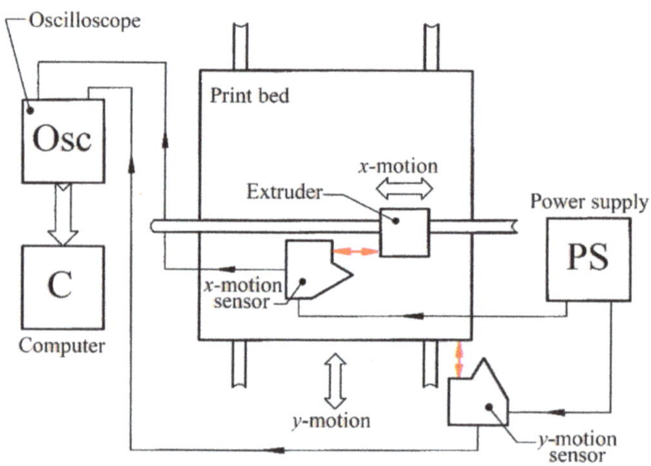

Figure 3. A scheme of the computer-assisted experimental setup.

Using an appropriate programming of the drive system written in G-code, the 3D printer was programmed to generate some identical repetitive circular trajectories of the extruder system related to the print bed (with 8.987 mm radius, each one covered in 8.987 s) for 50 s (for almost 5.5 complete trajectories). There are not special reasons to have this coincidence of values for radius and time, except the fact that these values should be accurately found by curve fitting (with the sine model) of the motions involved in the achieving of circular trajectory. However, these values assure a relatively small speed on the 2D circular trajectory. These trajectories are generated using two theoretical pure harmonic motions (x-motion accomplished by the extruder system, y-motion accomplished by the print bed) on the x and y-axes (both having a period T of 8.987 s), experimentally revealed by the computer measurement setup, as a detail in Figure 4 indicates. Here the blue-colored curve describes the motion on the x-axis; the one colored in red describes the motion on y-axis, both with 20,000 s^{-1} sampling rate. The choosing of this sampling rate is based on this argument: it should be at least two times bigger than the sampling rate of the sensors (10,000 s^{-1}). The sampling rate of the sensors acts as the Nyquist frequency for the sampling rate of the oscilloscope. It is not difficult to remark the resources of these two simultaneous evolutions for P&A evaluation. In Figure 5, a zoomed in detail in area A of Figure 4 proves that each of two real x and y-motions are not strictly pure, simple harmonic shapes (especially the y-motion). As a result, in a summary valuation, the trajectory of the extruder system related to the print bed does not have a strictly circular shape as expected.

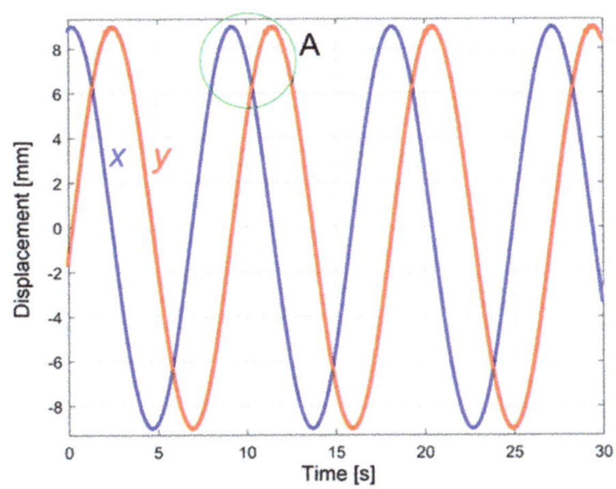

Figure 4. The evolution of x and y-motions during a repetitive circular trajectory.

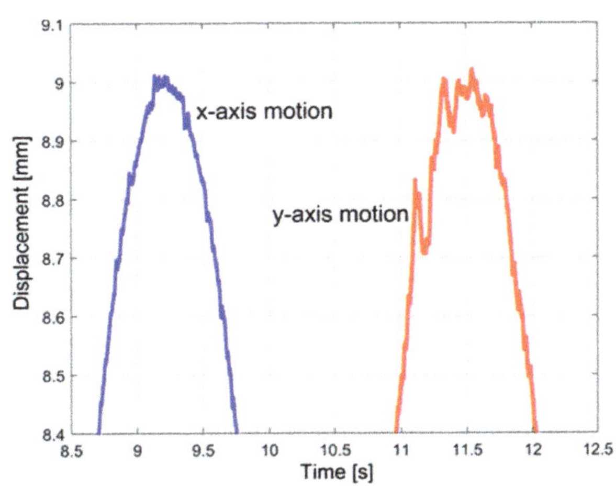

Figure 5. A zoomed in portion of Figure 4 (in area A).

These two evolutions are useful for P&A evaluation of the circular trajectories on a 3D printer in experimental terms. For this purpose, some different techniques of computer-aided signal processing will be applied (e.g., the recurrent periodical pattern detection on x and y-motion, curve fitting, circular fitting, low pass numerical filtering).

3. Experimental Results and Discussion

There are many important exploitable resources of x and y-motions description already partially revealed in Figures 4 and 5. As a first interesting approach, we propose the signals fitting, each one with a single harmonic function (fitting with a sine model). The Curve Fitting Tool from Matlab provides the best analytical approximation x_a and y_a of x and y-motion, as follows:

$$x_a(t) = 8.987 \cdot \sin(0.6991 \cdot t + 1.398) y_a(t) = 8.987 \cdot \sin(0.6991 \cdot t - 0.1883) \quad (1)$$

The fitting quality is confirmed for both motions by the same amplitude (8.987 mm, this being the radius of the circular trajectory) and angular frequency ω = 0.6991 rad/s (for a period $T = 2\pi/\omega$ = 8.987 s). These values found by curve fitting were already used in circular trajectory programming. However, there is a shift of phase by 1.5863 radians between motions ($1.5863 > \pi/2 = 1.5707$). This means that the x and y-axes are not rigorously perpendicular (there is an axes misalignment), as a first indicator for the lack of P&A. The printer works in a non-orthogonal $x0y$ system (with an angle of 90.893 degrees between axes). As a result, a programmed circular trajectory will be executed as an elliptical one. Nevertheless, this non-perpendicularity of x and y-axes revealed here can be compensated by programming. In a short definition, this compensation should solve this question: what kind of elliptical trajectory should be programmed in order to achieve a desired circular trajectory?

The evolution of residuals $x_r(t) = x(t) - x_a(t)$ and $y_r(t) = y(t) - y_a(t)$ from curve fitting of x and y-motions using a sine model are described in Figures 6 and 7 (with the same scale). The shape and the magnitude of residuals proves that x and y-motions are not perfectly harmonic (as expected), as a new indicator for the lack of P&A of circular trajectories.

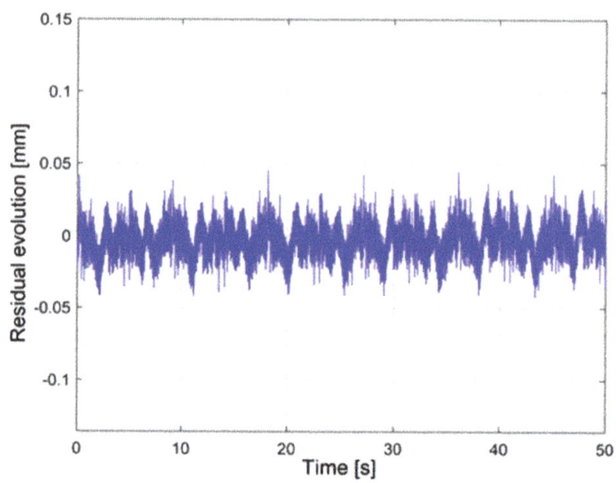

Figure 6. The evolution of residual x_r from the harmonic fitting model for x-motion.

The evolution of residuals from Figures 6 and 7 indicates that the negative effect on P&A of x-motion errors is smaller than of y-motion errors.

It is interesting to find out if there is a recurrent (or repeating)periodical pattern of the residual evolutions (x_r, y_r) on a complete period T of each motion (x, y). A simple way to check if there is a recurrent periodical pattern on x_r, and y_r evolutions is to build an average evolution of the residual (x_{Ar}, y_{Ar}) on a single period T with these definitions:

$$x_{Ar}(t) = \frac{1}{k}\sum_{i=0}^{k-1} x_r(t + i \cdot T) y_{Ar}(t) = \frac{1}{k}\sum_{i=0}^{k-1} y_r(t + i \cdot T) \quad (2)$$

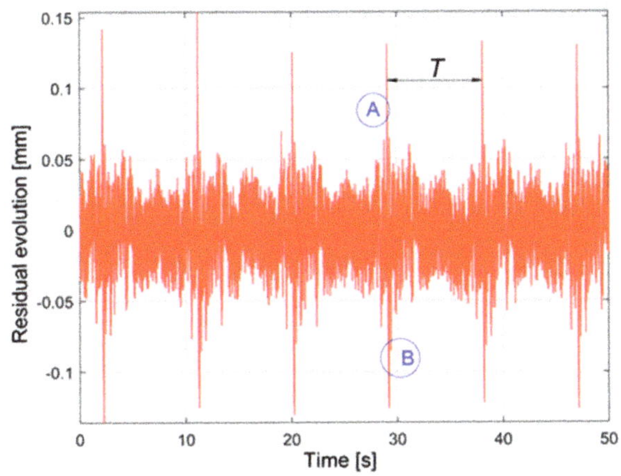

Figure 7. The evolution of residual y_r from the harmonic fitting model for y-motion.

Figure 8 presents the evolution of $x_{Ar}(t = 0 \div T$, with 179,740 samples) with $k = 5$ (for five completely circular trajectories), each sample of x_{Ar} being an average of k correlated samples of x_r. It is obvious that the x-motion has a well-defined repetitive periodical pattern, with systematic errors. In other words, x_r (and x-motion as well) is well correlated with itself.

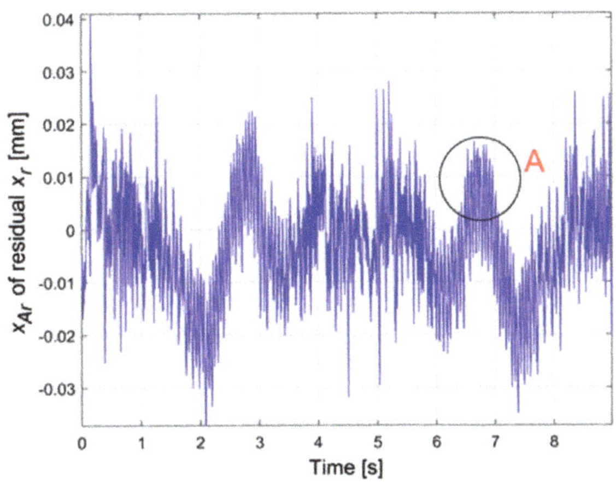

Figure 8. The average evolution x_{Ar} of the residual x_r for x-motion.

The curve fitting of x_{Ar} evolution using a sum of sine model delivers the analytical evolution x_{aAr} of x_{Ar} as Figure 9 indicates. In Equation (3) is depicted the analytical model for x_{aAr} as it follows:

$$x_{aAr}(t) = \sum_{j=1}^{n} a_{xj} \cdot \sin(b_{xj} \cdot t + c_{xj}) \quad (3)$$

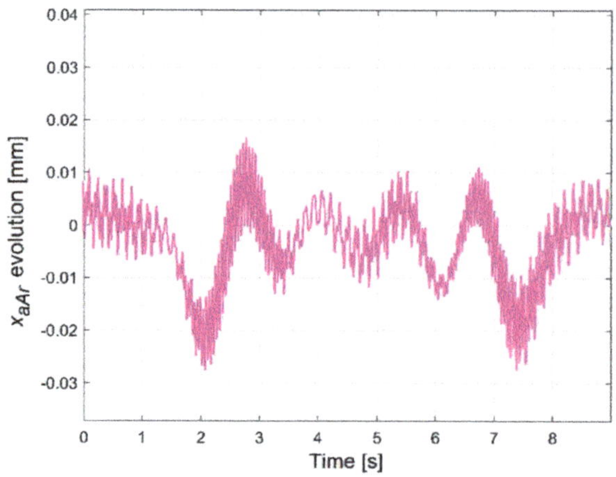

Figure 9. The evolution of an analytical model x_{aAr} (with n = 16) of the average residual x_{Ar}.

With n = 16, the values of a_{xj}, b_{xj} and c_{xj} involved in x_{aAr} model from Equation (3) are depicted in Table 1 (as results of x_{Ar} curve fitting).

Table 1. The values of a_{xj}, b_{xj} and c_{xj} involved in x_{aAr} model from Equation (3) with n = 16.

j	a_{xj}[mm]	b_{xj}[rad/s]	c_{xj}[rad]	j	a_{xj}[mm]	b_{xj}[rad/s]	c_{xj}[rad]
1	0.203	0.3379	1.969	**9**	0.001888	6.191	−2.542
2	0.004801	4.885	0.3713	**10**	0.002353	119.3	1.698
3	0.004398	1.383	1.758	**11**	0.1188	121.8	2.263
4	0.003755	2.083	1.377	**12**	0.001696	44.14	−3.301
5	0.004055	4.163	−2.608	**13**	0.00148	120.4	−0.3449
6	0.004279	3.47	−2.578	**14**	−0.1166	121.8	2.283
7	0.1981	0.3481	−1.225	**15**	0.001224	123.3	−2.064
8	0.001892	42.6	−2.541	**16**	0.0004954	118.6	−2.431

A better model x_{aAr} for x_{Ar} is available by increasing the value of n. There is not a total fit between x_{aAr} and x_{Ar}, mainly because the model is not able to describe the phenomena characterized by temporary variations of amplitudes.

On this subject, Figure 10 presents a detail with x_{aAr} and x_{Ar} evolutions located in the area A of Figure 8 with n = 105 (105 components in sum of sine model from Equation (3)). The strong variation of displacement depicted here is likely the result of structural vibrations of the 3D printer induced by the stepper motor.

The x_{Ar} and x_{aAr} evolutions are certain arguments that the experimental setup is able to describe the P&A of x-motion. Moreover, the analytical model from Equation (3) and Table 1 helps (at least in theoretical terms) to compensate for the errors of x-motion.

Figure 11 presents the evolution of y_{Ar} with k = 5 (for five completely circular trajectories, each sample of y_{Ar} being an average of k correlated samples of y_r). As with x-motion, it is obvious that the y-motion also has a well-defined periodic recurrent pattern (the same period T as x_{Ar}), having systematic errors. In other words, y_r (and y-motion as well) is well correlated with itself. Unfortunately, it was not possible to find an acceptable analytical model y_{aAr} with harmonic components (similar to the x_{aAr} model for x_{Ar} based on Equation (3)). A very high number of harmonic components (n) is necessary in this sum of sine model. A future approach intends to identify a more appropriate analytical model. There are strong repetitive irregularities on y-motion (and y_r and y_{Ar} as well) revealed in A, B areas of Figures 8 and 11. Likely they are irregularities generated by a suddenly releasing of a mechanical stress inside the toothed belt used to produce the y-motion.

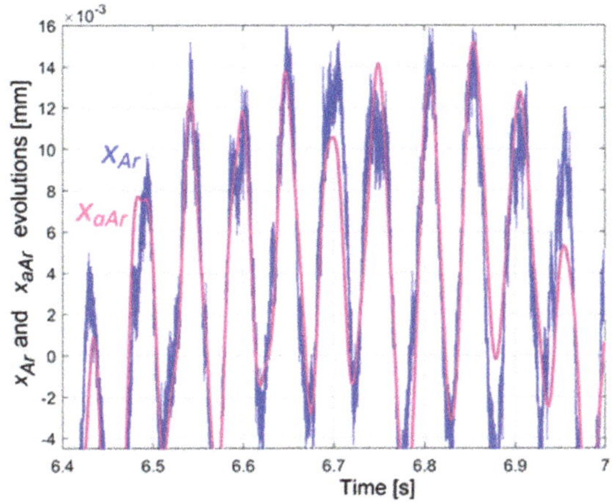

Figure 10. The evolution of x_{Ar} and an analytical model x_{aAr} (with n = 105) in area A of Figure 8.

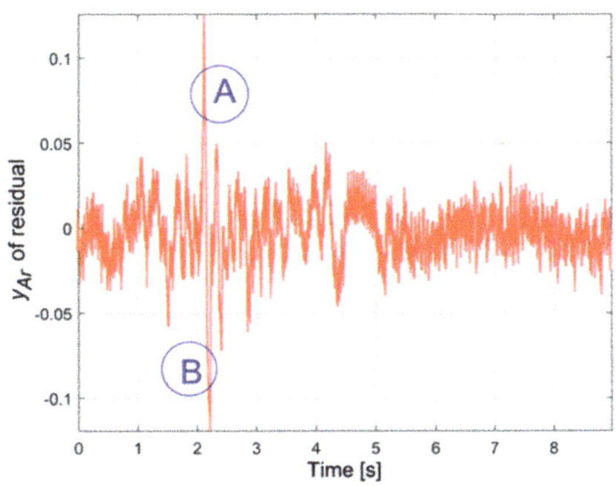

Figure 11. The average evolution y_{Ar} of the residual y_r for y-motion.

It is interesting now to examine the average trajectory generated by the 3D printer with $x_a(t) + x_{Ar}(t) = x_e(t)$ as x-motion and $y_a(t) + y_{Ar}(t) = y_e(t)$ as y-motion. It is obvious that this trajectory is not a perfect circle (at least because the wrong phase shift between x_a and y_a). A computer program was developed in order to find out the description of the least square circle (the coordinates x_c, y_c of center and the radius R_c as well) by circular fitting. This program is available for the fitting of any closed curve with known analytical description. The circular fitting supposes to find out the values x_c, y_c and R_c for which a fitting criterion ε described in Equation (4) reaches a minimum value.

$$\varepsilon = \sum_{i=1}^{N}\left[(x_{ei} - x_c)^2 + (y_{ei} - y_c)^2 - R_c^2\right] \quad (4)$$

In Equation (4), N is the number of samples x_{ei} or y_{ei} (N = 179,740) used for x_e-motion or y_e-motion description of the average trajectory. If the average trajectory is a perfect circle, then a perfect fitting produces a value ε = 0 for the fitting criterion. The circular fitting of average trajectory produces x_c = 0.00303 mm, y_c = 0.00252 mm and R_c = 8.9873 mm (this radius being very close to the amplitudes of x_a and y_a already revealed in Equation (1)). A first conventional graphical description of the circularity error of the average trajectory (as a first trajectory fitting residual, TFR_1) is available in polar coordinates (d_{i1}, α_{i1}), related to the least square circle, with d_{i1}, α_{i1} defined as:

$$d_{i1} = \left|\sqrt{(x_{ei} - x_c)^2 + (y_{ei} - y_c)^2} - R_c\right| \alpha_{i1} = \arctan^4\left(\frac{y_{ei} - y_c}{x_{ei} - x_c}\right) \quad (5)$$

Here d_{i1} is the distance from average trajectory to the least square circle, α_{i1} is the polar angle, with *arctan*^{4} the inverse of tangent function in four quadrants. This TFR_1 is also available in Cartesian coordinates as a curve described by a movable point having $d_{i1} \cdot cos(\alpha_{i1}) + x_c$ as abscissa and $d_{i1} \cdot sin(\alpha_{i1}) + y_c$ as ordinate. If the average trajectory is a perfect circle, then TFR_1 is a point placed in the center of the least square circle.

Figure 12 presents the TFR_1 of the average trajectory with a circular grid (with a 20 μm increment on radius). Here the maximum value of distance d_{i1} is 145.8 μm.

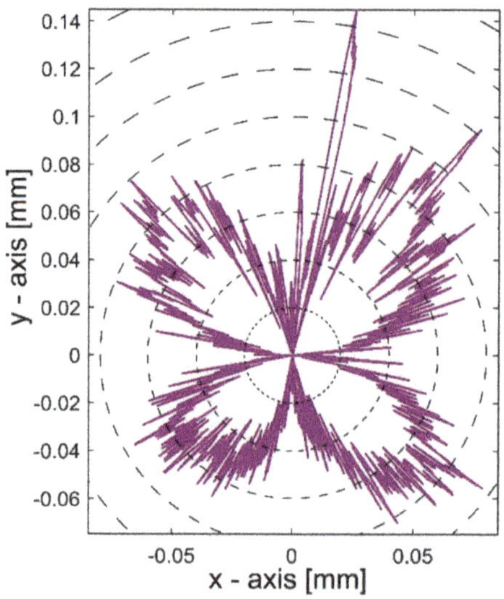

Figure 12. The evolution of TFR_1.

It is interesting to explain why TFR_1 from Figure 12 has four almost similar lobes. The dominant component (as amplitude) in x_e-motion is x_a, while the dominant component in y_e-motion is y_a (with x_a and y_a experimentally revealed by fitting and depicted in Equation (1) as pure harmonic motions).

As previously shown, these two components (having the same amplitude and angular frequency) are not rigorously shifted with $\pi/2$ (as expected).This means that the dominant part of the average trajectory (generated by x_a and y_a motions composition) is not a circle (as expected) but an ellipse. The circular fitting of an ellipse produces a least square circle which intersects the ellipse in four points and a TFR_1 with four lobes, according to the simulation results from Figure 13 (with an elliptical trajectory generated by two harmonic signals shifted with 2.1863 radians). A better explanation for these four lobes in Figure 12 is produced if over this figure is added the TFR_1 of the trajectory generated only by x_a and y_a-motions, with green color, as Figure 14 indicates (here both curves being traversed counter clockwise). A better approach to the shape of TFR_1 involved in Figure 14 is produced if the distances d_{i1} from Equation (5) are calculated related by a circle with smaller radius than the radius of the least square circle (R_c), as TFR_{1a}. As example, in TFR_{1a} from Figure 15, this radius is $R_c - 0.025$ μm.

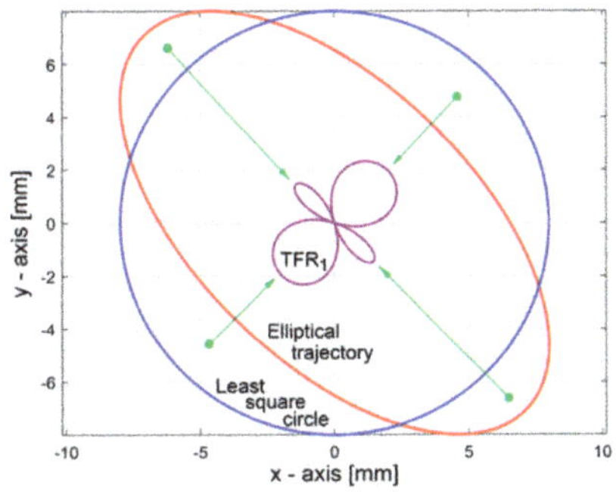

Figure 13. The evolution of TFR_1 generated by circular fitting on an elliptical trajectory (simulation).

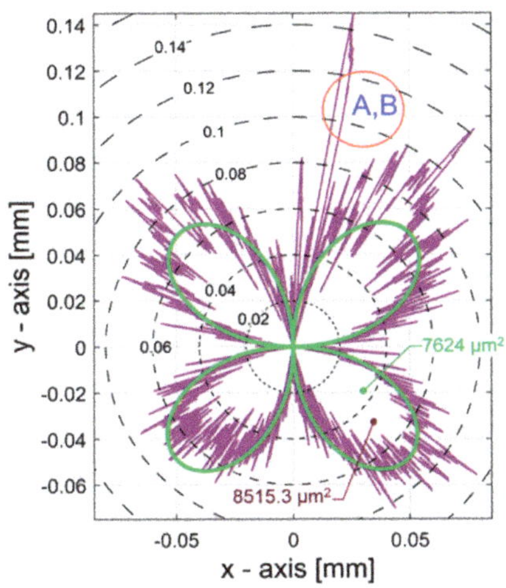

Figure 14. The evolution of TFR_1 generated by x_e and y_e and TFR_1 generated only by x_a and y_a (with green color).

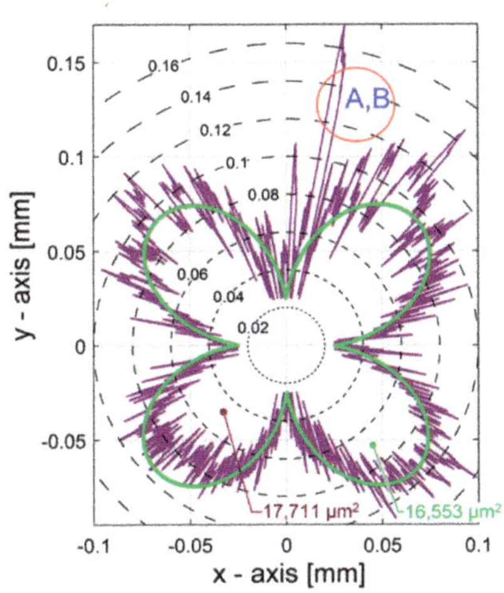

Figure 15. The evolutions of TFR_{1a}.

For P&A evaluation of the average circular trajectory (the result of x_e and y_e simultaneous motions), an important item is the size of the surface delimited by the trajectories fitting residuals (TFR_1 and TFR_{1a}). Each area is a sum of the areas of *N-1*neighboring triangles. All triangles share a common vertex placed in the origin of coordinate systems. The other two vertices are two successive points on TFR. With the area formula of a triangle from [30] (based on the vertices coordinates), the total area delimited by TFR_1 generated by x_e and y_e (Figure 12 or Figure 14) is calculated as 8515.3 μm^2, while the total area delimited by TFR_1 generated by x_a and y_a (Figure 14) is 7624 μm^2.

A second conventional graphical description of the circularity error of the average trajectory (as a second trajectory fitting residual, TFR_2) is available in polar coordinates (d_{i2}, α_{i2}) related to the minimum circumscribed circle (having the same center as the center of the least square circle), with d_{i2}, α_{i2} defined as:

$$d_{i2} = \sqrt{(x_{ei} - x_c)^2 + (y_{ei} - y_c)^2} - R_{cc}\alpha_{i2} = \arctan^4\left(\frac{y_{ei} - y_c}{x_{ei} - x_c}\right) \quad (6)$$

In the first Equation from (6), $R_{cc} = \min\left(\sqrt{(x_{ei} - x_c)^2 + (y_{ei} - y_c)^2}\right)$ is the radius of the minimum circumscribed circle.

Figure 16 presents the TFR_2 of the average trajectory generated by x_e and y_e (R_{cc} = 8.8434 mm) and TFR_2 generated by x_a and y_a (with green color, R_{cc} = 8.9171 mm), with circular grid (with a 50 μm increment on radius). Here the maximum value of distance d_{i2} is 246 μm.

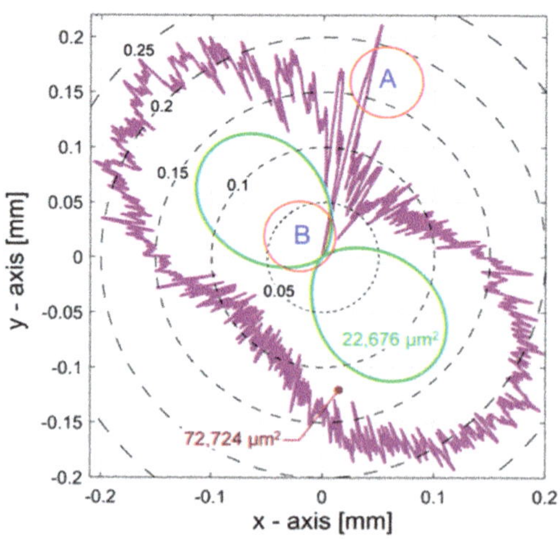

Figure 16. The evolution of TFR_2 generated by x_e and y_e and TFR_2 generated only by x_a and y_a (with green color).

A TFR_2 for a perfect circular average trajectory is a point placed in the origin of the least square circle.

The existence of these two lobes on TFR_2 in Figure 16 is explicable if, in addition to the comments and simulation done in Figure 13, we take into account that the minimum circumscribed circle touches the elliptical trajectory in two symmetrical points. The TFR_2 evolution of a pure elliptical trajectory related to the minimum circumscribed circle is depicted in Figure 17 (by simulation).

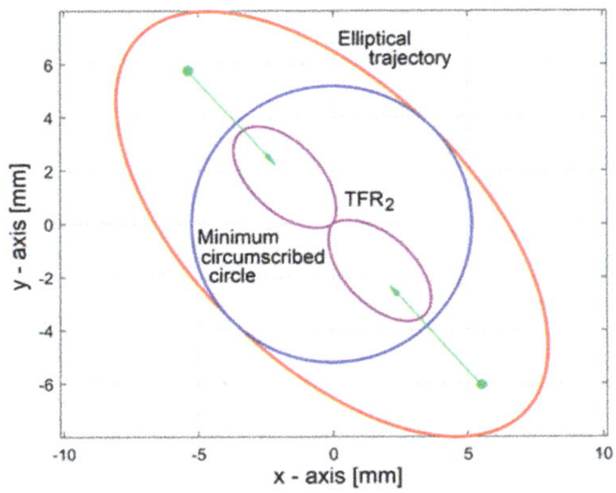

Figure 17. The evolution of TFR_2 generated by circular fitting on an elliptical trajectory (simulation).

Some supplementary resources on the P&A of circular trajectories are revealed by circular fitting of the 2D curve generated only by x_{Ar}-motion (already described in Figure 8) and y_{Ar}-motion (already described in Figure 11), as Figure 18 indicates.

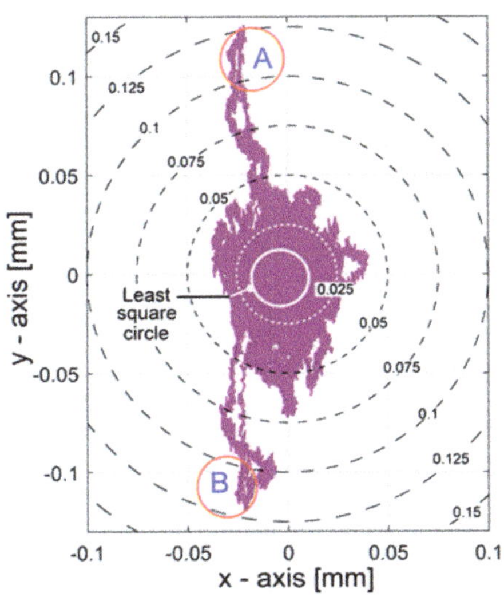

Figure 18. The evolution of the average trajectory generated by x_{Ar} and y_{Ar}.

The least square circle (14.2 μm radius, with center at $x_c = -3.3$ μm and $y_c = -1.9$ μm) should also be considered as an indicator for P&A of the average trajectory.

We should mention that the effect of strong repetitive irregularities on y-motion (y_r and y_{Ar}) already revealed in A, B areas on Figures 8 and 11 are also well described in Figures 14–16 and Figure 18. Moreover, the mirroring of these events A, B in Figure 14 or Figure 15 confirms the previously formulated hypothesis (the comments in Figure 11) that they are related by a sudden release of a mechanical stress inside the toothed belt used for y-motion. In Figure 11, a maximum positive peak from A is immediately followed by a minimum negative peak from B. Therefore, in Figure 14, these two peaks A, B are described as a single peak because of modulus in definition of d_{i1} (Equation (5)).

As it is clearly indicated in Figures 8 and 11, there are strong vibrations on both motions (on x and y), with a negative effect on the P&A of circular trajectories. There are two different strategies available to reduce or to eliminate these vibrations.

The first strategy (probably as the better approach) is to use each stepper motor also as an actuator inside an open-loop active vibration suppression system. The second strategy is to use passive dynamic vibration absorbers or tune mass dampers as well [31] placed on the print bed and on the extruder system. The effect of vibration suppression on TFR_1 or TFR_{1a} shapes should be similar to the effect of a low pass numerical filtering of x_e and y_e. Figure 19 presents the new shape of TFR_{1a} (as TFR_{1af}) if x_e and y_e motions are filtered with a moving average numerical filter (with 1000 samples in the average). As expected, the size of the surface delimited by the TFR_{1af} is not significantly changed (17,399 μm^2 here by comparison with 17,711 μm^2 on TFR_{1a} from Figure 15). The evolution of TFR_{1af} from Figure 19 is also useful when only the influence of the low frequency variable components from x_e and y_e-motions on P&A is investigated and used for errors compensation. The compensation is a feasible option with an appropriate control of stepper motors since they are operated using the microstepping drive technique [32].

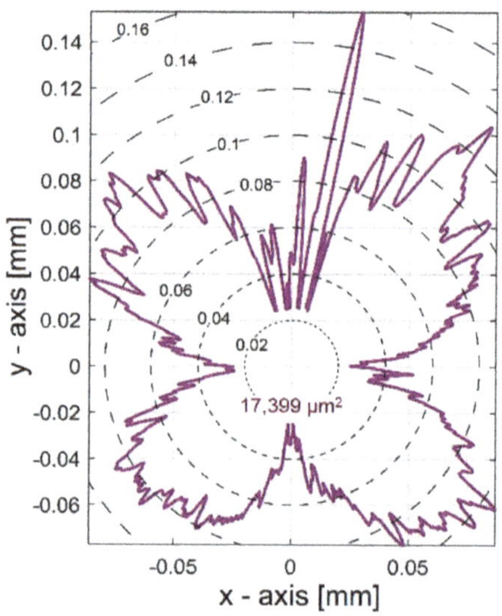

Figure 19. The evolution of TFR_{1a} from Figure 15 with a low pass filtering of x_e and y_e (as TFR_{1af}).

If the accuracy describes how close the real trajectory is to a desired circle (or how close the shapes of TFR_1, TFR_{1a} and TFR_{1af} by a point are), the precision describes the repeatability of real trajectories, each trajectory being generated using a complete cycle (period) of x and y-motions, partially described in Figure 4. The best way to compare these real trajectories is to use the comparison between trajectories fitting residuals related to a circle with a smaller radius than the radius of least square circle (defined similarly to TFR_{1a}) but using low pass filtered x and y-motions (as TFR_{3af}). A perfect coincidence of real trajectories should produce a perfect coincidence of TFR_{3af} trajectories. Figure 20 presents the evolution of TFR_{3af} for five successive real trajectories ($TFR_{3af1} \div TFR_{3af5}$) and the evolution of TFR_{1a}.

Figure 21 presents a zoomed in detail of Figure 20 in area B. As expected, there is not a perfect coincidence of TFR_{3af} trajectories, despite some certain shape similarities (except in area A on Figure 20 where the trajectory TFR_{3af2} is extremely different). It is obvious that the difference between trajectories is less than 10 μm (except in area A). Without any improvement of the 3D printer structure and kinematics, this should also be the theoretical precision after an eventual compensation of the errors (using an appropriate control of the stepper motors). With ideal errors compensations, the trajectories TFR_{1f} should be bordered outside by a circle with 10 μm radius (or 35 μm radius for the trajectories TFR_{3af}).

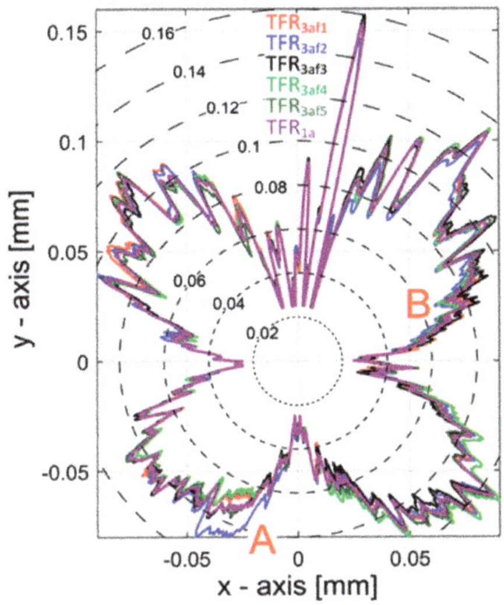

Figure 20. The evolution of TFR_{3af} trajectory for five successive real trajectories ($TFR_{3af1} \div TFR_{3af5}$) and the evolution of TFR_{1a}.

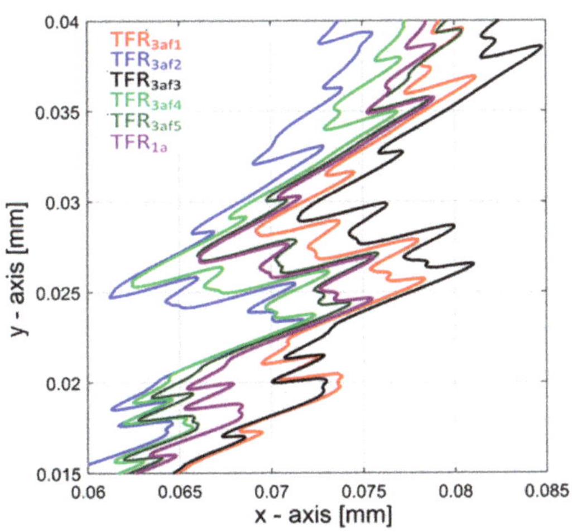

Figure 21. A zoomed in detail of Figure 20 (in B area).

A complete estimation of 3D printer P&A should consider specific trajectories (e.g., circles as in this study) placed in different positions in different 2D locations of the printing volume travelled with different speeds, clockwise and counter clockwise.

4. Conclusions and Future Work

Some theoretical and experimental approaches related to the precision and accuracy (P&A) of a 3D printer, particularly for 2D circular trajectories, were achieved in this paper. The choosing of 2D circular trajectories was inspired from ISO standards methods for P&A evaluation of CNC manufacturing systems (ISO 230-4 [27]) due to some similarities in terms of motion control. The evolution of the simultaneous displacement on two theoretically orthogonal axes (x and y) during a repetitive 2D circular trajectory of the extruder system related to the print bed was simultaneously and continuously measured using a computer-assisted setup based with two contactless optical sensors and a numerical oscilloscope (for sampling and data acquisition). The signals of description for x and y-motions were numerically processed in order to find out some motions characteristics involved in the evaluation of P&A for circular trajectories.

First, a non-perpendicularity of x and y-axes (or an axis misalignment) were experimentally detected (with 0.893 degrees error) by means of the curve fitting (using a pure sine model) of the dominant harmonic components of each signal (x and y-motions signal). Because of this error, a circular programmed trajectory is executed as an elliptical one (typically for a non-orthogonal $x0y$ coordinate system), a topic also confirmed by some supplementary signal processing results.

Second, it was established that the x and y-motions are not simple pure harmonic motions. The residuals of previous curve fitting on each axis movement describe the deviation from pure harmonic shape. A procedure for finding a repetitive periodical pattern in the evolution of these residuals was established and applied with good results. The model of each repetitive pattern is useful in the amelioration of P&A by correction and compensation. The analytical description of the x-motion residual pattern was already established (by curve fitting with a sum of sine model); a future approach intends to do the same for y-motion residual.

Third, a procedure of the description for an average 2D trajectory (an average of several successive theoretically identical circular trajectories) was established. A computer-aided procedure of fitting for closed curves (particularly a circular trajectory) was developed. The circular fitting of the 2D average trajectory was done related to the least square circle. Two conventional graphical descriptions of the circularity errors of the average trajectory were proposed (as trajectory fitting residuals TFR_1 and TFR_2): first description being related to the least square circle, second related to the minimum circumscribed circle.

The shape of these trajectories fitting residuals and the size of the surface delimited by them are useful in P&A evaluation of circular trajectories in order to verify that the 3D printer works properly and, especially, for systematic errors compensation purposes. For example, the non-perpendicularity of x and y-axes previously detected is mirrored in the shape of the average trajectory and, finally, in the shape of these two trajectory fitting residuals (with four similar lobes on TFR_1 and two similar lobes on TFR_2). The deviation from the harmonic shape for x and y-motions is described on these trajectories.

A future approach will be focused on finding a complete procedure of experimental research of P&A using high range/resolution non-contact displacement sensors placed on each of three axes. Some complex 3D curves will be used as test trajectories. A numerical interface between the experimental setup and the 3D printer will be developed in order to perform automated testing and errors compensation.

These signal processing procedures are available to verify the P&A of 2Dcircular trajectories on any other similar equipment (e.g., a 3D CNC manufacturing system).

Author Contributions: Conceptualization, M.H. and C.-G.D.; methodology, M.H. and C.-G.D.; software, A.M.; validation, A.M., M.H. and D.-F.C.; formal analysis, D.-F.C.; investigation, F.N. and C.-G.D.; resources, D.-F.C.; data curation, A.S. and F.C.; writing—original draft preparation, M.H.; writing—review and editing, M.H. and D.-F.C.; visualization, A.S. and A.M. supervision, M.H.; project administration, D.-F.C.; funding acquisition, D.-F.C. All authors have read and agreed to the published version of the manuscript.

Funding: This work was supported by the Technical University Gheorghe Asachi din Iasi, Romania, through the grant GI/P29/2021.

Institutional Review Board Statement: Not applicable.

Informed Consent Statement: Not applicable.

Data Availability Statement: Not applicable.

Conflicts of Interest: The authors declare no conflict of interest.

References

1. Tiwari, K.; Kumar, S. Analysis of the factors affecting the dimensional accuracy of 3D printed products. *Mater. Today Proc.* **2018**, *5*, 18674–18680. [CrossRef]
2. Hällgren, S.S.; Pejryd, L.; Ekengren, J. 3D Data Export for Additive Manufacturing—Improving Geometric Accuracy, 26th CIRP Design Conference. *Procedia CIRP* **2016**, *50*, 518–523. [CrossRef]
3. Lianghua, Z.; Xinfeng, Z. Error Analysis and Experimental Research on 3D Printing. *IOP Conf. Ser. Mater. Sci. Eng.* **2019**, *592*, 012150.
4. Baumann, F.; Bugdayci, H.; Grunert, J.; Keller, F.; Roller, D. Influence of slicing tools on quality of 3D printed parts. *Comput. Aided Des. Appl.* **2016**, *13*, 14–31. [CrossRef]
5. Hernandez, D.D. Factors Affecting Dimensional Precision of Consumer 3D Printing. *J. Aviat. Aeronaut. Aerosp.* **2015**, *2*, art.2. [CrossRef]
6. George, E.; Liacouras, P.; Rybicki, F.J.; Mitsouras, D. Measuring and Establishing the Accuracy and Reproducibility of 3D Printed Medical Models. *Radiographics* **2017**, *37*, 1424–1450. [CrossRef] [PubMed]
7. Chae, M.P.; Chung, R.D.; Smith, J.A.; Hunter-Smith, D.J.; Rozen, W.M. The accuracy of clinical 3D printing in reconstructive surgery: Literature review and in vivo validation study. *Gland Surg.* **2021**, *10*, 2293–2303. [CrossRef]
8. Abdelmomen, M.; Dengiz, F.O.; Tamre, M. Survey on 3D Technologies: Case Study on 3D Scanning, Processing and Printing with a Model. In Proceedings of the 2020 21st International Conference on Research and Education in Mechatronics (REM), Cracow, Poland, 9–11 December 2020; pp. 1–6.
9. Xu, J.; Ding, L.; Love, P.D. Digital reproduction of historical building ornamental components: From 3D scanning to 3D printing. *Autom. Constr.* **2017**, *78*, 85–96. [CrossRef]
10. Favero, C.S.; English, J.D.; Cozad, B.E.; Wirthlin, J.O.; Short, M.M.; Kasperc, F.K. Effect of print layer height and printer type on the accuracy of 3-dimensional printed orthodontic models. *Am. J. Orthod. Dentofac. Orthop.* **2017**, *152*, 557–565. [CrossRef]
11. Vitolo, F.; Martorelli, M.; Gerbino, S.; Patalano, S.; Lanzotti, A. Controlling form errors in 3D printed models associated to size and position on the working plane. *Int. J. Interact. Des. Manuf.* **2018**, *12*, 969–977. [CrossRef]
12. Tanoto, Y.Y.; Anggono, J.; Siahaan, I.H.; Budiman, W. The Effect of Orientation Difference in Fused Deposition Modelling of ABS Polymer on the Processing Time, Dimension Accuracy, and Strength, International Conference on Engineering, Science and Nanotechnology 2016 (ICESNANO 2016). *AIP Conf. Proc.* **2016**, *1788*, 030051-1-030051-7.
13. Divyathej, M.V.; Varun, M.; Rajeev, P. Analysis of mechanical behavior of 3D printed ABS parts by experiments. *Int. J. Sci. Eng. Res.* **2016**, *7*, 116–124.
14. Msallem, B.; Sharma, N.; Cao, S.; Halbeisen, F.S.; Zeilhofer, H.F.; Thieringer, F. Evaluation of the Dimensional Accuracy of 3D-Printed Anatomical Mandibular Models Using FFF, SLA, SLS, MJ, and BJ Printing Technology. *J. Clin. Med.* **2020**, *9*, 817. [CrossRef] [PubMed]
15. Alsoufi, M.S.; Elsayed, A.E. Surface Roughness Quality and Dimensional Accuracy—A Comprehensive Analysis of 100% Infill Printed Parts Fabricated by a Personal/Desktop Cost-Effective FDM 3D Printer. *Mater. Sci. Appl.* **2018**, *9*, 11–40. [CrossRef]
16. Mantada, P.; Mendricky, R.; Safka, J. Parameters Influencing the Precision of Various 3D Printing Technologies. *MM Sci. J.* **2017**, *5*, 2004–2012. [CrossRef]
17. Li, L.; McGuan, R.; Isaac, R.; Kavehpour, P.; Candler, R. Improving precision of material extrusion 3D printing by in-situ monitoring & predicting 3D geometric deviation using conditional adversarial networks. *Addit. Manuf.* **2021**, *38*, 101695.
18. Salmi, M.; Paloheimo, K.S.; Tuomi, J.; Wolff, J.; Mäkitie, A. Accuracy of medical models made by additive manufacturing (rapid manufacturing). *J. Cranio-Maxillo-Facial Surg.* **2013**, *41*, 603–609. [CrossRef]
19. Kacmarcik, J.; Spahic, D.; Varda, K.; Porca, E.; Zaimovic-Uzunovic, N. An investigation of geometrical accuracy of desktop 3D printers using CMM. *IOP Conf. Ser. Mater. Sci. Eng.* **2018**, *393*, 012085. [CrossRef]
20. Jadayel, M.; Khameneifar, F. Improving Geometric Accuracy of 3D Printed Parts Using 3D Metrology Feedback and Mesh Morphing. *J. Manuf. Mater. Process.* **2020**, *4*, 112. [CrossRef]
21. Majarena, A.C.; Aguilar, J.J.; Santolaria, J. Development of an error compensation case study for 3D printers. *Procedia Manuf.* **2017**, *13*, 864–871. [CrossRef]
22. Keaveney, S.; Connolly, P.; O'Cearbhaill, E.D. Kinematic error modeling and error compensation of desktop 3D printer. *Nanotechnol. Precis. Eng.* **2018**, *1*, 180–186. [CrossRef]
23. Qian, S.; Bao, K.; Zi, B.; Wang, N. Kinematic Calibration of a Cable-Driven Parallel Robot for 3D Printing. *Sensors* **2018**, *18*, 2898. [CrossRef]
24. Kochetkov, A.V.; Ivanova, T.N.; Seliverstova, L.V.; Zakharov, O.V. *Kinematic Error Modelling of Delta 3D Printer*; Materials Science Forum, Trans Tech Publications, Ltd.: Zurich, Switzerland, 2021; Volume 1037, pp. 77–83.
25. Cajal, C.; Santolaria, J.; Velazquez, J.; Aguado, S.; Albajez, J. Volumetric error compensation technique for 3D printers. *Procedia Eng.* **2013**, *63*, 642–649. [CrossRef]
26. Duan, M.; Yoon, D.; Okwudire, C.E. A limited-preview filtered B-spline approach to tracking control—With application to vibration-induced error compensation of a 3D printer. *Mechatronics* **2018**, *56*, 287–296. [CrossRef]
27. *Test Code for Machine Tools—Part 4: Circular Tests for Numerically Controlled Machine Tools*; ISO230-4; ISO: Geneva, Switzerland, 2005.
28. Anet. Available online: https://anet3d.com/pages/a8 (accessed on 7 November 2021).
29. MICRO-EPSILON. Available online: http://www.img.ufl.edu/wiki/images/ILD_2000_Datasheet.pdf (accessed on 7 November 2021).

30. Math Open Reference. Available online: https://www.mathopenref.com/coordtrianglearea.html (accessed on 7 November 2021).
31. Yang, F.; Sedaghati, R.; Esmailzadeh, E. Vibration suppression of structures using tuned mass damper technology: A state-of-the-art review. *J. Vib. Control* **2021**, in press. [CrossRef]
32. Barabas, Z.A.; Morar, A. High performance microstepping driver system based on five-phase stepper motor (sine wave drive). *Procedia Technol.* **2014**, *12*, 90–97. [CrossRef]

applied sciences

MDPI

Article

An Experimental Approach on Beating in Vibration Due to Rotational Unbalance

Dragos-Florin Chitariu *, Florin Negoescu, Mihaita Horodinca, Catalin-Gabriel Dumitras, Gures Dogan and Mehmet Ilhan

"Gheorghe Asachi" Technical University of Iasi, 700050 Iasi, Romania; florin.negoescu@academic.tuiasi.ro (F.N.); mihaita.horodinca@academic.tuiasi.ro (M.H.); catalin-gabriel.dumitras@academic.tuiasi.ro (C.-G.D.); gures.dogan@student.tuiasi.ro (G.D.); ilhan.mehmet@student.tuiasi.ro (M.I.)

* Correspondence: dragos.chitariu@tuiasi.ro; Tel.: +40-741-269-491

Received: 2 September 2020; Accepted: 28 September 2020; Published: 1 October 2020

Abstract: This paper proposes a study in theoretical and experimental terms focused on the vibration beating phenomenon produced in particular circumstances: the addition of vibrations generated by two rotating unbalanced shafts placed inside a lathe headstock, with a flat friction belt transmission between the shafts. The study was done on a simple computer-assisted experimental setup for absolute vibration velocity signal acquisition, signal processing and simulation. The input signal is generated by a horizontal geophone as the sensor, placed on a headstock. By numerical integration (using an original antiderivative calculus and signal correction method) a vibration velocity signal was converted into a vibration displacement signal. In this way, an absolute velocity vibration sensor was transformed into an absolute displacement vibration sensor. An important accomplishment in the evolution of the resultant vibration frequency (or combination frequency as well) of the beating vibration displacement signal was revealed by numerical simulation, which was fully confirmed by experiments. In opposition to some previously reported research results, it was discovered that the combination frequency is slightly variable (tens of millihertz variation over the full frequency range) and it has a periodic pattern. This pattern has negative or positive peaks (depending on the relationship of amplitudes and frequencies of vibrations involved in the beating) placed systematically in the nodes of the beating phenomena. Some other achievements on issues involved in the beating phenomenon description were also accomplished. A study on a simulated signal proves the high theoretical accuracy of the method used for combination frequency measurement, with less than 3 microhertz full frequency range error. Furthermore, a study on the experimental determination of the dynamic amplification factor of the combination vibration (5.824) due to the resonant behaviour of the headstock and lathe on its foundation was performed, based on computer-aided analysis (curve fitting) of the free damped response. These achievements ensure a better approach on vibration beating phenomenon and dynamic balancing conditions and requirements.

Keywords: lathe headstock; rotational unbalance; vibration; beating; signal processing

---

1. Introduction

Rotating unbalance is a topic frequently mentioned in the analysis of the dynamics of rotary bodies (rotordynamics [1]). The rotating unbalance occurs due to an asymmetry of mass distribution (in some different regions of the rotary body, the center of mass is not placed on the axis of rotation). Centrifugal forces occur in these unbalanced regions of the rotary body. The resultant of these centrifugal forces is transmitted through the bearings to the structure where the rotary body is placed. In each bearing the resultant of centrifugal forces has two components at orthogonal directions. Each component acts as a harmonic excitation force against the structure, thus generating vibrations.

For some specific appliances these vibrations are desirable (e. g. in vibration shakers used also as mechanical vibration exciters [2,3], vibration alert systems in mobile phones, electronic vibrating bracelets [4], and haptic feedback devices with vibrations [5]). Generally, the vibrations due to the rotating body unbalances have highly undesirable effects (e.g., bad surface quality in the grinding process [6,7], premature bearing destruction [8,9]), and human body discomfort [10]). In order to measure [11] and to eliminate the unbalance of rotary bodies [12–14] (using additional balancing masses and additional inertia [15]), several special requirements must be met and specialized equipment must be used [16,17]. Actually, one of the best ways to balance rotary unbalanced bodies is through the use of self-balancing systems [18,19].

Sometimes two rotary unbalanced bodies having almost the same angular speed (or rotational frequency) which rotate in the same structure (e.g., in centerless grinding machines, [20]) produce a vibration beating phenomenon [10,21,22].

Each rotary unbalanced body generates a vibration. The addition of two vibrations, having slightly different frequencies, produces the aforementioned beating phenomenon. This is a resultant vibration with periodical variation of amplitude, with nodes (where the amplitude has a minimum value, the addition of the two vibrations produces destructive interference, 180 degrees out of phase between the constituents of the resultant vibration) and anti-nodes (a maximum amplitude, where the addition of the two vibrations produces constructive interference, with zero degrees shift of phase between constituents) [23]. Obviously the beating phenomenon in mechanics is not solely related to the vibrations produced by rotary unbalanced bodies, it also occurs when two vibration modes with almost the same modal frequency are excited [24,25], and it occurs as well when a system vibrates simultaneously due to forced sinusoidal excitation close to a resonant frequency and due to a free response [26–29].

Some specific appliances use the vibration beating phenomenon to monitor the condition of mechanical systems, e.g., monitoring the adhesion integrity of single lap joints [30], monitoring the structural integrity of helicopter rotor blades [31], and for seismic vibration testing [32]. A vibration beating mechanism in piezoelectric energy harvesting systems is proposed in [33].

In machine tools, in addition to a critical source of vibrations (self-excited vibrations in turning [34], milling [35] or grinding [36] processes), the vibrations produced by rotary unbalance (generated by tools [37], shafts [38] or work pieces [36]) and particularly the beating phenomenon [6] created by rotary unbalanced bodies, are important items. This paper proposes some approaches, in theoretical and experimental terms, to address the vibration beating phenomenon produced inside a Romanian lathe headstock SNA 360, by two inner unbalanced rotary shafts, rotating with very close angular speeds. The main achievements of this paper are producing results in the areas of: vibration beating monitoring, conversion of a velocity vibration signal into a displacement signal (by antiderivative calculus), and the evolution of beating vibration signal frequency (pattern, simulation, measurement and accuracy measurement), as well as producing a study of the influence of headstock and lathe foundation dynamics on vibration amplitudes.

2. A Theoretical Approach

Assume that the rotary unbalance of each shaft (each of which rotates at angular speeds of ω_1 and ω_2, respectively) is reducible at the asymmetry of mass distribution with unbalance masses m_1 and m_2, respectively, placed in a single plane at distances r_1 and r_2, respectively, to the rotation axis. The horizontal projection of the centrifugal forces of each rotary unbalance ($F_1 = m_1\omega_1^2 r_1$ and $F_2 = m_2\omega_2^2 r_2$) generates vibration displacements $y_1 = kD_{af1}F_1 cos(\theta_1)$ and $y_2 = kD_{af2}F_2 cos(\theta_2)$, where k is the stiffness of the headstock and the lathe foundation, D_{af1} and D_{af2} are the dynamic amplification factors and $\theta_1 = \omega_1 t + \varphi_1$ and $\theta_2 = \omega_2 t + \varphi_2$ are the instantaneous values of the angle of the centrifugal forces with respect to the horizontal direction (φ_1 and φ_2 being the instantaneous values of these angles at $t = 0$). With these considerations, a complete description of y_1 and y_2 of the vibrations is given below:

$$y_1 = kD_{af1}m_1\omega_1^2 r_1 cos(\omega_1 t + \varphi_1) = A_1 cos(\omega_1 t + \varphi_1) \tag{1}$$

$$y_2 = kD_{af2}m_2\omega_2^2 r_2 cos(\omega_2 t + \varphi_2) = A_2 cos(\omega_2 t + \varphi_2) \tag{2}$$

Here $A_1 = kD_{af1}m_1\omega_1^2 r_1$ and $A_2 = kD_{af2}m_2\omega_2^2 r_2$ are the vibration amplitudes of the two shafts, respectively. The headstock and the lathe vibrate as a single body on its foundation (as a mass–spring–damper system) with a vibratory motion which is the result of the addition $y_1 + y_2$ of these two vibrations, a periodical non-harmonic motion that presents itself as a beating phenomenon [23], with nodes and anti-nodes (as the simulation from Figure 1 proves). According to Figure 1, if the period of vibration y_1 is $T_1 = 2\pi/\omega_1$ and $T_2 = 2\pi/\omega_2$ is the period of vibration y_2, then the period T_b of the beating phenomenon (the beat period being the time between two anti-nodes or between two nodes, as well) and the periods T_1, T_2 (with $T_2 < T_1$) should fulfill this obvious condition:

$$T_b = nT_1 = (n+1)T_2 \tag{3}$$

with n being a natural number, defined from Equation (3) as:

$$n = T_2/(T_1 - T_2) \tag{4}$$

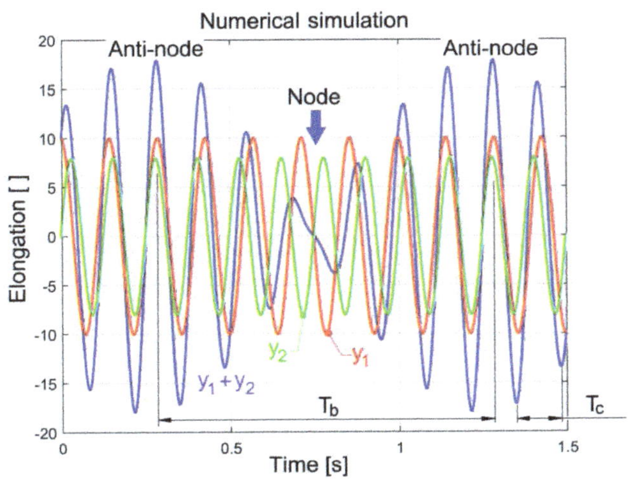

Figure 1. A simulation of the beating phenomenon, where: y_1 is the vibration of shaft 1; y_2 is the vibration of shaft 2, T_b is the beat period; and T_c is the period of the resultant vibration $y_1 + y_2$.

In Figure 1 $n = 7$. If in Equations (3) and (4) the periods are replaced by frequencies ($T_b = 1/f_b$, $T_1 = 1/f_1$, $T_2 = 1/f_2$), then the resulting frequency f_b of the beating phenomenon (beat frequency or the number of nodes per second, as well) is:

$$f_b = f_2 - f_1 \tag{5}$$

In Figure 1 $f_2 = 8$ Hz and $f_1 = 7$ Hz; these generate $f_b = 1$ Hz with $T_b = 1$s ($A_1 = 10$, $A_2 = 8$, $\varphi_1 = 0$, $\varphi_2 = -\pi/2$).

According to [39], the resultant waveform of the vibration addition $y_1 + y_2$ has the frequency $f_c = 1/T_c$ (as a combination frequency or modulation frequency, with the period T_c highlighted on Figure 1) defined as:

$$f_c = (f_1 + f_2)/2 \tag{6}$$

This paper will prove by simulations and experiments that this definition is not accurate, especially when $A_1 \neq A_2$.

These theoretical considerations and some other supplementary issues and procedures will be confirmed by experimental approach in this paper.

3. Experimental Setup

Figure 2a presents a lateral view of both shafts (1 and 2, placed in the headstock) involved in the beating phenomenon due to rotary unbalance.

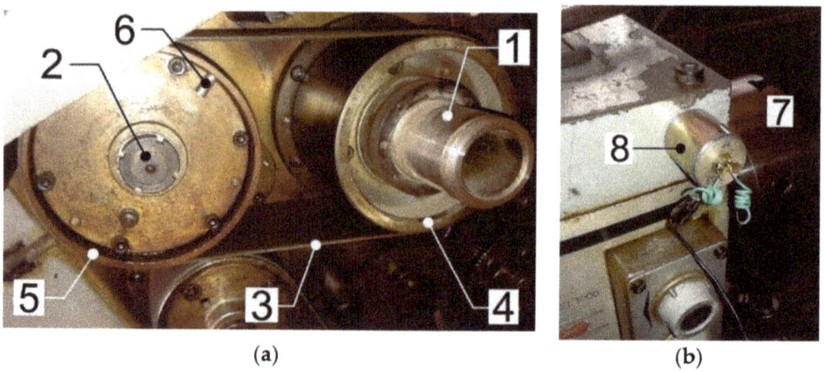

Figure 2. (**a**) A lateral view of the shafts involved in the beating phenomenon (with a flat belt transmission between the shafts); (**b**) A front view of the lathe headstock with the vibration sensor.

A friction belt transmission with a flat drive belt 3, a pulley 4 (on shaft 1) and a pulley 5 (on shaft 2) synchronously rotates both shafts (the theoretical value of transmission speed ratio is 1:1). Here 6 depicts an additional mass (10.8 g, a permanent magnet) placed in different angular positions on pulley 5 and used to change the internal unbalancing of the shaft 2.

The shaft 1 is also the lathe main spindle (with the jaw chuck labelled with 7 on Figure 2b placed on the opposite side of Figure 2a). Figure 2b shows an absolute velocity vibration sensor 8 (an electrodynamic seismic geophone Geo Space GS 11D, now HGS Products HG4 as described in [40]), placed on the headstock. The geophone corner frequency (8 Hz) is smaller than the minimum frequency of the headstock vibration (17 Hz). No significant resonant amplification at the corner frequency can be identified (the open circuit damping being 34% of critical damping). The geophone sensitivity is 31.89 V/m/s.

The signal delivered by the vibration sensor (a voltage proportional to the vibration velocity of the headstock) is numerically acquired by a personal computer via a computer-assisted numerical oscilloscope PicoScope 4424 from Pico Technology, Saint Neots, UK (USB powered, four channels, 12 bits resolution, 80 MS/s maximum sampling, 32 MS memory) [41]. Due to the high sensitivity of the geophone and oscilloscope features (e.g., a numerically controlled internal amplifier) the supplementary amplification of the signal delivered by vibration sensor is not necessary.

The computer-aided processing of this signal was done in Matlab. Firstly, the signal delivered by sensor must be mathematically divided by the sensitivity of the sensor in order to obtain the vibration velocity evolution. Secondly, the velocity of vibration must be numerically integrated (by antiderivative calculus) in order to obtain the vibration displacement evolution. The description and analysis of the evolution of some of the other vibration features (e.g., the combination frequency f_c, the beat vibration amplitude and the free vibrations of the lathe on foundation) were also performed.

In order to measure the average value of the instantaneous angular speed (IAS) ω_1 of the main spindle (shaft 1) rigorously, the technique described in our previous work [42] was used (with a two phase multi-pole AC generator placed in the jaw chuck, as an IAS sensor). The same technique (which refers only to signal processing, briefly described later on) was used for an accurate measurement of the combination frequency f_c.

4. Experimental Results and Discussion

4.1. A Beating Phenomenon Described in Vibration Velocity

Figure 3 presents the evolution of the headstock vibration velocity during a time interval of 200 s, described with 1 MS (or 1,000,000 samples as well) so a sampling interval of Δt = 200 μs, when the main spindle (shaft 1) rotates (and shaft 2, as well) in the steady-state regime, with constant IAS, with an average value of ω_1 = 109.2369 rad/s (for f_1 = 17.3856 Hz average rotation frequency, or 1043.1 revolutions per minute on average).

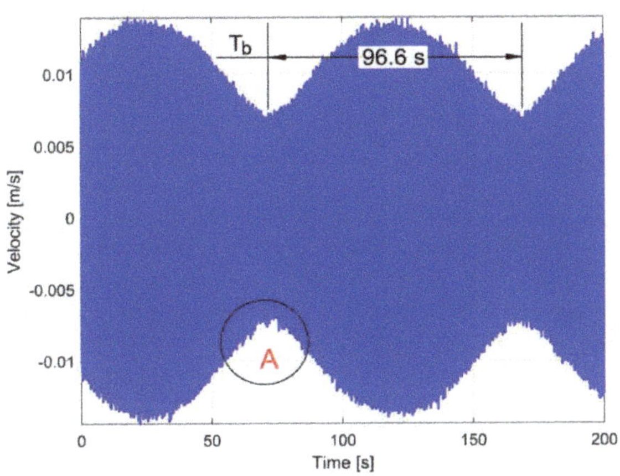

Figure 3. The evolution of the velocity of headstock vibrations with a beating phenomenon due to rotary unbalanced shafts; here T_b is the beat period, A is a label for a future comment on signal evolution.

It is obvious that Figure 3 depicts a vibration beating phenomenon with nodes and anti-nodes, with a very high value of the period T_b (96.6 s) and consequently with a very small value of beat frequency $f_b = 1/T_b$ = 1/96.6 Hz. The beating phenomenon proves that the IASs ω_1 and ω_2 and also rotation frequencies f_1 and f_2 as well, are slightly different because the diameters of pulleys 4 and 5 involved in belt transmission are not strictly the same. With $T_1 = 1/f_1$ and the relationship between T_1 and T_b from Equation (3) there are $n \approx T_b{\cdot}f_1 \approx 1679$ periods T_1 between nodes (and between anti-nodes as well). This is an approximated value of n because the frequency f_1 is not rigorously constant (as is proved, later in this paper). According to Equation (3), at each n complete rotations of shaft 1, the shaft 2 makes $n + 1$ or $n - 1$ rotations (one rotation difference), which means that the experimentally revealed value of the belt transmission speed ratio is $\omega_2/\omega_1 = T_1/T_2 = n/(n \pm 1) \approx 1679/(1679 \pm 1)$. This is also the ratio between pulleys diameters: the diameter of pulley 4 divided by the diameter of pulley 5 (assuming that there is no slipping between the belt and the pulleys).

The additional mass 6 (Figure 2a) was placed in a certain angular position on pulley 5, in order to obtain a maximum value of the amplitude A_2, and a maximum difference between amplitudes in anti-nodes and nodes as well. It is obvious that the main spindle (shaft 1) is also unbalanced; otherwise the beating phenomenon does not occur.

Figure 4 presents a zoomed-in detail in the area labelled A in Figure 3. Here the dominant component (≈6 mm/s amplitude) is the sum of two vibrations created by rotary unbalances; the other low amplitude (and high frequency) components are related by vibrations generated by some other headstock rotary components.

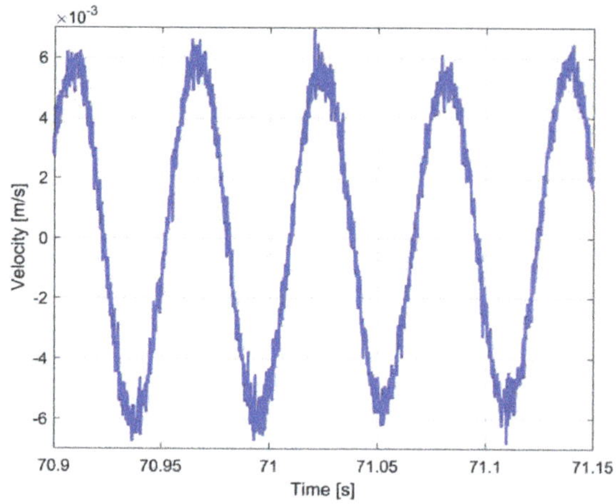

Figure 4. A zoomed-in detail of Figure 3 in the area labelled as A.

With the values for f_1 and f_b revealed before, the frequency f_2 is a result of Equation (5), with two possible values ($f_2 = f_1 - f_b = 17.3752$ Hz or $f_2 = f_1 + f_b = 17.3959$ Hz). Because in Equation (5) the notations f_1 and f_2 are arbitrary, this equation should be reconsidered as $f_b = |f_2 - f_1|$.

As a consequence, the angular speed $\omega_2 = 2\pi f_2$ has two possible values ($\omega_2 = 109.1716$ rad/s or $\omega_2 = 109.3016$ rad/s), as does the speed ratio (ω_2/ω_1) of the driving belt (with $\omega_1 = 109.2369$ rad/s). To find the right value of f_2 (and ω_2 as well) the technique described in [42] should be used (with an IAS sensor placed on shaft 2).

The evolution from Figure 3 is an addition of vibrations velocities ($v = dy_1/dt + dy_2/dt$) generated by both of the unbalanced shafts (1 and 2). It is expected that the beating phenomenon keeps the main characteristics (e.g., T_b, T_c values or f_b, f_c values, as well) if it is described using the addition of vibration displacements ($s = y_1 + y_2$), except for the amplitudes in nodes and anti-nodes which significantly decreases.

4.2. The Description of the Beating Phenomenon in Vibration Displacement by Numerical Integration

The vibration displacement evolution can be obtained from vibration velocity evolution by numerical integration (antiderivative calculus). Based on the approximate definition of velocity (derivative of displacement) $v = ds/dt \approx \Delta s/\Delta t$, a current sample of velocity v_i is defined using two successive samples of displacement s_i, s_{i-1} (in the displacement interval $\Delta s = s_i - s_{i-1}$) and the values of time t_i, t_{i-1} for these samples, in the sampling interval $\Delta t = t_i - t_{i-1}$ (usually this is a constant value) as:

$$v_i = \frac{s_i - s_{i-1}}{\Delta t} \tag{7}$$

This is an approximation of the first derivative of displacement as backward finite difference, with $i > 1$ [43]. The current sample of displacement s_i can be simply mathematically extracted from Equation (7) as:

$$s_i = v_i \Delta t + s_{i-1} \tag{8}$$

Equation (8) describes a sample s_i of displacement related to velocity, this also being our proposal for a description of numerical integration of velocity (antiderivative calculus). According to Equation (8) the sample s_i depends on sampling interval Δt (here Δt is the time $t_i - t_{i-1}$ between two consecutive samples of velocity v_i and v_{i-1} or two consecutive samples of displacement s_i and s_{i-1} as well), the velocity sample v_i and the previous sample of displacement s_{i-1}, as the result of a previous step of numerical integration. The numerical integration from Equation (8) is available for $i > 1$. Of course, it is mandatory to know the value of the first sample of displacement s_1, this being an indefinite value because $i > 1$. This is exactly the constant C of integration (usually an arbitrary value). Pure harmonic signals are numerically integrated, in that case, evidently $C = 0$.

Figure 5a describes the graphical result of numerical integration of vibration elongation evolution from Figure 3 using Equation (8), with $C = s_1 = 0$. Certainly this evolution is not strictly related to the vibration displacement from the beating phenomenon.

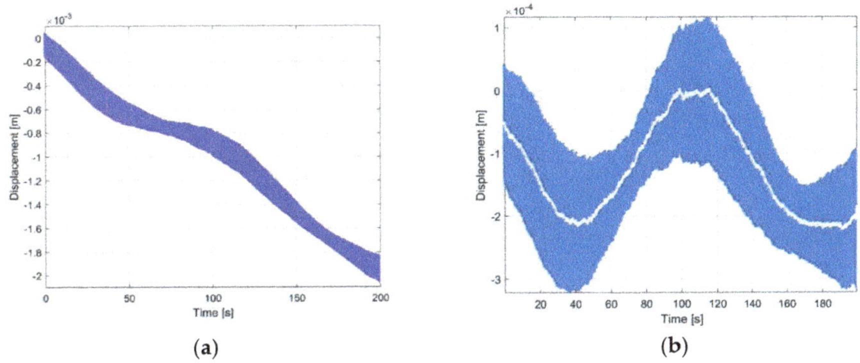

Figure 5. (**a**) The result of numerical integration of velocity evolution from Figure 3; and (**b**) the result of removing the zero-offset influence on numerical integration of velocity from Figure 5a.

We found that the oscilloscope generates a very small negative constant zero offset. As the theory of integration establishes, the numerical integration of this constant zero offset produces a component with linear evolution, experimentally confirmed in Figure 5a by the evolution with negative slope. The removal of this linear component produces the result from Figure 5b (the evolution emphasized in blue).

It is evident that Figure 5b is not the expected evolution of the vibration displacement in beating phenomenon. Surely, there is not a mistake in the numerical integration proposal in Equation (8) because the numerical derivative of the evolution from Figure 5b using Equation (7) produces exactly the evolution of velocity, as Figure 6 indicates (by comparison with Figure 3).

Our first attempt to explain this deficiency in this result of numerical integration is related by the constant of integration C.

Intuitively it is supposed that somehow the hypothesis that $C = 0$ is wrong. Perhaps the effect of this wrong hypothesis is mirrored in the evolution from Figure 5b and its effect should be removed (as the influence of negative zero offset was removed before).

It was discovered by numerical simulation that the numerical integration of a computer-generated beating vibration velocity signal (similarly to those depicted in Figure 3) using Equation (8), with $C = 0$, produces a vertically shifted evolution with a constant nonzero value, which should be mathematically removed.

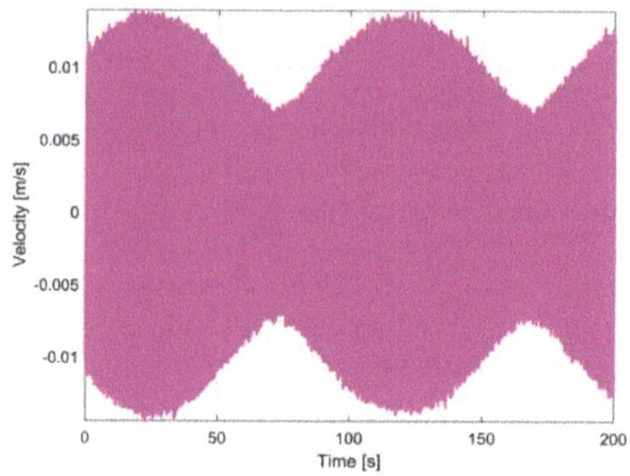

Figure 6. The result of the numerical derivative of the evolution from Figure 5b (vibration velocity, practically similar with Figure 3).

This approach assumed that in the result of numerical integration of vibration velocity depicted in Figure 5b, a supplementary low frequency component was generated and should be removed. For the time being we unfortunately do not have a consistent explanation for the appearance of this low frequency component. This low frequency component (depicted in Figure 5b in white) was detected by low-pass numerical filtering of the vibration displacement signal (the evolution emphasised in blue).

A computer-generated moving average filter [43] was used, with the first notch frequency equal to the combination frequency $f_c = (f_1 + f_2)/2$ (assuming that this definition from Equation (6) is accurate), in order to completely remove the variable component from Figure 5b having the resultant vibration frequency, and in order to obtain the low frequency component. The number of points in the average of the filter is defined as integer of the ratio $1/(f_c \Delta t)$. The removal of this low frequency component from the result of numerical integration (the vibration displacement signal) depicted in Figure 5b is shown in Figure 7. It is obvious that this evolution properly describes the resultant vibration displacement during the beating phenomenon, previously described in Figure 3, by the velocity of the resultant vibration.

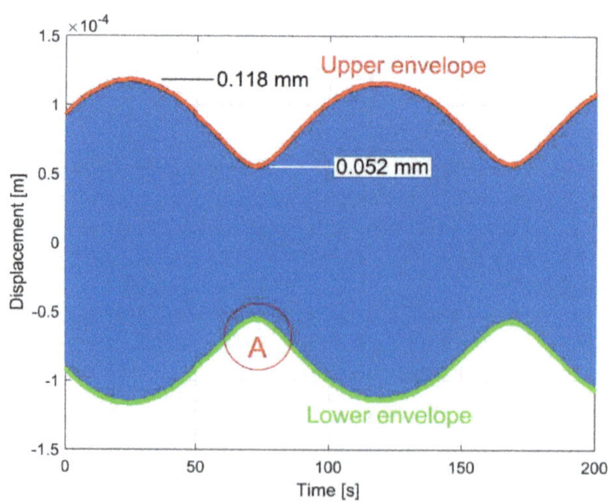

Figure 7. The headstock vibration displacement evolution during the beating phenomenon, deduced by numerical integration and correction of the signal depicted in Figure 3.

There is a supplementary confirmation of this result: the numerical differentiation of the vibration displacement signal from Figure 7 (using Equation (7)) fits very well with the vibration velocity signal from Figure 3, as a very short detail (15 ms duration) of both evolutions (given in Figure 8a) from the area labelled as A (Figures 3 and 7) indicates. Thus, the absolute velocity vibration sensor together with the proposed numerical signal integration method acts as an absolute displacement vibration sensor.

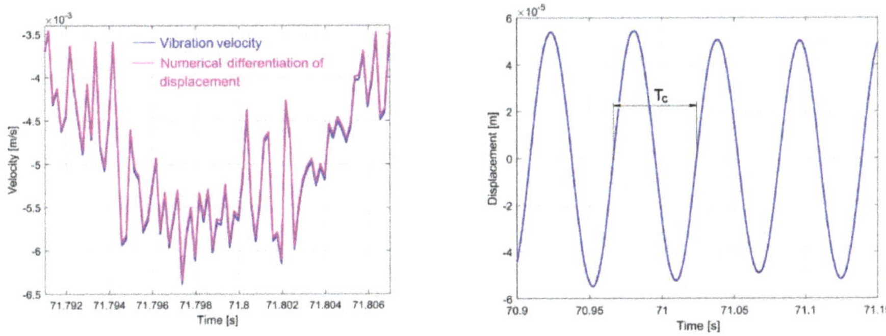

(a) **(b)**

Figure 8. (**a**) A detail concerning the evolution of velocity (Figure 3) overlaid on the numerical differentiation of the displacement depicted in Figure 7; and (**b**) a detail of area A of Figure 7 with T_c, the period of the resultant vibration.

Figure 8b presents a short detail of the vibration displacement evolution in the area labelled with A in Figure 7. This figure has the same size on the abscissa as Figure 4. By comparison with Figure 4, the evolution is much smoother here, as a consequence of numerical integration, which drastically reduces the amplitudes of high frequency components. The integration acts as a low-pass filter.

In Figure 7 two relationships between the vibrations amplitudes A_1 and A_2 are available (from Equations (1) and (2)) due to the constructive interference in anti-nodes ($A_1 + A_2 = 118$ μm) and destructive interference in nodes ($A_1 - A_2 = 52$ μm), so $A_1 = 85$ μm and $A_2 = 33$ μm. For the time being A_1 does not necessarily refer to vibration amplitude generated by the main spindle or shaft 1.

4.3. The Evolution of Frequency for Resultant Vibration in Beating Phenomenon

An interesting item in the beating phenomenon is the evolution of frequency of the resultant vibration f_c (also known as combination frequency or modulation frequency, $f_c = 1/T_c$, with T_c highlighted in Figure 8b). In [39] this frequency is defined as the average of both frequencies (f_1, f_2) involved in the beating (Equation (6)).

A beating phenomenon was simulated using the sum of two harmonic vibrations displacements $y_{1s}(A_1,f_1)$ and $y_{2s}(A_2,f_2)$—already described in Equations (1) and (2) with different values of amplitudes ($A_2 > A_1$) and frequencies f_1 and f_2, close to those from the experiment described in Figures 3 and 7 (with $f_1 = 17.3856$ Hz and $f_2 = 17.3959$ Hz), for a duration equal to T_b (placed between two anti-nodes).

For six different values of amplitudes (A_1 increases and A_2 decreases), the evolution of the combination frequency f_c and its average value was determined on the vibration beating simulated signal, as Figure 9 indicates, using a high accuracy measurement technique from a previous work [42]. Here each peak describes the value of frequency f_c in vibration beating node. It is obvious that f_c is not constant and, in contradiction with Equation (6) and [39], $f_c \neq (f_1 + f_2)/2$. Here, with $A_1 > A_2$ the average f_c is very close to f_2, with $f_c > f_2$.

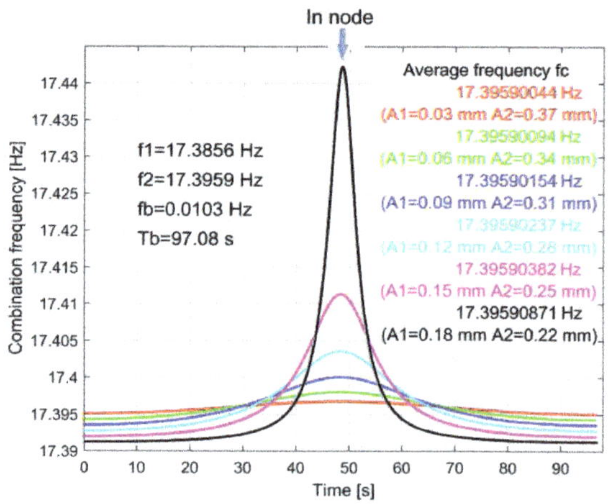

Figure 9. The evolution of the instantaneous combination frequency f_c on the simulated vibration beating during a beat period T_b (with $A_2 > A_1$ and $f_2 > f_1$).

A similar simulation was done in the same conditions, now with $A_1 > A_2$, as Figure 10 indicates (here A_1 decreases and A_2 increases). Similar to the simulation given in Figure 9, it is obvious that f_c is not constant and again, in contradiction with Equation (6) and [39], $f_c \neq (f_1 + f_2)/2$. The average frequency f_c is very close to f_1, with $f_c < f_1$.

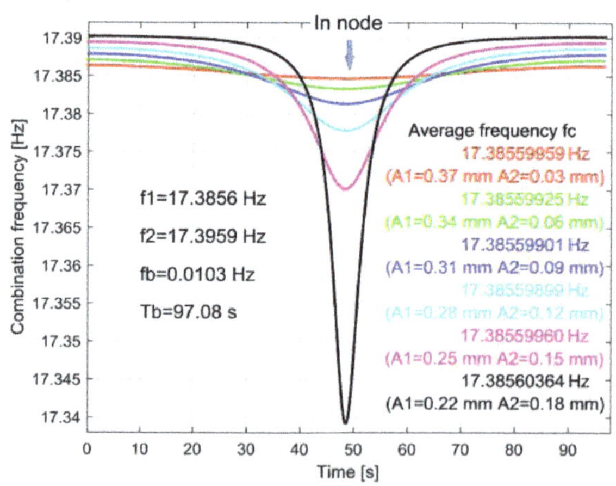

Figure 10. The evolution of the instantaneous combination frequency on simulated vibration beating during a beat period (the same condition as in Figure 9, except for the amplitudes relationship: $A_1 > A_2$.

There are two conclusions here, in contradiction with the literature [39]:

- The combination frequency f_c is not constant over a period T_b (even if its variation is not significant);
- The average value of the combination frequency f_c over a period T_b is practically the same as the frequency of the input vibration in the beating phenomenon ($y_{1s}(A_1,f_1)$ or $y_{2s}(A_2,f_2)$), whose amplitudes are higher (e.g., if $A_2 > A_1$ then the average $f_c \approx f_2$).

Some supplementary simulations for many other values of frequencies f_1 and f_2 (and consequently T_b) completely confirm these conclusions.

Figure 11 presents the evolution of the instantaneous combination frequency f_c during the vibration beating phenomenon (displacement of headstock) experimentally described in Figure 7.

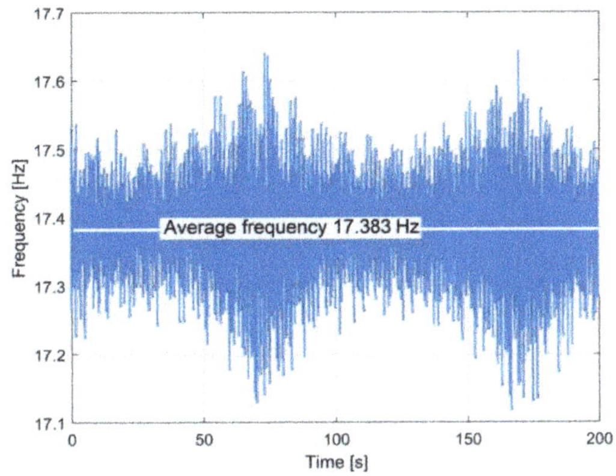

Figure 11. The evolution of the instantaneous combination frequency f_c during the vibration beating phenomenon (displacement of headstock) described in Figure 7.

Apparently, this is a very noisy evolution. The dominant component of the signal from Figure 7 is the displacement of resultant vibration $y_1 + y_2$. The frequency measurement method [42] is based on detection of zero-crossing moments of this signal (a topic discussed later on). It is obvious that many other additional vibrations of the lathe headstock (some of them with high frequency) disturb the accuracy of the zero-crossing detections, as a main reason for the noisy evolution from Figure 11.

The best information available in Figure 11 is the average value of the combination frequency f_c ($\overline{f}_c$ = 17.3830 Hz, very close to the rotational frequency of the main spindle, f_1 = 17.3856 Hz). Based on the previous conclusions from Figures 9 and 10 it is evident that the amplitude A_1 of unbalanced vibration generated by the main spindle (shaft 1) is higher than the amplitude A_2 of shaft 2 (so A_1 = 85 μm and A_2 = 33 μm, an item analysed before). As previously mentioned, the rotational frequency of shaft 2 is $f_2 = f_1 - f_b$ = 17.3752 Hz or $f_2 = f_1 + f_b$ = 17.3959 Hz.

Figure 12 presents a low-pass filtered evolution of the instantaneous combination frequency from Figure 11, (using a multiple moving average filter [43], as a well-known method to attenuate the signal noise).

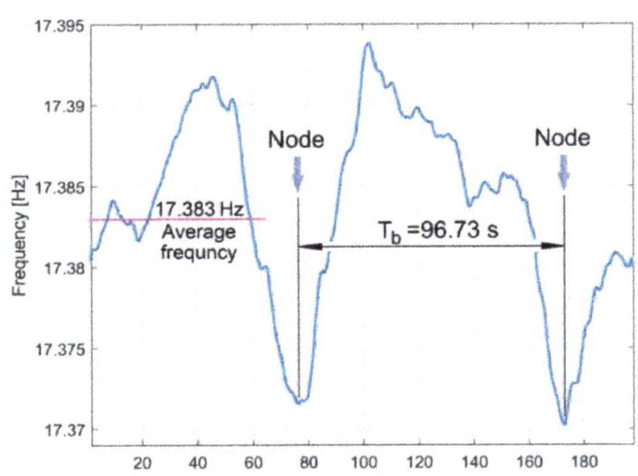

Figure 12. The evolution of the low pass filtered instantaneous combination frequency f_c during the vibration beating phenomenon described in Figure 7 (a low pass filtering of Figure 11).

Despite a relatively strong irregular variation of the combination frequency f_c (due to the variation of experimental conditions: e.g., the small variation of rotational speeds of shafts 1 and 2, caused mainly by the variation of frequency of the supplying voltages applied to the asynchronous driving motor, around a theoretical value of 50 Hz, as Figure 13 clearly indicates), the previous simulations and conclusions are fully experimentally confirmed. Three supplementary identical experiments confirm the evolution presented in Figure 12.

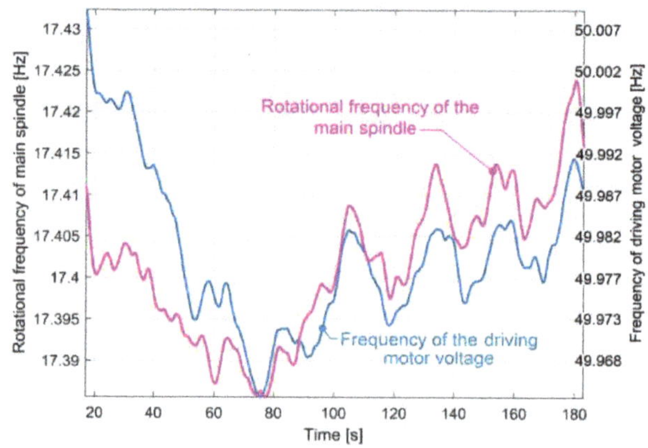

Figure 13. Low-pass filtered rotational frequency evolution of the main spindle and supplying voltage frequency evolution of the driving motor.

Firstly, in Figure 12 there are two negative peaks (for the two nodes in Figure 3 or Figure 7; each node produces a negative peak on f_c evolution, an item already discussed in the simulation from Figure 10) at a time interval very close to the beat period value T_b, already defined in Figure 3 (96.73 s here, compared with 96.6 s in Figure 3).

Secondly, as shown in Figure 14, a superposition of filtered frequency f_c evolution from Figure 12 (here in a conventional blue coloured description) over the experimental envelopes of vibration displacement in beating (the same as those depicted in Figure 7) indicates that the negative peaks of f_c are placed, as expected, in nodes.

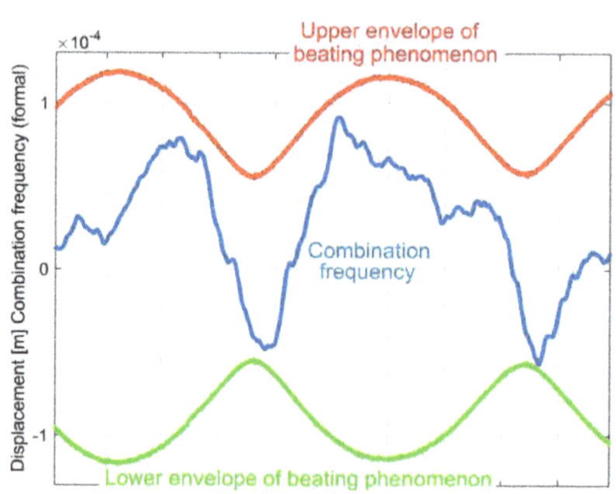

Figure 14. The position of negative peaks on the combination frequency (formally represented) relative to the position of nodes on the vibration beating phenomenon.

The small displacement to the right of the negative peaks of the filtered combination frequency evolution (as against the nodes on Figure 14) is not related to the numerical filtering. This is proved by the result of the simulation of Figure 14, as given in Figure 15 (with addition of pure harmonic signals y_{1s} and y_{2s} in vibration beating simulation).

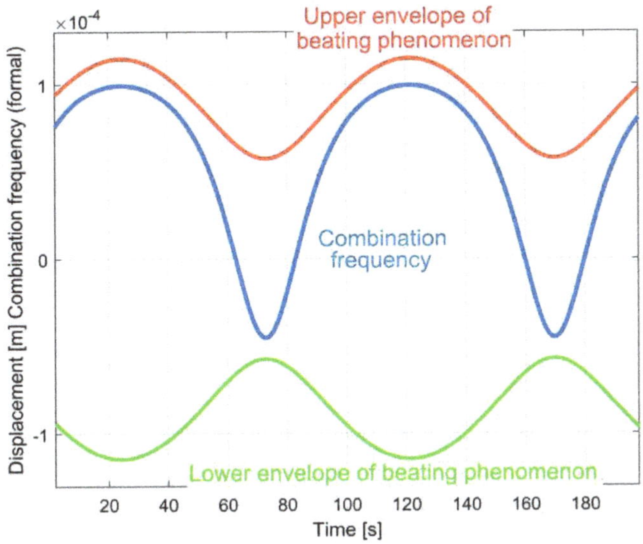

Figure 15. The result of a numerical simulation for the evolutions described in Figure 14.

This periodic pattern of filtered combination frequency f_c evolution experimentally revealed in Figures 12 and 14 (according to the simulations from Figures 10 and 15) is strongly attenuated if the amplitude A_2 becomes significantly lower than A_1 (and vice versa).

An important question is in order here due to a very small variation of filtered frequencies revealed before (less than 50 mHz full scale evolution in Figures 9, 10, 12, 13 and 15): how accurate is this frequency measurement method [42]?

In this measurement method (e.g., the measurement of the combination frequency f_c of vibration displacement signal from Figure 7), the computer-aided detection of the time interval between each two consecutive zero-crossing moments (t_{zcj} and t_{zcj+1}) of a periodical signal is used. This time interval defines a semi-period $T_c/2 = t_{zcj+1} - t_{zcj}$ as $T_c/2 = 1/2f_c$, or a value $f_c = 1/T_c$. When the result of multiplication of two successive displacements samples s_i and s_{i-1} (having the sampling times t_i and t_{i-1}, with $i > 1$) is negative or zero ($s_i \cdot s_{i-1} < 0$ or $s_i \cdot s_{i-1} = 0$) a zero-crossing moment is detected (e.g., t_{zcj}) and calculable as the abscissa of the intersection of a line segment defined by the points of coordinates (t_i, s_i) and (t_{i-1},s_{i-1}) on the t-axis (as x-axis in Figure 8b). The main reason for frequency measurement error $\varepsilon_f \neq 0$ is a consequence of calculation errors for two successive zero-crossing moments $\varepsilon_j \neq 0$ (for t_{zcj}) and $\varepsilon_{j+1} \neq 0$ (for t_{zcj+1}). These ε_j and ε_{j+1} errors are caused by the replacement of a harmonic evolution with a linear evolution between those two successive displacement samples involved in each zero-crossing moment definition. With $t_{i-1} - t_i = \Delta t$ (Δt being the sampling interval) the error $\varepsilon_j = 0$ only in three situations: (1) if $s_i = 0$ (the end of the line segment is placed on the t-axis, with $t_j = t_i$), (2) if $s_{i-1} = 0$ (the start of line segment is placed on t-axis, with $t_j = t_{i-1}$) and (3) if $-s_i = s_{i-1}$ (the middle of the line segment is placed on t-axis, with $t_j = t_{i-1} + \Delta t/2$). A similar approach is available for the next two successive samples (s_{i+h} and s_{i+h-1}) involved in the definition of t_{zcj+1} moment and ε_{j+1} error (with h as the integer part of the ratio $T_c/\Delta t$). If simultaneously ε_j and $\varepsilon_{j+1} = 0$ then $\varepsilon_f = 0$. Any other definition of sampling times generates frequency measurement errors $\varepsilon_f \neq 0$.

A computer-aided calculus was performed for frequency measurement error ε_f of a harmonic simulated signal with frequency 17.383 Hz (the average value of the combination frequency f_c) during a semi-period. Here 10,000 different values of sampling time t_1 (between 0 and Δt, with $\Delta t = 200$ μs, the same sampling interval as in Figures 3 and 7) and $t_2 = \Delta t - t_1$ (between Δt and 0) for the first two successive displacement samples involved in the calculus of the first zero-crossing moment t_{zc1} was used. Figure 16 describes the evolution of the frequency measurement error $\varepsilon_f(t_1)$.

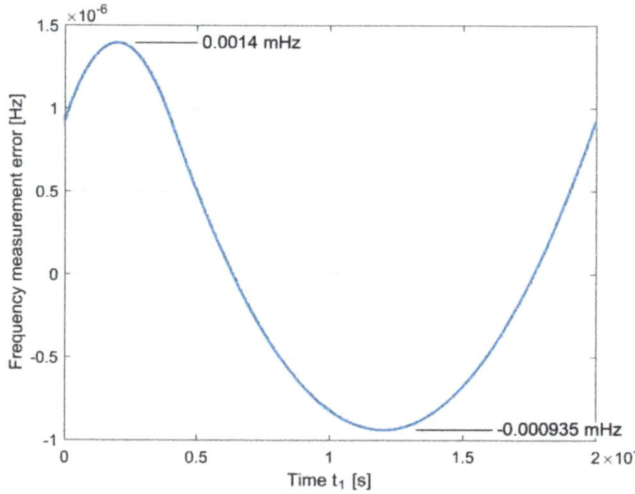

Figure 16. The evolution of the frequency measurement error ε_f (for a simulated harmonic signal with combination frequency $f_c = 17.383$ Hz) versus the evolution of the first sampling time ($t_1 = 0 \div \Delta t$, or $t_1 = 0 \div 200$ μs) involved in the first zero-crossing time (t_{zc1}) calculus.

As Figure 16 clearly indicates, the frequency measurement error ε_f of the combination frequency f_c is variable and placed between −0.000935 and +0.0014 mHz. The result of the measured frequency of the simulated signal is $17.383^{+1.4\ \mu\text{Hz}}_{-0.935\ \mu\text{Hz}}$ as a description of the accuracy measurement. Very similar limits for the ε_f error are calculated for a harmonic signal with frequency f_1. If the value of the frequency $f_c = 1/T_c$ used in simulation accomplishes the condition $T_c = h\Delta t$ (with h being an integer), then $\varepsilon_f = 0$ for any value $t_1 = 0 \div \Delta t$.

4.4. *The Influence of the Lathe Suspension Dynamics on Beating Vibrations Amplitude*

The relative high vibration displacement amplitude of the headstock during the beating phenomenon (as shown in Figure 7) has an evident explanation: the vibration frequencies f_1, f_2 and f_c as well, are close to the first resonant frequency (vibration mode) of the headstock and lathe on its foundation (as a single body mass–spring–damper vibratory system). This means that the dynamic amplification factors D_{af1} and D_{af2}, (involved in Equations (1) and (2)) are significantly higher than 1 (because of resonant amplification). In order to prove that, the resources of a very simple experiment performed with the same experimental setup are available: the evolution of headstock vibration velocity after an impulse excitation produced with a rubber mallet (hammer) in the same direction with y_1 and y_2 vibrations (as Figure 17 describes).

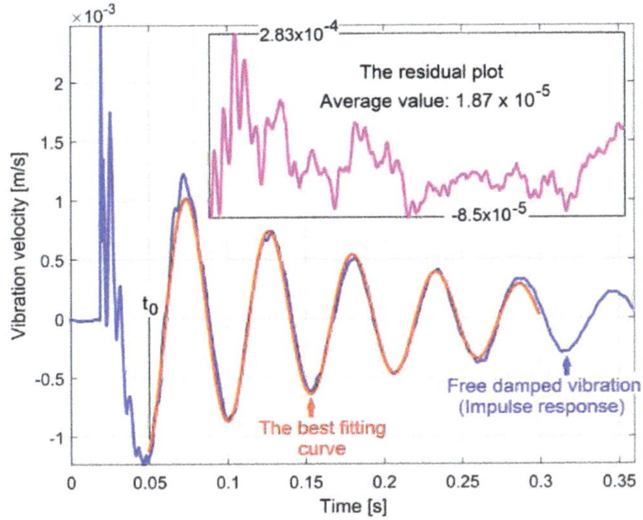

Figure 17. Some experimental results on signal processing related to free damped vibration of headstock after an impulse excitation (with a rubber mallet).

Here the blue curve partially depicts the free damped vibration velocity v_{fd} (acquired with the geophone sensor); the red coloured one depicts the best fitting curve of a part of the free response (with 25,000 samples and 500 ns sampling time). The curve fitting [44] was done in Matlab, with an adequate computer program specially designed for this paper, based on a known theoretical model of free viscous damped vibration velocity response [45]:

$$v_{fd}(t) = a \cdot e^{-bt} sin(p_1 t + \alpha) \tag{9}$$

The best fitting curve (in red in Figure 17) is described with $a = 1.170 \cdot 10^{-3}$ m/s, $b = 5.899$ s^{-1} (as damping constant), $p_1 = 117.952$ rad/s (as angular frequency of damped harmonic vibration) and $\alpha = 4.986$ rad (as phase angle at the origin of time t_0 on Figure 17). The angular natural frequency ($p = \sqrt{p_1^2 + n^2} = 118.099$ rad/s) and the damping constant b are useful in the definition [45] of dimensionless dynamic amplification factor D_{af} from forced vibrations of harmonic excitation (as happens during the beating phenomenon, assuming that the combination frequency is approximately constant):

$$D_{af} = \frac{1}{\sqrt{\left[1 - \left(\frac{\omega}{p}\right)^2\right]^2 + \left(2\frac{\omega}{p} \cdot \frac{b}{p}\right)^2}} \tag{10}$$

Here $\omega = 2\pi f$ is the angular frequency of harmonic excitation on frequency f. Based on previous experimental results of curve fitting (with b and p values in Equation (10)) Figure 18 presents the simulated evolution of D_{af} related to the frequency of excitation (1 ÷ 35 Hz range). Because of a low damping constant b, the system presents resonant amplification, with a maximum value $D_{af} = 10.01$ on $f = 18.749$ Hz frequency.

Based on the previous experimentally determined frequencies f_1 and f_2, with $f = f_1 = 17.3856$ Hz gives the result $D_{af1} = 5.831$ and with $f = f_2 = 17.3752$ Hz (or $f = f_2 = 17.3959$ Hz) the result is $D_{af2} = 5.803$ (or $D_{af2} = 5.859$). For $f = \overline{f}_c = 17.383$ Hz (Figure 12) the result is $D_{afc} = 5.824$ (the coordinates of point A on Figure 18). This means that, because of mechanical resonance, the vibration amplitude generated by the beating phenomenon of the headstock and the lathe on its foundation (already revealed in Figure 7) is amplified on average by 5.824 times.

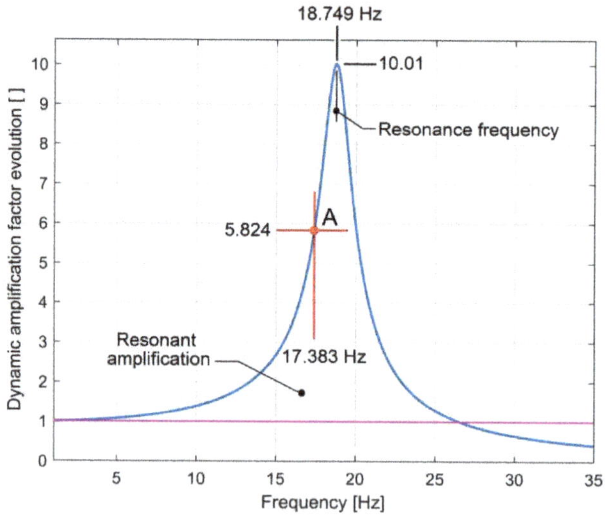

Figure 18. The evolution of the dynamic amplification factor D_{af} generated by the headstock foundation in the resonance area, based on Equation (10) and experimental free damped response analysis.

Besides the amplification of the vibration, the resonant behaviour also introduces a significant shift of phase γ between the excitation (unbalancing) force and the vibration displacement, theoretically described [45] as depending on ω (and excitation frequency f as well) with the equation:

$$\gamma = arctan[\frac{2\frac{b}{p}\frac{\omega}{p}}{1-\left(\frac{\omega}{p}\right)^2}] \tag{11}$$

With the b and p values previously determined, the values of shift of phase calculated for each frequency are: $\gamma_1 = 0.5691$ rad for $f = f_1 = 17.3856$ Hz and $\gamma_2 = 0.5656$ rad (or $\gamma_2 = 0.5725$ rad) for $f = f_2 = 17.3752$ Hz (or $f_2 = 17.3959$ Hz). For $f = \overline{f}_c = 17.383$ Hz the result is $\gamma_c = 0.5682$ rad.

The knowledge of both of these resonant characteristics (the value of the dynamic amplification factor D_{af} and especially the phase shift γ) is important for a next approach of the dynamic balancing of these two shafts placed inside the headstock.

As a general comment, we should mention that the resonance behaviour of this low damped vibratory system—as previously mentioned—is a consequence of the disponibility of this system to absorb modal mechanical energy. The system works as a narrow-band modal energy absorber [46].

5. Conclusions and Future Work

Some specific features of the beating vibration phenomenon discovered on a headstock lathe have been revealed in this paper.

An experimental description (with theoretical approaches based on simulations) of this beating vibration phenomenon with very low beat frequency (1/96.6 Hz) was performed. The beating phenomenon occurs due to the addition of vibrations produced by two unbalanced shafts, rotating with very close instantaneous angular speeds (rotating frequencies), with constructive interference in anti-nodes and destructive interference in nodes.

The absolute velocity signal of vibration beating (delivered by a vibration electro-dynamic sensor placed on the headstock) was converted into a displacement signal. For this purpose, a fully confirmed method of numerical integration (antiderivative calculus), with theoretical and experimental approaches was applied. This method is deduced from the approximation of the formula for the first derivative of displacement, as a backward finite difference [43]). An appropriate technique of correction of this numerical antiderivative calculus method was also introduced (mainly by removing the low frequency displacement signal component generated by numerical integration). Thus, an absolute velocity vibration sensor together with a numerical integration procedure plays the role of an absolute vibration displacement sensor.

A consistent part of the research was focused on the resultant vibration displacement signal, mainly on the evolution of frequency (or the combination frequency f_c) related to the nodes and anti-nodes position. It was theoretically discovered (by simulation) and was experimentally proved that, in opposition to the literature reports, the combination frequency is not constant, and the definition of its average value is wrong. The evolution of the combination frequency has a specific periodic pattern (having the same frequency as the beat frequency) with small variation (tens of millihertz) and negative or positive peaks placed in beating nodes. The appearance of these peaks (negative or positive) depends on the relationship between amplitudes and frequencies of vibrations involved in the beating phenomenon. The small variation of frequency inside the pattern and the correlation between the frequencies of different experimental signals (the combination frequency f_c, the rotation frequency f_1 of the main spindle, and the supply voltage frequency of the driving motor) have been correctly described as a result of a high accuracy procedure of frequency measurement, developed in a previous work [42] and successfully applied here. It was proved on a simulated signal (having a frequency of 17.383 Hz, equal to the average value of the combination frequency in vibration beating phenomenon) that this procedure has less than ±1.5 μHz measurement error.

The influence of the behaviour of the headstock and lathe foundation dynamics (as a rigid body placed on a spring–damper system) on the vibration induced by unbalanced rotors and the beating phenomenon was also investigated. Based on computer-aided analysis of free damped viscous response (by curve fitting), the characteristics of foundation dynamics were experimentally revealed (mainly the values of natural angular frequency and the damping constant). Considering these values, the dynamic amplification factors of vibrations (mainly of the resultant vibration) and phase shift between centrifugal forces (as excitation forces produced by unbalanced rotary shafts) and the vibrations generated by these forces were calculated.

For each experiment, numerical simulation and signal processing procedures, several computer programs written in Matlab were successfully used.

In the future, the theoretical and experimental approaches will be focused on the influence of dynamic unbalancing and vibration beating on the active and instantaneous electrical power absorbed by the driving motor of the headstock. There is a logical reasoning for these approaches: the headstock vibration motion (especially during the resonant amplification behaviour revealed in Figure 18) should be mechanically powered. Of course, the instantaneous and active mechanical power (difficult to measure) is delivered by the driving motor as an equivalent of instantaneous and active electric power absorbed from the electrical supply network (easier to measure).

Several theoretical and experimental studies on computer-aided balancing of each rotary shaft inside the lathe headstock will be performed (using two absolute velocity sensors and an appropriate method of computer-assisted experimental balancing). A study on the vibration beating phenomenon produced by more than two unbalanced rotary bodies will be done.

Author Contributions: Conceptualization, D.-F.C. and M.H.; methodology, D.-F.C. and M.H.; software, F.N. and C.-G.D.; validation, M.H. and C.-G.D.; formal analysis, D.-F.C. and M.H.; investigation, D.-F.C., C.-G.D. and M.H.; resources, M.H.; data curation, M.H.; numerical integration method, M.H.; writing—original draft preparation D.-F.C. and M.H.; writing—review and editing, G.D., M.I.; visualization, M.H. and F.N.; supervision, M.H. and C.-G.D.; project administration, M.H. and C.-G.D.; funding acquisition, D.-F.C., G.D. and M.I. All authors have read and agreed to the published version of the manuscript.

Funding: This work was accomplished with the support of the COMPETE project nr. 9PFE/2018, funded by the Romanian Government.

Acknowledgments: This work was accomplished with the support of COMPETE project nr. 9PFE/2018, funded by the Romanian Government.

Conflicts of Interest: The authors declare no conflict of interest.

Nomenclature

A_1, A_2	The amplitudes of vibrations y_1, y_2 [m]
ae^{-bt}	The envelope of free viscous damped vibration velocity response [m/s]
b	The damping constant [s^{-1}]
C	The constant of velocity signal integration [m]
D_{af}	Theoretical dynamic amplification factors of vibrations [ ]
D_{af1},D_{af2}	Dynamic amplification factors of vibrations y_1, y_2 produced by shafts 1, 2 [ ]
D_{afc}	Dynamic amplification factor of resultant vibration $y_1 + y_2$ at average frequency f_c [ ]
dy_1/dt, dy_2/dt	The derivative of vibration displacements y_1, y_2 (vibration velocities) [m/s]
f	The frequency of harmonic excitation of the lathe headstock [Hz]
F_1, F_2	The horizontal projection of the rotary unbalance forces generated by shafts 1 and 2 [N]
f_1, f_2	The frequency of vibrations y_1, y_2 [Hz]
f_b	The beat frequency [Hz]
f_c	The frequency of the resultant vibration $y_1 + y_2$, or combination frequency [Hz]
IAS	Instantaneous angular speed [rad/s]
k	The stiffness of headstock and lathe foundation [N/m]
m_1, m_2	Unbalance mass on rotary shafts 1, 2 [Kg]
n	A natural number involved in the definition of the beat period T_b
p	The natural angular frequency [rad/s]
p_1	The angular frequency of damped harmonic vibration [rad/s]
r_1, r_2	The distance between the center of the unbalance mass and the rotation axis on shafts 1, 2 [m]
s	The addition of vibration displacements $s = y_1 + y_2$ [m]
s_i, s_{i+1}	Two successive displacement samples of vibration [m]
s_{i+h}, s_{i+h-1}	Two successive displacement samples of vibration [m]
t	Time [s]
t_0	The origin of time for the theoretical model of free damped vibration velocity [s]
t_{zcj}, t_{zcj+1}	Two successive zero-crossing moments of the displacement vibration signal involved in frequency measurement [s]
Δt	Sampling interval for a numerically described signal [s]
T_1, T_2	The periods of vibrations y_1, y_2 [s]
T_b	The beat period, with $T_b = 1/f_b$ [s]
T_c	The period of the resultant vibration $y_1 + y_2$, with $T_c = 1/f_c$ [s]
v	The velocity of the resultant vibration in beating [m/s]
v_{fd}	The vibration velocity of the headstock during a free damped response [m/s]
v_i	A sample of the vibration velocity [m/s]
y_1, y_2	The vibration displacement generated by shafts 1, 2 [m]
y_{1s}, y_{2s}	Simulated vibration displacement signals [m]
α	The phase angle at the origin of time t_0 for a theoretical model of free damped vibration velocity [rad/s]
ε_f	The error in the frequency measurement [Hz]
ε_j, ε_{j+1}	The calculus errors for two successive zero-crossing moments [s]
γ	The shift of phase between the excitation force and the vibration displacement in the free damped response [rad]
θ_1, θ_2	The instantaneous value of the angle of centrifugal forces to the horizontal direction [rad]
φ_1, φ_2	The values of θ_1 and θ_2 at the origin of time, $t = 0$ [rad]
ω	The angular frequency of harmonic excitation of the lathe headstock [rad/s]
ω_1, ω_2	The instantaneous angular speed of the rotary shafts 1, 2 [rad/s]

References

1. Muszyńska, A. *Rotordynamics*; CRC Press: Boca Raton, FL, USA, 2005; ISBN 978-0-8247-2399-6.

2. Wang, Y.; Yuan, X.; Sun, R. Critical techniques of design for large scale centrifugal shakers. *World Inf. Earthq. Eng.* **2011**, *27*, 113–123.
3. Anekar, N. Design and testing of unbalanced mass mechanical vibration exciter. *IJRET* **2014**, *3*, 107–112.
4. Conley, K.; Foyer, A.; Hara, P.; Janik, T.; Reichard, J.; D'Souza, J.; Tamma, C.; Ababei, C. Vibration alert bracelet for notification of the visually and hearing impaired. *J. Open Hardw.* **2019**, *3*, 4. [CrossRef]
5. Fontana, F.; Papetti, S.; Jarvelainen, H.; Avanzini, F. Detection of keyboard vibrations and effects on perceived piano quality. *J. Acoust. Soc. Am.* **2017**, *142*, 2953–2967. [CrossRef]
6. Ippolito, R.; Settineri, L.; Sciamanda, M. Vibration monitoring and classification in centerless grinding. In *International Centre for Mechanical Sciences (Courses and Lectures)*; Kuljanic, E., Ed.; Springer: Vienna, Austria, 1999; p. 406. [CrossRef]
7. Merino, R.; Barrenetxea, D.; Munoa, J.; Dombovari, Z. Analysis of the beating frequencies in dressing and its effect in surface waviness. *CIRP Annal. Manuf. Technol.* **2019**, *68*, 353–356. [CrossRef]
8. Mayoof, F.N. Beating phenomenon of multi-harmonics defect frequencies in a rolling element bearing: Case study from water pumping station. *World Acad. Sci. Eng. Technol.* **2009**, *57*, 327–331.
9. Pham, M.T.; Kim, J.-M.; Kim, C.H. Accurate bearing fault diagnosis under variable shaft speed using convolutional neural networks and vibration spectrogram. *Appl. Sci.* **2020**, *10*, 6385. [CrossRef]
10. Park, J.; Lee, J.; Ahn, S.; Jeong, W. Reduced ride comfort caused by beating idle vibrations in passenger vehicles. *Int. J. Ind. Ergon.* **2017**, *57*, 74–79. [CrossRef]
11. Huang, P.; Lee, W.B.; Chan, C.Y. Investigation of the effects of spindle unbalance induced error motion on machining accuracy in ultra-precision diamond turning. *IJMTM* **2015**, *94*, 48–56. [CrossRef]
12. Norfield, D. *Practical Balancing of Rotating Machinery*; Elsevier Science: Amsterdam, The Netherlands, 2005.
13. Thearle, E.L.; Schenectady, N.Y. Dynamic balancing of rotating machinery in the field. *ASME Trans.* **1934**, *56*, 745–775.
14. Kelm, R.; Kelm, W.; Pavelek, D. Rotor Balancing Tutorial. In Proceedings of the 45th Turbomachinery and 32th Pump Symposia, Houston, TX, USA, 12–16 September 2016.
15. Wijk, V.; Herder, J.L.; Demeulenaere, B. Comparison of various dynamic balancing principles regarding additional mass and additional inertia. *J. Mech. Robot.* **2009**, *1*, 4. [CrossRef]
16. Isavand, J.; Kasaei, A.; Peplow, A.; Afzali, B.; Shirzadi, E. Comparison of vibration and acoustic responses in a rotary machine balancing process. *Appl. Acoust.* **2020**, *164*, 107258. [CrossRef]
17. Bertoneri, M.; Forte, P. Turbomachinery high speed modal balancing: Modelling and testing of scale rotors. In Proceedings of the 9th IFToMM International Conference on Rotor Dynamics. Mechanisms and Machine Science, Milan, Italy, 22–25 September 2017; Pennacchi, P., Ed.; Springer: Cham, Switzerland, 2017. [CrossRef]
18. Caoa, H.; Dörgelohb, T.; Riemerb, O.; Brinksmeier, E. Adaptive separation of unbalance vibration in air bearing spindles. *Proc. CIRP* **2017**, *62*, 357–362. [CrossRef]
19. Majewski, T.; Szwedowicz, D.; Melo, M.A.M. Self-balancing system of the disk on an elastic shaft. *J. Sound Vib.* **2015**, *359*, 2–20. [CrossRef]
20. Hashimoto, F.; Gallego, I.; Oliveira, J.F.G.; Barrenetxea, D.; Takahashi, M.; Sakakibara, K.; Stalfelt, H.O.; Staadt, G.; Ogawa, K. Advances in centerless grinding technology. *CIRP Annal.* **2012**, *61*, 747–770. [CrossRef]
21. Zhang, Z.X.; Wang, L.Z.; Jin, Z.J.; Zhang, Q.; Li, X.L. Non-whole beat correlation method for the identification of an unbalance response of a dual-rotor system with a slight rotating speed difference. *MSSP* **2013**, *39*, 452–460. [CrossRef]
22. Zhang, Z.X.; Zhang, Q.; Li, X.L.; Qian, T.L. The whole-beat correlation method for the identification of an unbalance response of a dual-rotor system with a slight rotating speed difference. *MSSP* **2011**, *25*, 1667–1673. [CrossRef]
23. Lee, Y.J. 8.03SC Physics III: Vibrations and Waves. Fall 2016. Massachusetts Institute of Technology: MIT Open CourseWare. 2016. Available online: https://ocw.mit.edu (accessed on 17 September 2020).
24. Kim, S.H.; Lee, C.W.; Lee, J.M. Beat characteristics and beat maps of the King Seong-deok Divine Bell. *JSV* **2005**, *281*, 21–44. [CrossRef]
25. Yuan, L.; Järvenpää, V.M. On paper machine roll contact with beating vibrations. *Appl. Math. Mech.* **2006**, *6*, 343–344. [CrossRef]
26. Gao, H.; Meng, X.; Qian, K. The Impact analysis of Beating vibration for active magnetic bearing. *IEEE Access* **2019**, *7*, 134104–134112. [CrossRef]
27. Preumont, A. *Twelve lectures on Structural Dynamics*; Université Libre de Bruxelles: Brussels, Belgium, 2012.

28. Horodinca, M.; Seghedin, N.; Carata, E.; Boca, M.; Filipoaia, C.; Chitariu, D. Dynamic characterization of a piezoelectric actuated cantilever beam using energetic parameters. *Mech. Adv. Mater. Struct.* **2014**, *21*, 154–164. [CrossRef]
29. Endo, H.; Suzuki, H. Beating vibration phenomenon of a very large floating structure. *J. Mar. Sci. Technol.* **2018**, *23*, 662–677. [CrossRef]
30. Carrino, S.; Nicassio, F.; Scarselli, G. Subharmonics and beating: A new approach to local defect resonance for bonded single lap joints. *J. Sound Vib.* **2019**, *456*, 289–305. [CrossRef]
31. Cattarius, J.; Inman, D.J. Experimental verification of intelligent fault detection in rotor blades. *Int. J. Syst. Sci.* **2000**, *31*, 1375–1379. [CrossRef]
32. Morrone, A. Seismic vibration testing with sine beats. *Nucl. Eng. Des.* **1973**, *24*, 344–356. [CrossRef]
33. Huang, H.H.; Chen, K.S. Design, analysis, and experimental studies of a novel PVDF–based piezoelectric energy harvester with beating mechanisms. *Sens. Actuators A Phys.* **2016**, *238*, 317–328. [CrossRef]
34. Basovich, S.; Arogeti, S. Identification and robust control for regenerative chatter in internal turning with simultaneous compensation of machining error. *Mech. Syst. Sign. Process.* **2021**, *149*, 107208. [CrossRef]
35. Li, D.; Cao, H.; Liu, J.; Zhang, X.; Chen, X. Milling chatter control based on asymmetric stiffness. *Int. J. Mach. Tools Manuf.* **2019**, *147*, 103458. [CrossRef]
36. Yana, Y.; Xub, J.; Wiercigrochc, M. Influence of work piece imbalance on regenerative and frictional grinding chatters. *Proc. IUTAM* **2017**, *22*, 146–153. [CrossRef]
37. Kimmelmann, M.; Stehle, T. Measuring unbalance-induced vibrations in rotating tools. *MATEC Web Conf.* **2017**, *121*, 03012. [CrossRef]
38. Gohari, M.; Eydi, A.M. Modelling of shaft unbalance: Modelling a multi discs rotor usingK-nearest neighbor and decision tree algorithms. *Measurement* **2020**, *151*, 107253. [CrossRef]
39. Hui, S.; Xiaoqiang, H.; Zhiqiang, D.; Lichang, X.; Haishui, J. Research on Low Frequency Noise Caused by Beat Vibration of Rotary Compressor. In *Compressor Engineering Conference, Proceedings of the 24th International Compressor Engineering Conference, Purdue, Italy, 9–12 July 2018*; Purdue University: Purdue, Italy, 2018; p. 2528.
40. Available online: http://hgsindia.com/www.hgsproducts.nl/Pdf/196216HG-4%20V%201.1.pdf (accessed on 17 September 2020).
41. Available online: https://www.picotech.com/oscilloscope/4224-4424/picoscope-4224-4424-overview (accessed on 17 September 2020).
42. Horodinca, M.; Ciurdea, I.; Chitariu, D.F.; Munteanu, A.; Boca, M. Some approaches on instantaneous angular speed measurement using a two-phase n poles AC generator as sensor. *Measurement* **2020**, *157*, 107636. [CrossRef]
43. Chapra, S.C.; Canale, R.P. *Numerical Methods for Engineers*, 7th ed.; McGraw Hill Education: New York, NY, USA, 2015.
44. Arlinghaus, S. *Practical Handbook of Curve Fitting*, 1st ed.; CRC Press: Boca Raton, FL, USA, 2020.
45. Shabanna, A.A. *Theory of Vibrations. An introduction*; Springer: New York, NY, USA, 1991.
46. Schmitz, T. Modal interactions for spindle, holders, and tools. *Proc. Manuf.* **2020**, *48*, 457–465. [CrossRef]

applied sciences

MDPI

Article

In Vivo Evaluation of Biocompatibility of Three Biomaterials Used in Endodontics for Prosthetic Purposes in Complex Rehabilitation Treatment

Liana Aminov [1,†], Eusebiu Viorel Sindilar [2,†], Aurelian Sorin Pasca [3,*,†], Cristina Antohi [1,†], Yllka Decolli [1,†], Ovidiu Stamatin [4,*,†], Lupu Iulian Costin [4,†], Bogdan Petru Bulancea [4,†], Laurian Francu [5,†], Eugen Mihalas [6,†] and Laura Elisabeta Checherita [1,†]

1 2nd Dental Medicine Department, "Gr. T. Popa" University of Medicine and Pharmacy, 700115 Iasi, Romania; lianaaminov@yahoo.com (L.A.); cristinaantohi_med@yahoo.com (C.A.); ydecolli@yahoo.com (Y.D.); checherita.laura@gmail.com (L.E.C.)
2 Department of Surgery, Faculty of Veterinary Medicine, "Ion Ionescu de la Brad" University of Life Sciences, 700490 Iași, Romania; esindilar@yahoo.com
3 Department of Pathology, Faculty of Veterinary Medicine, "Ion Ionescu de la Brad" University of Life Sciences, 700490 Iași, Romania
4 3rd Dental Medicine Department, "Gr. T. Popa" University of Medicine and Pharmacy, 700115 Iasi, Romania; lupu_costin@yahoo.com (L.I.C.); bulanceabogdan20@yahoo.com (B.P.B.)
5 Anatomy Medical Department, "Gr. T. Popa" University of Medicine and Pharmacy, 700115 Iasi, Romania; laurianfrancu@yahoo.com
6 1st Dental Medicine Department, "Gr. T. Popa" University of Medicine and Pharmacy, 700115 Iasi, Romania; mihalas.eugen@umfiasi.ro
* Correspondence: passorin@yahoo.com (A.S.P.); ovidiustamatin@yahoo.com (O.S.)
† All authors have equal scientifical contribution to this article.

Citation: Aminov, L.; Sindilar, E.V.; Pasca, A.S.; Antohi, C.; Decolli, Y.; Stamatin, O.; Costin, L.I.; Bulancea, B.P.; Francu, L.; Mihalas, E.; et al. In Vivo Evaluation of Biocompatibility of Three Biomaterials Used in Endodontics for Prosthetic Purposes in Complex Rehabilitation Treatment. *Appl. Sci.* **2021**, *11*, 6519. https://doi.org/10.3390/app11146519

Academic Editor: Andrei Victor Sandu

Received: 28 May 2021
Accepted: 13 July 2021
Published: 15 July 2021

Abstract: The ideal biomaterial used in endodontics in the process of sealing the radicular canals should possess a group of qualities for a predictable outcome: biocompatibility, initiation of ontogenesis and cementogenesis, ease of handling, sufficient manipulation time, and convenient price. For a perfect sealing, the root canal treatment can be followed by prosthetic restoration. This study of biocompatibility aims to determine the quantification of the local reaction following the implantation of three biomaterials in the rabbit subcutaneous connective tissue. The used biomaterials with particular reparative properties are: MTA (Mineral Trioxide Aggregate, Dentsply, Tulsa Dental, Johnson City, TN, USA), Sealapex (Kerr, Switzerland), and DiaRoot BioAggregate (Innovative BioCaramix Inc, Vancouver, BC, Canada). The first two biomaterials (MTA, Sealapex) are already being used in endodontic treatments, and the latter was newly introduced during the concrete development of the study. This is an experimental study focused on qualitative and quantitative analysis based on histopathological examination and underlined by the positive result of the study undertaken of the applicability of oral rehabilitation treatments, increasing patients' quality of life by a significant proportion of 95%, and generating optimal functionality of the stomatognathic system with prosthetic devices as well as accomplishing the objectives of homeostasis.

Keywords: biomaterials; subcutaneous implantation; biocompatibility; prosthetic restoration; oral rehabilitation homeostasis

1. Introduction

An important objective in endodontic therapy is to induce periapical bone repair and to stimulate cementogenesis. The biomaterials used are placed in close contact with both the soft and hard periodontal tissues [1]. The biomaterial may cause local or systemic side effects due to direct contact or through the leakage of the substances released from the material into the periodontal tissue or alveolar bone [2,3].

The biological compatibility of the root sealant is of key importance, because in clinical conditions these materials are put in direct contact with vital tissues, and the tissue response to these materials can influence the outcome of endodontic treatment [3,4].

Considering the above-mentioned aspects, the ideal endodontic repair and sealing materials should possess some of these properties: adherent to the canal walls, in order to ensure a tight seal of the root canal system; nontoxic; easy to manipulate; well tolerated by the periradicular tissues; promoting bone healing, non-resorbable; radiopaque; unaffected by the presence of moisture; and not staining the surrounding tissues [5,6].

Several methods have been used to evaluate the biocompatibility of endodontic cements. One of the most practical and widely used methods is the implantation of the material in the subcutaneous connective tissue in rats [7,8]. The irritating effect of the materials can be evaluated by histopathological examination of the tissue response around the implants.

Ongoing studies aim to create a well-adapted dental structure as a support for further biocompatible restorations in the context of complex oral rehabilitation treatments, so as to reduce the occlusal forces that can be exerted through these materials to keep the oral homeostasis in normal parameters [9].

Aim of Study

Our biocompatibility experimental study aims to evaluate the local reaction following the implantation of three biomaterials in the rabbit subcutaneous connective tissue.

The main objective is to make a comparative assessment to understand the response of living tissues in direct contact to the biomaterials, guided by the answer suggested by the next part of the practical applicability study, by using prosthetic and complex rehabilitation means in the context of dysfunctional syndrome, which recorded a positive result followed over time by their response to applied treatments, following the principles of homeostasis.

2. Materials and Methods

The materials used were as follows:

- MTA (Mineral Trioxide Aggregate, Dentsply, Tulsa Dental, Johnson City, TN, USA) is a material with a highly efficient antibacterial effect and is alkaline, made of calcium hydroxide, bismuth oxide (Bi_2O_3), calcium sulfate ($CaSO_4$), tricalcium silicate ($(CaO)_3{\cdot}SiO_2$), dicalcium silicate ($(CaO)_2{\cdot}SiO_2$), tricalcium aluminate ($(CaO)_3{\cdot}Al_2O_3$).
- Sealapex (Kerr, Switzerland)—used for root canal sealing—has the following chemical composition: barium sulfate, titanium dioxide, zinc oxide, calcium hydroxide, butilbenzen, sulfonamide, zinc stearate.
- DiaRoot BioAggregate (Innovative BioCeramix, Inc., Vancouver, BC, Canada) is a material similar in structure to MTA that additionally contains ceramic nanoparticles. It has proven antiseptic proprieties and at the same time stimulates cementogenesis. The chemical composition includes: calcium silicate, calcium hydroxide, hydroxyapatite, tantalum oxygen (Ta_2O_5).

The used endodontic materials were prepared according to the instructions of the producer and then were introduced in polyethylene tubes 10 mm in length and 1.5 mm in diameter.

2.1. In Vivo Experiment

Twenty-one Belgian Giant rabbits, aged 4 months and weighing 3.5 kg ($\pm$50 g), raised and fed in identical conditions (food and water ad libitum) were used.

The rabbits were divided into 4 groups:

- Group A—6 rabbits—receiving MTA implants;
- Group B—6 rabbits—receiving Sealapex implants;
- Group C—6 rabbits—receiving DiaRoot implants;
- Group D (control)—3 rabbits—receiving empty polyethylene tube implants.

The experimental period lasted 60 days, while the rabbits were kept in similar conditions and were fed identical food, except 24 h prior to the dental implant surgery when they received water but no food. All rabbits were fed a normal diet until the end of the study.

2.2. Surgical and Post-Operative Protocol

The surgical interventions were made under general anesthesia and aseptic conditions. Prior to anesthesia, rabbits received atropine premedication (0.02 mg/kg; Atropina, Pasteur Institute, Bucharest, Romania).

Anesthesia was induced with xylazine (0.1 mg/kg i.m, Xylazine Bio 2%, Bioveta, Czech Republic) and ketamine (10 mg/kg i.m., Ketaminol® 10, Intervet International GmbH, Neufahrn bei Freising, Germany).

Preoperatively, the lateral thoracic regions were shaved and disinfected with antiseptic solution 96% (Videne; Adams Healthcare Ltd., Birmingham, UK).

The rabbits were first placed in the left and then in the right lateral decubitus position.

Skin incisions were made for implanting the tubes (Figure 1).

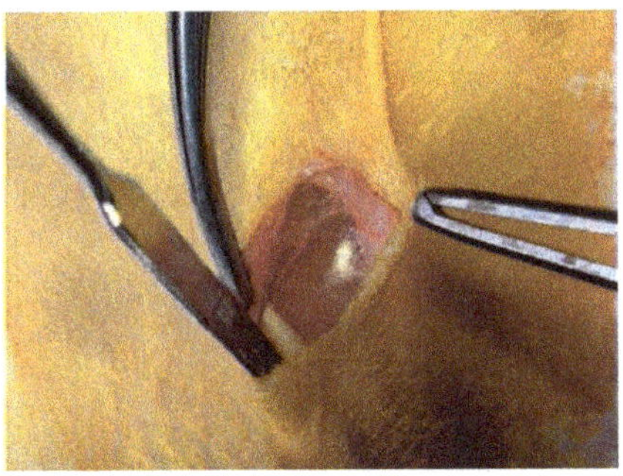

Figure 1. Subcutaneous technique for implantation of biomaterials.

Tubes were inserted as deep as the created tissue pocket allowed. In the end, the surgical wound was sutured with non-absorbable suture thread (Figure 2).

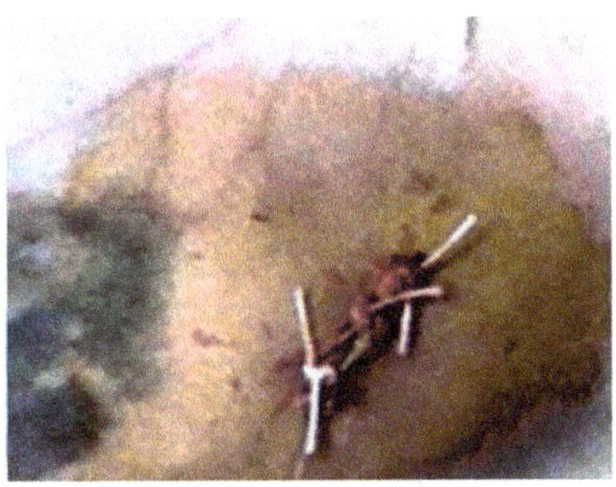

Figure 2. Wound suture after implant.

For 3 postoperative days, each rabbit was given an analgesic (carprofen 4 mg kg^{-1} i.sc; Rimadyl®, Pfizer, Tadworth, UK) and prophylactic antibiotic (7.5 mg kg^{-1} amoxicillin; VEYX® YL LA 200, Veyx-Pharma GmbH, Schwarzenborn, Germany).

At 7, 30, and 60 days after implantation, 2 rabbits in groups A, B, and C and one rabbit in the control group (D) were killed using T61 solution (2 mL/kg, i.p., MSD Animal Health GmbH, Cuxhaven, Germany).

2.3. Histopathological Protocol

Immediately after animals were sacrificed [1], fragments of subcutaneous connective tissue were collected from biomaterial implant sites (Figure 3).

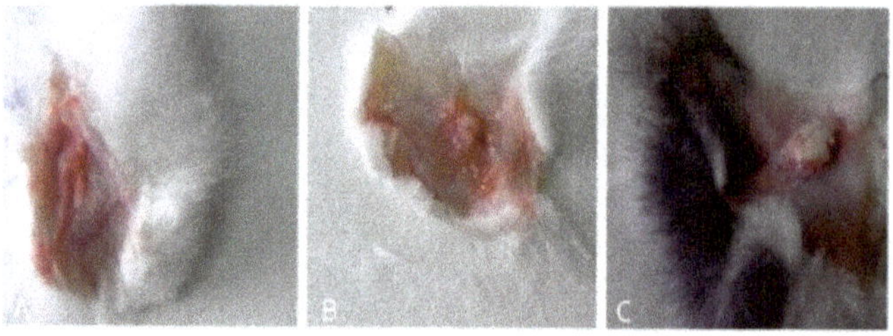

Figure 3. Samples selected for sampling: (**A–C**) lax subcutaneous tissue after implantation of three biomaterials.

Briefly after sampling, all specimens were fixed in 10% buffered formalin and embedded in parafine with a tissue processor Leica TP1020 (Leica Microsystems GmbH, Wetzlar, Germany). Sections of 5μm thickness were obtained with a Microtome SLEE CUT 6062 (SLEE Medical GmbH, Mainz, Germany) and then de-paraffinized and stained by the Masson trichrome techniques. The qualitative histology was performed from stained sections using a light microscope Leica DM 750 (Leica Microsystems GmbH, Germany) with an attached digital camera Leica ICC50 HD (Leica Microsystems GmbH, Germany) Germany). The photographs were taken with Leica Application Suit Software (LAS) version 4.2.

3. Results

When the tissue sample was collected, 7 days after implantation, a well-defined fibrous capsule of large size (on average = 3.5/14.2 mm in group B; close values in group A, on average = 3.1/13.6 mm; and the lowest size in group C, on average = 1.9/10.5 mm) was formed (Figure 3A–C).

Seven days after the implantation, the implant sites of the three biomaterials in the four groups (A, B, C, and D) were histologically examined.

In group A (implanted with MTA) we noticed a well-defined area of peripheral necrosis (surrounding the biomaterial) in which incompletely resorbed MTA fragments and an intense influx of leukocytes consisting of macrophages, histiocytic cells, and neutrophils were present, together with a high number of fibroblasts and collagen fibers (Figures 4 and 5).

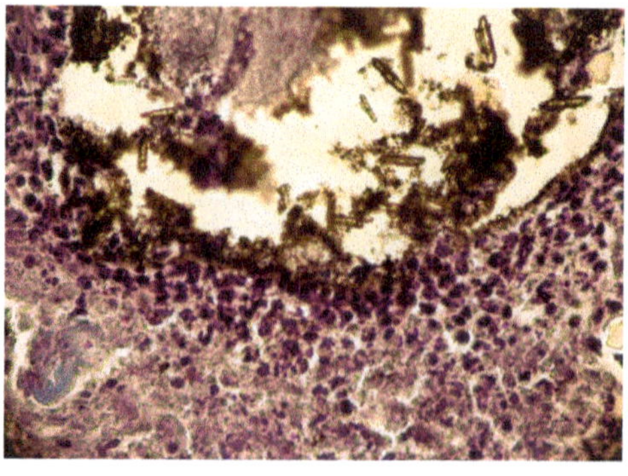

Figure 4. Group A. Subcutaneous conjunctive tissue. Area of necrosis, intense leukocyte infiltration.

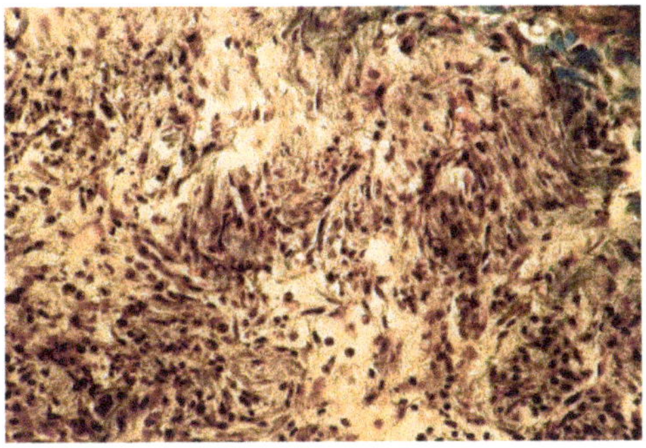

Figure 5. Group A. Buffer zone, consisting of fibroblast proliferation and collagen fiber synthesis. Col. Trichrome Masson, ×200.

The implanted material was partially reabsorbed. Local mineralization, probably due to the diffusion of released calcium ions from the implant material, was seen in some muscle fibers of the skin (Figure 6).

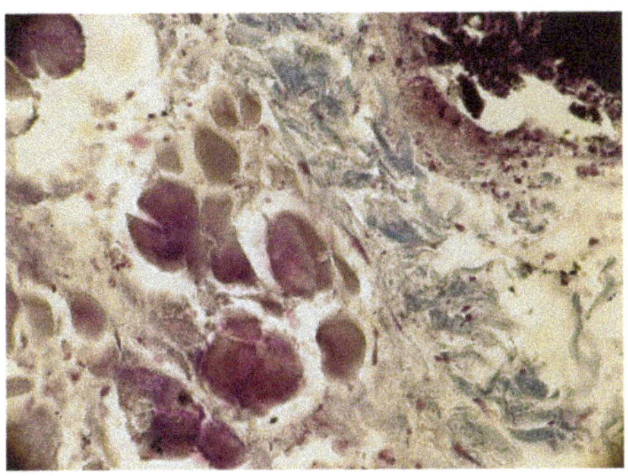

Figure 6. Calcification of muscular fibers from the skin. Col. Trichrome Masson, ×200.

In group B, the local reaction was slightly reduced in intensity, with the necrotic area being much smaller compared to group A. The present cells, neutrophils, macrophages, and fibroblasts were fewer in number. The implanted material was partially reabsorbing, as shown by the migration of neutrophils and macrophages to the implantation site (Figures 7 and 8).

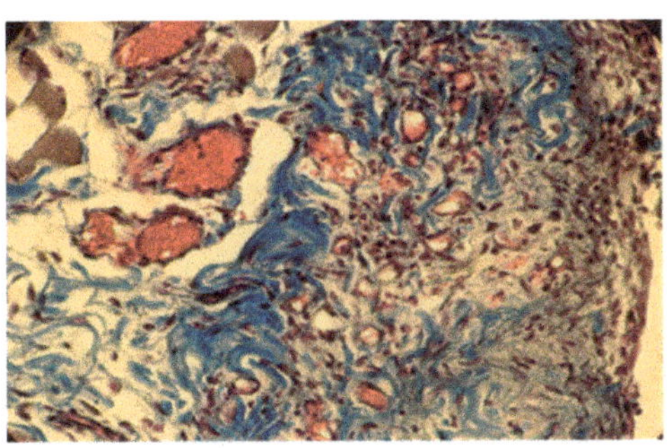

Figure 7. Group B. Moderate inflammation and mild congestive subcutaneous connective tissue in rabbits at 7 days after implantation. Col. Trichrome Masson, ×400.

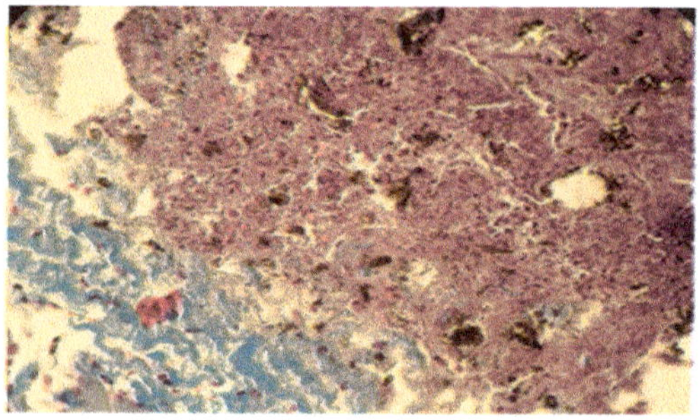

Figure 8. Moderate necrosis area with pockets of implanted material and moderate leukocyte influx. Partial resorption of the implant material. Subcutaneous tissue, 7 days after implantation. Col. Trichrome Masson, ×200.

In group C, it was noticed that the local inflammatory reaction induced by the implantation of DiaRoot BioAggregate was much less intense than in groups A and B. The area of necrosis was a small band surrounding the implanted material, with fewer neutrophils and macrophages. Fibroblast differentiation was also reduced (Figures 9 and 10).

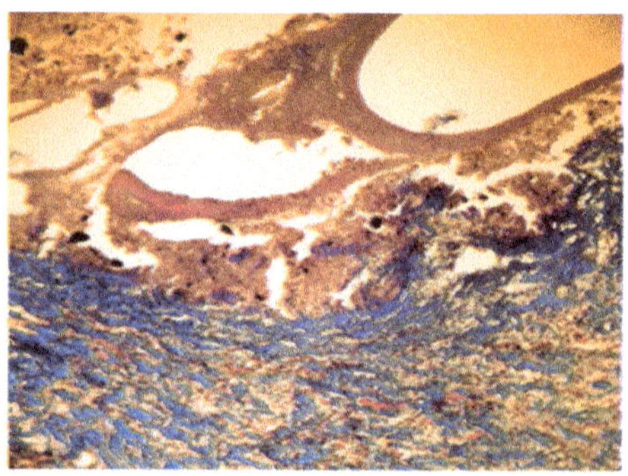

Figure 9. Group C. Moderate necrosis area with pieces of implanted material and moderate leukocyte influx. Partial resorption of the implant material. Subcutaneous tissue—7 days after implant. Col. Trichrome Masson, ×100.

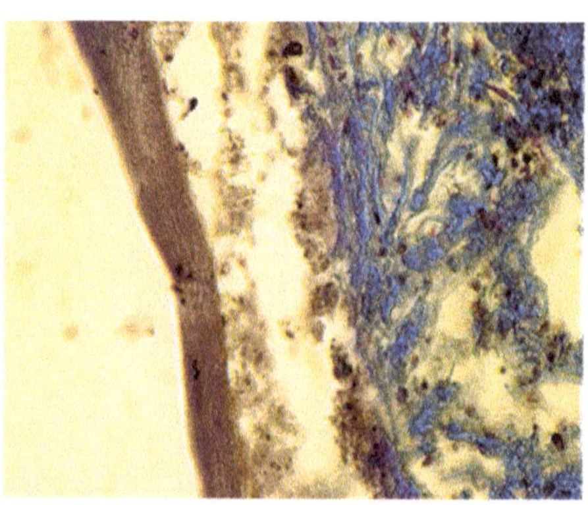

Figure 10. Group C. Moderate necrosis area with pieces of implanted material and moderate leukocyte influx. Partial resorption of the implanted material. Subcutaneous tissue—7 days after implant. Col. Trichrome Masson, ×400.

In samples collected 30 days after implantation, there was a significant decrease in local cellular reactions in all groups (A, B, and C), and inflammation was almost nonexistent, with rare inflammatory infiltrate cells being observed only in groups B and A (Figures 11 and 12).

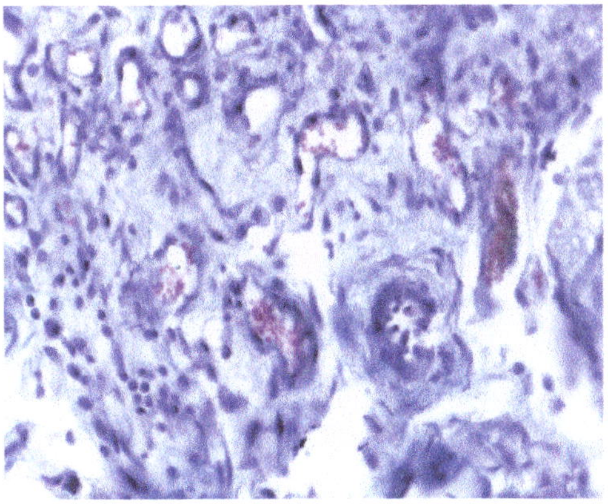

Figure 11. Group A. Scar tissue and rare inflammatory cells at the implant site with MTA after 30 days. Rare inflammatory cells. Col. Trichrome Masson, ×400.

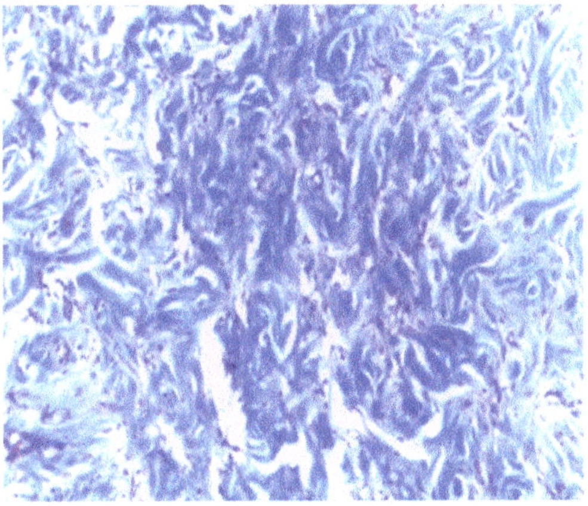

Figure 12. Group B. Dense subcutaneous tissue over the implant with MTA after 30 days. Rare inflammatory cells. Col. Trichrome Masson, ×400.

4. Discussion

Mineral Trioxide Aggregate (MTA) was recommended in the past as a repair material for root perforations. It was developed by Loma Linda University (in 1993) and was originally introduced as a retrograde filling material [2]. It was shown to result in a much lower percolation in contact with the surrounding tissues in comparison with the most commonly used materials such as amalgam, intermediate restorative material (IRM), and Super-EBA. For this reason, MTA was considered the material of choice for the repair of root perforation [3] since it was shown to be biocompatible with the periradicular tissue (minimal inflammatory response, ability to allow regeneration of hard tissue structures such as bone and cementum) [4], thus facilitating the regeneration of the periodontal supporting apparatus [5,6]. In research on human osteoblast models, it was found that MTA stimulated the production of cytokines, such as interleukin-1α, interleukin-1β, and interleukin-6, which are involved in bone turnover [10].

Sealapex is a calcium hydroxide-based sealer commonly used in endodontics, with both its biostimulating properties and cytotoxicity in contact with the periradicular tissues being extensively discussed in the literature [10–12]. In general, Sealapex demonstrated reduced inflammatory reaction as compared to other endodontic sealers, showing moderate inflammation at 48 h that became mild in later periods. Other zinc oxide-eugenol-based sealers (Endoflas, Tubliseal) were toxic after 48 h and 7 days. The toxicity decreased gradually in later periods [13].

The cytotoxicity of BioAggregate (BA), the third material used in the study, was also evaluated, as compared to that of MTA, on mesenchymal cell cultures, with the results showing that there was no statistically significant difference between MTA and BA in any of the experimental time periods. DiaRoot BioAggregate displayed in vitro compatibility similar to MTA [14].

In the present study, 7 days after implantation, the presence of more pronounced inflammatory reactions and the presence of multinucleated giant cells were histologically noted in contact with the remaining enclaves of Sealapex and MTA, while in the BA samples, the inflammatory reaction was considerably reduced, as demonstrated by the absence of these cells and reduced inflammatory cell infiltrate. This indicates a biocompatibility of BA with the living subcutaneous tissue, compared to the first two materials. The inflammation [12] and the presence of thicker bands of necrotic tissue in contact with the implanted materials (thicker in MTA compared to Sealapex and BA) after the first 7 days may be accounted for by both the different chemical composition of the cements and the high alkaline pH of MTA: after manipulation, the pH of MTA was 10.2 initially and rose to 12.5, 3 h after mixing, remaining constant thereafter. The maximum pH of Sealapex is 9.1 and that of BA is 12 after setting [15].

At 30 days after implantation, histological analysis [14] of the samples showed no significant differences between the three studied materials [15], with several isolated groups of inflammation cells [16] being found only in one Sealapex sample and one MTA sample.

The fact that no inflammation reaction was histologically detected 3 months after subcutaneous implantation with any of the used materials [17] is due to the phagocytic cell activity and tissue repair capacity [18] characteristic to these animals.

All kinds of specialized research on types of biomaterials used for endodontic restorations consider in the end the benefit of the patients, ensuring good biocompatibility, integration, stability, and functionality so that the predictable prosthetic rehabilitation increases quality of life by a proportion of 95% for the patients. Unidental restoration and Richmond crown result in an increase of 45%, and double prosthetic pieces elements result in an increase of 32% in a multiple construction bridge. A special importance can be observed in the approach of patients who presented clinical signs of dysfunctional syndrome in 25% who experienced pain, 15% with clinical signs of joint cracks and crepitation, and muscle dysfunctions in 35%, all due to occlusal imbalances, which require treatment with mouth guards and balneal–physical–kinetic therapy rehabilitation complex treatment in order to accomplish the principles of homeostasis and its stability [18–20] (Figure 13).

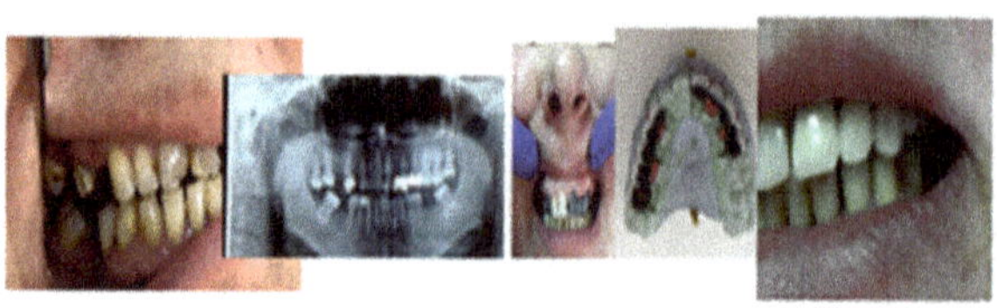

Figure 13. Images of the initial clinical situation, paraclinical orthopantomography examination, and the occlusal rehabilitation for the necessary foundation of the stomatognathic system homeostasis.

We consider the possibility of prosthetic rehabilitation an important stage in which from the start we observe the efficiency of those biomaterials with which we experimented in our study. Functionality and integration can be ways to quantify the study's goal.

In this case, increased efficiency and optimization of the functions of the stomatognathic system are observed resulting from the type of prosthesis and the other treatments related to the rehabilitation.

5. Conclusions

– Based on the reaction of the three materials, it can be assumed that the most biocompatible material is DiaRoot BioAggregate, followed by Sealapex and MTA.
– The smallest necrotic area and lowest cellularity were seen around the DiaRoot BioAggregate. Seven days after implantation, the material best preserved in tissues was BA, meaning that it was the least reabsorbing.
– The other two implanted materials significantly induced tissue irritation, histologically reflected by a thicker necrotic area and a considerable neutrophils and macrophages influx.
– MTA was found to have the highest local mineralization effect, based on the mineralization of adjacent muscle fibers.
– The fact that fibroblast differentiation and synthesis of connective tissue fibers were observed around the implant site 7 days after the MTA implantation, a reaction absent in the case of the other two materials, proves a slightly lower biocompatibility of MTA compared to BioAggregate.
– In the end, the advantage for patients is in creating better comfort under the premises of a good biocompatibility and integration in order to sustain all types of further individual prosthetic rehabilitation treatments and to ultimately obtain the attributes of homeostasis.

Author Contributions: Conceptualization, L.A. and L.E.C.; methodology, E.V.S. and A.S.P.; software, L.F.; validation, O.S. and B.P.B.; formal analysis, C.A., E.M. and Y.D.; investigation, L.I.C.; resources, E.V.S., A.S.P., O.S. and B.P.B.; data curation, L.I.C.; writing—original draft preparation, L.F.; writing—review and editing, L.A. and L.E.C.; visualization, E.V.S.; supervision, A.S.P.; project administration, C.A., E.M. and Y.D. All authors have read and agreed to the published version of the manuscript.

Funding: This research received no external funding.

Institutional Review Board Statement: The study was conducted according with national and international regulations on animal welfare, identification, control andelimination of factors causing physiological and behavioral disorders: Directive EC86/609 EU; Government Ordinance no.37/2002, approved by Law no. 471/2002; Law 205/2004 on animal protection, amended and supplemented by Law no. 9/2008. The study was approved by the Ethics Committee of Faculty of Veterinary Medicine Iasi (Protocol Code No. 533 / 22.04.2020).

Conflicts of Interest: The authors declare no conflict of interest.

References

1. European Parliament, Council of the European Union. Directive 2010/63/EU of the European Parliament and of the council of 22 September 2010 on the protection of animals used for scientific purposes. *Off. J. Eur. Union* **2010**, *L276*, 33–79.
2. Torabinejad, M.; Parirokh, M.; Dummer, P.M.H. Mineral trioxide aggregate and other bioactive endodontic cements: An up-dated overview—part II: Other clinical applications and complications. *Int. Endod. J.* **2018**, *51*, 284–317. [CrossRef] [PubMed]
3. Mathew, L.A.; Kini, S.; Acharya, S.R.; Kamath, S.; Menezes, N.D.; Rao, S. A comparative evaluation of the microleakage of blood-contaminated mineral trioxide aggregate and Biodentine as rot-end filling materials: An in vitro study. *J. Interdiscip. Dent.* **2016**, *6*, 19–24. [CrossRef]
4. Asokan, S.; Samuel, A.; Priya, P.R.G.; Thomas, S. Evaluation of sealing ability of Biodentine™ and mineral trioxide aggregate in primary molars using scanning electron microscope: A randomized controlled in vitro trial. *Contemp. Clin. Dent.* **2016**, *7*, 322–325. [CrossRef] [PubMed]
5. Upadhyay, Y. Mineral Trioxide Aggregate Repair of Perforated Internal Resorption: A Case Report. *JOHCD* **2012**, *6*, 149–150. [CrossRef]
6. Sinkar, R.C.; Patil, S.S.; Jogad, N.P.; Gade, V.J. Comparison of sealing ability of ProRoot MTA, RetroMTA, and Biodentine as furcation repair materials: An ultraviolet spectrophotometric analysis. *J. Conserv. Dent.* **2015**, *18*, 445–448. [CrossRef] [PubMed]
7. Torabinejad, M.; Parirokh, M. Mineral trioxide aggregate: A comprehensive literature review—Part II: Leakage and biocompatibility investigations. *J. Endod.* **2010**, *36*, 190–202. [CrossRef] [PubMed]
8. Jafarnia, B.; Jiang, J.; He, J.; Wang, Y.-H.; Safavi, K.E.; Zhu, Q. Evaluation of cytotoxicity of MTA employing various additives. *Oral Surg. Oral Med. Oral Pathol. Oral Radiol. Endodontol.* **2009**, *107*, 739–744. [CrossRef] [PubMed]

9. Rodriguez-Lozano, F.J.; García-Bernal, D.; Onate-Sanchez, R.E.; Ortolani-Seltenerich, P.S.; Forner, L.; Moraleda, J.M. Evaluation of cytocompatibility of calcium silicate-based endodontic sealers and their efects on the biological responses of mesenchymal dental stem cells. *Int. Endod. J.* **2017**, *50*, 67–76. [CrossRef] [PubMed]
10. Scelza, M.Z.; Linhares, A.B.; Silva, L.E.; Granjeiro, J.M.; Alves, G. A multiparametric assay to compare the cytotoxicity of endodontic sealers with primary human osteoblasts. *Int. Endod. J.* **2011**, *45*, 12–18. [CrossRef] [PubMed]
11. De-Deus, G.; Canabarro, A.; Alves, G.; Linhares, A.; Senne, M.I.; Granjeiro, J.M. Optimal cyto-compatibility of a bioceramic nanoparticulate cement in primary human mesenchymal cells. *J. Endod.* **2009**, *10*, 1387–1390. [CrossRef] [PubMed]
12. Klasser, G.D.; Green, C.S. The changing field of Temporomandibular disorders: What dentists need to know. *JCDA* **2009**, *5*, 49–53.
13. Estrela, C.; Bammann, L.L.; Estrela, C.R.D.A.; Silva, R.S.; Pécora, J.D. Antimicrobial and Chemical Study of MTA, Portland Cement, Calcium Hydroxide Paste, Sealapex and Dycal. *Braz. Dent. J.* **2000**, *11*, 3–9. [PubMed]
14. Simsek, N.; Bulut, E.T.; Ahmetoğlu, F.; Alan, H. Determination of trace elements in rat organs implanted with endodontic repair materials by ICP-MS. *J. Mater. Sci. Mater. Med.* **2016**, *27*, 1–6. [CrossRef] [PubMed]
15. Carausu, E.M.; Trandafir, V.; Ghibu, L.; Stamatin, O.; Checherita, L.E. Study of Electrolyte Serum Disturbances and Acid-base Status in Patients with Oral-maxillofacial and Dental Sepsis. *Rev. Chim.* **2017**, *68*, 1552–1556. [CrossRef]
16. Kim, S.G.; Malek, M.; Sigurdsson, A.; Lin, L.M.; Kahler, B. Regenerative endodontics: A comprehensive review. *Int. Endod. J.* **2018**, *51*, 1367–1388. [CrossRef] [PubMed]
17. Pereira, R.F.; Cintra, L.; Tessarin, G.W.L.; Chiba, F.Y.; Mattera, M.S.D.L.C.; Scaramele, N.; Tsosura, T.V.S.; Ervolino, E.; Oliveira, S.; Sumida, D.H. Periapical Lesions Increase Macrophage Infiltration and Inflammatory Signaling in Muscle Tissue of Rats. *J. Endod.* **2017**, *43*, 982–988. [CrossRef] [PubMed]
18. Totu, E.E.; Cristache, C.M.; Isildak, I.; Yildirim, R.; Burlibasa, M.; Nigde, M.; Burlbasa, L. Preliminary Studies on Citotoxicity and Genotoxicity Assessment of the PMMA-TiO2 Nanocompozites for Stereolithographic Complete Dentures Manufacturing. *Rev. Chim.* **2018**, *69*, 1160–1165. [CrossRef]
19. Ogunlewe, M.O.; Agbelusi, G.A.; Gbotolorun, O.M.; James, O. A review of temporomandibular joint disorders (TMD's) presenting at the Lagos University Teaching Hospital. *Nig. Q. J. Hosp. Med.* **2008**, *18*, 57–60. [PubMed]
20. Chisnoiu, A.M.; Picos, A.M.; Popa, S.; Chisnoiu, P.D.; Lascu, L.; Picos, A.; Chisnoiu, R. Factors involved in the etiology of temporomandibular disorders-a literature review. *Clujul Med.* **2015**, *88*, 473. [CrossRef] [PubMed]

applied sciences

MDPI

Article

Nano-Architectonics of Antibiotic-Loaded Polymer Particles as Vehicles for Active Molecules

Norin Forna [1,2,†], Daniela Damir [1,*,†], Letitia Doina Duceac [3,4,†], Marius Gabriel Dabija [1,3,*,†], Gabriela Calin [4,*,†], Daniela Luminita Ichim [4,5,*,†], Cristian Gutu [6,7,†], Carmen Grierosu [2,4,†], Lucian Eva [3,4,†], Mădălina Irina Ciuhodaru [1,8,†], Elena Roxana Bogdan Goroftei [7,9,†], Elena Ariela Banu [7,9,†], Liviu Stafie [4,10,†], Ciolpan Gabriela [11,12,†], Geta Mitrea [7,13,†] and Constantin Marcu [7,14,†]

1 "Grigore T. Popa" University of Medicine and Pharmacy Iasi, 16 Universitatii Str., 700115 Iasi, Romania; norin.forna@gmail.com (N.F.); mciuhodaru@yahoo.com (M.I.C.)
2 Orthopaedic Trauma Surgery Clinic, Clinical Rehabilitation Hospital, 14 Pantelimon Halipa Str., 700661 Iasi, Romania; grierosucarmen@yahoo.com
3 "Prof. Dr. Nicolae Oblu" Neurosurg Hospital Iasi, 2 Ateneului, 700309 Iasi, Romania; letimedr@yahoo.com (L.D.D.); lucianeva74@yahoo.com (L.E.)
4 Faculty of Dental Medicine, "Apollonia" University of Iasi, 11 Pacurari Str., 700511 Iasi, Romania; dr_liviustafie@yahoo.com
5 "Dr. Iacob Czihac" Military Emergency Clinical Hospital, 9 Henri Berthelot, 700483 Iasi, Romania
6 "Dr. Aristide Serfioti" Emergency Military Hospital, 199 Traian Str., 800150 Galati, Romania; dr.c.gutu@gmail.com
7 Faculty of Medicine and Pharmacy, University Dunarea de Jos, 47 Domneasca Str., 800008 Galati, Romania; elenamed84@yahoo.com (E.R.B.G.); banuariela@yahoo.com (E.A.B.); getamitrea@yahoo.com (G.M.); marcu_saar@yahoo.de (C.M.)
8 "Elena Doamna" Obstetrics and Gynecology Hospital, 29 Elena Doamna Str., 700398 Iasi, Romania
9 "Sf Ioan" Emergency Clinical Hospital, 2 Gheorghe Asachi Str., 800494 Galati, Romania
10 Public Health Directorate of Iasi, 2-4 Vasile Conta, 7001016 Iasi, Romania
11 Doctoral School of Economics and Business Administration-Management Specialization, Alexandru Ioan Cuza University, 700506 Iasi, Romania; gabriela.ciolpan@gmail.com
12 "Sf. Maria" Emergency Clinical Hospital Children Iasi, 62 Vasile Lupu Str., 700309 Iasi, Romania
13 "Sf. Ap. Andrei" Emergency Clinical Hospital, 177 Brailei Str., 800578 Galati, Romania
14 Saarbrucken-Caritasklink St. Theresia University Hospital, 66113 Saarbrücken, Germany
* Correspondence: danadamir@yahoo.com (D.D.); mariusdabija.md@gmail.com (M.G.D.); m_gabriela2004@yahoo.com (G.C.); danielaluminitaichim@yahoo.com (D.L.I.)
† These authors contributed equally to this work.

Citation: Forna, N.; Damir, D.; Duceac, L.D.; Dabija, M.G.; Calin, G.; Ichim, D.L.; Gutu, C.; Grierosu, C.; Eva, L.; Ciuhodaru, M.I.; et al. Nano-Architectonics of Antibiotic-Loaded Polymer Particles as Vehicles for Active Molecules. *Appl. Sci.* **2022**, *12*, 1998. https://doi.org/10.3390/app12041998

Academic Editor: Domenico Lombardo

Received: 22 December 2021
Accepted: 30 January 2022
Published: 14 February 2022

Abstract: Recently, nanotechnology research studies have been proven that use of various nanoparticles as drug delivery systems to target and to annihilate pathogenic microorganisms may be a good solution for prevention and treatment of severe infection. In the last few years, antimicrobial drug encapsulation into nano-sized systems has materialized as a promising alternative that increased drug efficacy and minimized adverse effects. Physicochemical properties of erythromycin-loaded polymer nanoparticles were assessed using particle size distribution, HPLC, FTIR, TG/DTA, and SEM characterization techniques. The as-prepared samples exhibited an average particle size of 340 and 270 nm, respectively, with erythromycin content of 99.7% in both samples. From the release profile of erythromycin from PLA/PLGA, a prolonged drug release can be observed from both Ery-PLA and Ery-PLGA nanostructures. Morphology images exhibited spherical, rigid, and ring-shaped nanoparticles. Thermal analytical study in the case of Ery-PLA and Ery-PLGA samples showed that pure drug has an endothermic peak at around 150 °C assigned to a melting point. The antibiotic melting peak disappeared for both antibiotic-loaded PLA and PLGA nanoparticles thermographs, denoting the presence of erythromycin. This indicates that the antibiotic is uniformly dispensed throughout the host polymer matrix at nanometer scale. FTIR spectra of Ery-PLA and Ery-PLGA nano-architectures with almost similar peaks indicated no alteration in chemical structure of drug-loaded polymer nanoparticles.

Keywords: PLA; PLGA; erythromycin; polymer nano-particles

1. Introduction

Bacterial infections happen when microorganisms exhibit its pathogenicity thus inducing disease and causing damage of the host area. Whilst various drugs were developed to prevent and to treat infections, drug resistance and drug side effects still remain an important global health problem. Although many procedures for prevention and treatment of several infections already exist, there is an essential need for improved or novel manners for bacterial annihilation. An insufficient therapy could be due to the inefficient drug delivery, despite a drug's therapeutic efficacy having been demonstrated for years. The occurrence of drug resistance caused by the excess and misuse of the antimicrobial agents, thus limiting their efficacy, represents a major interest for public health today. This issue determined multidisciplinary research to find improved therapies to defeat antibiotic resistance. An important approach to prevent and to treat several infections caused by drug resisting pathogens is the use of drug-loaded nanoparticles in order to deliver the antimicrobial agent to the targeted area. The advantages of the nano-medicine include the improvement of drug efficiency and stability, decrease of drug toxicity and prolongation of drug release, and precise-targeting to in infected area [1,2]. The development of polymeric nanoparticles loaded with antimicrobial drugs enhanced the therapeutic action and the pharmacokinetics of antimicrobials. Several drugs can be encapsulated into the submicron-sized polymeric matrix or absorbed on the surface [3–6].

Erythromycin (Figure 1), a macrolide antibiotic, is currently used for several bacterial infections treatment, often prescribed as an alternative to penicillin allergic patients. It is an effective drug administered as a prophylactic treatment or for various bacterial infections in several medical specialties, such as cardiology, neurosurgery, pediatrics, pediatric surgery, pulmonology, orthopedics, and obstetrics/gynecology. Its therapeutic actions include respiratory tracts infections, chlamydia infections, syphilis, pelvic inflammatory disease, skin infections, and digestive infections, as well as prevention of bacterial endocarditis and relapses of acute rheumatoid arthritis in patients allergic to beta-lactams, chlamydial conjunctivitis, listeriosis, legionella pneumophila infections, and whooping cough [7,8].

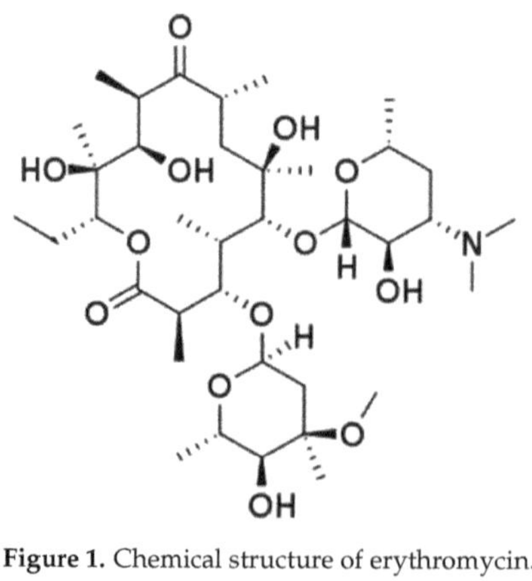

Figure 1. Chemical structure of erythromycin.

Erythromycin limits the development of bacteria by interfering with protein synthesis. Erythromycin binds irreversibly to the 50S ribosomal subunit (23S sequence), thus preventing the synthesis of bacterial proteins. The action spectrum of erythromycin is similar to that of penicillin, which is also active against some strains of penicillin-resistant staphylococci, being generally bacteriostatic or bactericidal, depending on the dose. It acts, especially, on gram-positive germs (streptococci, pneumococci), as well as on gram-negative germs (neisserii, haemophili), spirochetes, rickettsia, and mycoplasmas. Adverse reactions include gastrointestinal disorders (nausea, vomiting, diarrhea), which disappear after discontinuation of treatment or after dose reduction, or as hepatotoxic phenomena with long-term treatment [9–13].

Nanotechnology implied in medical field brings out considerable therapeutic benefits, especially in public health. An excellent approach in treating and prevention infections consist of using intercalated nanomaterials loaded with specific drugs as carriers to deliver the API at the affected site. Targeted nanoparticles transport has the potential to carry various drugs to diagnose, to treat, and to prevent diseases at cellular, and even molecular, level [14–16].

This study refers to synthesis and physico-chemical characterization of erythromycin-loaded poly (lactic acid) (PLA—Figure 2) and poly (lactic-co-glycolytic acid) (PLGA—Figure 3), as polymers chains are partitioned by hydrolysis into lactic and glycolytic acid natural metabolites, being then eliminated by citric acid cycle metabolic pathway. In the last few decades, both PLA and PLGA biopolymers have been studied for various medical and pharmaceutical applications, especially as drug delivery systems, as they proved high safety and excellent biocompatibility. An important issue of this work is the encapsulation of erythromycin active molecules into PLA and PLGA nanoparticles and their evaluation for further delivery to targeted site of infections [17–24]. These antibiotic-loaded polymeric nanoparticle delivery systems improve the prevention and/or treatment of severe infections caused by antibiotic pathogen resistance by improving preparation methods, thus ensuring a drug release profile according to its action. Moreover, this approach includes erythromycin as anti-infective agent knowing that, today, the medical field around the world is facing drug-resistant microorganisms. Prolonged and controlled release of encapsulated erythromycin consists of longer contact time in the circulatory system to prevent the multiplication of bacteria, especially multi-drug resistant ones, and to limit the side effects of the drug.

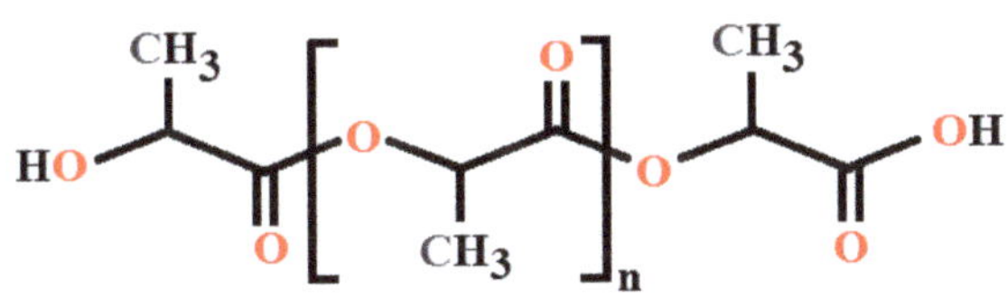

Figure 2. Chemical structure of poly (lactic acid), where n represents the number of chains.

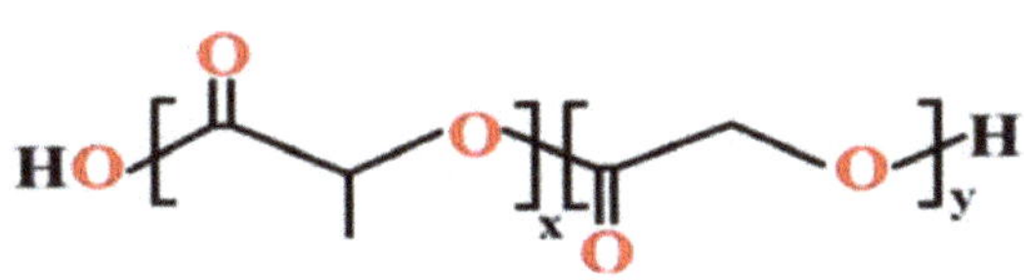

Figure 3. Chemical structure of poly (lactic-co-glycolic acid), where x represents the number of lactic acid units, and y represents the number of glycolic acid units.

The characterization of as synthesized samples was performed in this study to provide useful formulate for a prolonged and controlled drug release system in prevention and treatment of several infections.

In addition to the obtained samples type drug delivery systems used in medicine, medical teams must have key communication skills effective in improving patient care. An extended approach for standardizing and improving communication, especially between disciplines, is the "SBAR" method (Situation, Background, Assessment, and Recommendation) [25,26].

Interdisciplinary cooperation is important for obtaining new drug formulations in the context of reducing morbidity and mortality in healthcare facilities. An important goal of this research is to obtain a new polymer nanostructure loaded with antibiotics for the prevention and treatment of healthcare-associated infections. PLA/PLGA polymers are good candidates for drug carriers due to their biodegradability [27,28]. A significant feature of these delivery systems is the disintegration kinetics, which can be prolonged for the sustained release of therapeutically agents, thus ensuring enough time for the drug to cause its effect [29–32]. An important aspect in nanomedicine is the incorporation of the antibiotic in the polymeric structure of PLA and PLGA to limit the action of infectious agents. This work focused on preparing erythromycin-loaded poly (lactic acid) (PLA) and poly (lactic-co-glycolic acid) (PLGA) and further evaluation of this broad spectrum curative agent encapsulated into polymer matrix.

2. Materials and Methods

2.1. Materials

PLA and PLGA polymers 50:50 H with an average molecular weight of 40,000–65,000 Da and 40,000–75,000 Da, respectively, were purchased from Sigma Aldrich (Darmstadt, Germany). PVA, as a surface active agent, and the other reagents used for samples preparation were also purchased from Sigma Aldrich, Germany. Ultrapure water from a Milli-Q water equipment was used.

2.2. Preparation Methods

Double emulsion solvent evaporation technique was used to prepare erythromycin-PLA and erythromycin-PLGA nanoparticles. The organic phase contained 0.5 g of PLA and PLGA, each dissolved in 10 mL mixture of dichloromethane and acetone in a volume ratio of 85:15. The internal aqueous phase contained 40 mg erythromycin (in a weight ratio of 1:10 drug-polymer) dissolved in phosphate buffer solution of pH 6.0. Afterward, organic phase and internal aqueous phase were mixed together by ultra-sonication for half a minute under cooling atmosphere to form W1/O emulsion, which was slowly added to an aqueous solution containing 100 mL of 1% PVA under vigorous stirring at 7000 rpm about 10 min. Then, the obtained W1/O/W2 mixture was agitated at 500 rpm for 24 h, for almost entirely organic solvent evaporation. The resulting products were vigorously cleaned using ultrapure water and centrifuged for 20 min at 10,000 rpm, until the nanoparticles formed. Finally, the antibiotic-polymer nanohybrids were lyophilized for further analysis, and the supernatant was stored for assessment of drug content.

The as-prepared erythromycin-polymer nano-scaled particles-type drug delivery systems (Figure 4) were denoted as Ery-PLA and Ery-PLGA, respectively.

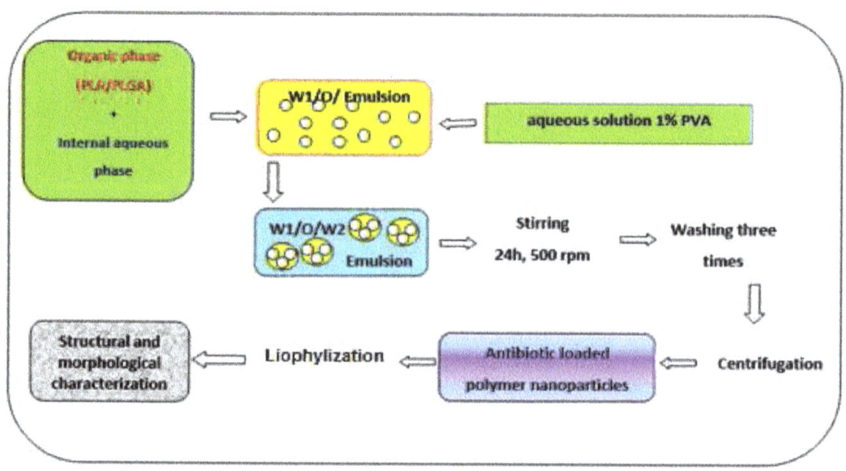

Figure 4. General scheme for erythromycin-loaded PLA/PLGA nanoparticles preparation.

2.3. Characterization Methods

ZetaSizer Nano ZS was used for the detection of aggregates and measurement of small samples permitting the evaluation of particle and molecule size, translational diffusion, electrophoretic mobility, and zeta potential of particles at high and low concentrations. The zeta potential analyzer uses electrophoretic light scattering for particles, molecules and surfaces, measuring range for sizes from 0.3 nm to 10 μm.

The HPLC (high pressure liquid chromatography) method was used in order to determine the amount of the erythromycin from the obtained samples. The stationary phase used a column at 70 °C, and the mobile phase consisted of a mixture of acetonitrile-0.2 M ammonium acetate-methanol-water, in a 450:100:100:350 volume ratio, adjusted at pH 7.0. The UV detection of column effluent was registered at λ = 215 nm. The injected volume was 100 μL for the test solutions consisting of erythromycin propionate-loaded polymers, the reference solution containing standard erythromycin A, and the reference solution containing erythromycin B CRS and erythromycin C CRS, in equal amounts, diluted to 50 mL with hydrolysis solution of dibasic potassium hydrogen phosphate (pH 8.0).

In vitro release study of erythromycin from PLA/PLGA nanoparticles was carried out using a Cary 60 UV-Vis—Agilent spectrometer (Agilent, Santa Clara, CA, USA), into a phosphate buffer solution at a pH value of 7.5. A well-established amount of each sample was suspended in 5 mL PBS in a centrifuge tube and ultrasound for a short time. The samples were then incubated at 37 °C, and, at set times, the samples were centrifuged. The supernatant was then collected, and the nano-powders were reconstituted by the addition of fresh PBS. The percentage of antibiotic release was represented as a function of contact time.

Morphological features of erythromycin-loaded both PLA and PLGA nanoparticles were analyzed by using a TESCAN scanning electron microscope (Brno, Czech Republic).

Another important characterization of drug-loaded polymer is represented by thermal analytical study. The Thermogravimetric Analyzer, Discovery TGA 5500 (TAInstruments, New Castle, DE, USA), monitors the stability of a sample (polymers, organic and inorganic compounds) in terms of weight variation versus time/temperature, in a controlled atmosphere, giving information about temperature (range: room temperature ÷ 1000 °C), evaluation of the degradation temperatures, steps of degradation, and determination of weight loss at a certain temperature under specific conditions.

An Alpha Bruker FT-IR spectrometer (Bremen, Germany), spectral range 4000–400 cm^{-1}, and 4 cm^{-1} resolution, permits evaluation of interaction ways between the antibiotic and the two polymers.

3. Results

Physical properties analysis was performed by determination of particle size distribution and zeta potential, as shown in Figures 5 and 6.

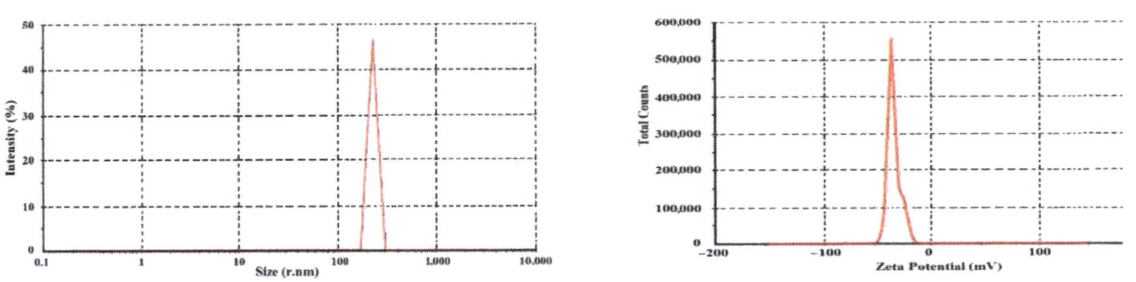

Figure 5. Distribution pattern of erythromycin-loaded PLA nanoparticles.

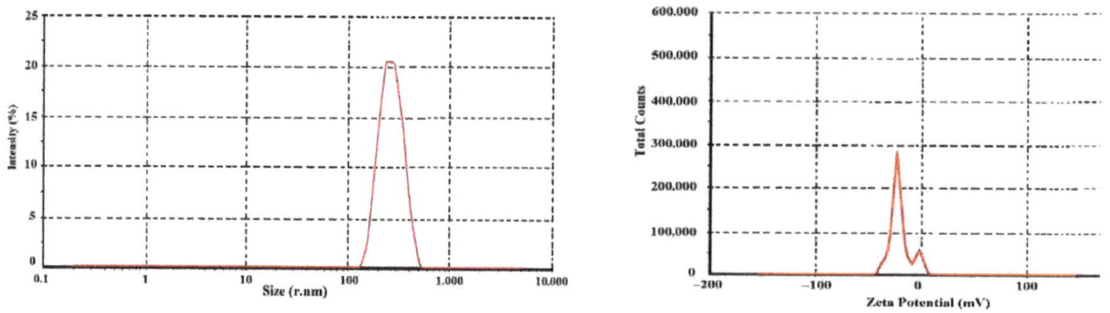

Figure 6. Distribution pattern of erythromycin-loaded PLGA nanoparticles.

Particle size distribution of PLA and PLGA nanoparticles indicated a mean particle size of approximately 44 nm and −25.2 mv zeta potential, and mean particle size of 180 nm and −33.1 mv for PLGA, respectively.

After loading the antibiotic onto PLA/PLGA matrix, erythromycin-loaded PLA nanoparticles were recorded around 325 nm in size, while drug-loaded PLGA nanoparticles were recorded approximately 258 nm in size, depending on the initial antibiotic concentration. Electro-kinetic potential measured at the exterior side of the colloidal nanostructures indicated zeta potential values of drug-loaded polymer nanoparticles of −11.3 mv for PLA and −17.3mv for PLGA, respectively, which denoted the degree of rejection of similarly charged particles in dispersion. A higher zeta potential for nanoparticles indicates the stability of colloidal dispersions resisting nanoparticles aggregation.

HPLC chromatograms are presented in Figure 7. It can be seen from the chromatogram of the samples that the retention times of erythromycin are approximately the same as the retention times of erythromycin in the reference chromatogram.

The percentage content of erythromycin A was calculated using the chromatogram shown in Figure 7a, as a reference, and the percentage content of erythromycin B using the chromatogram shown in Figure 7b, as a reference. The amount of erythromycin propionate in both samples was expressed as the sum of erythromycin A and erythromycin B and was 99.7% of the initial amount used to prepare antibiotic-loaded nanoparticles for both samples.

We observed (Figure 8) that drug release from Ery-PLA was slowly more increased than from Ery-PLGA nanostructures. In the case of Ery-PLA, almost 55% of the antibiotic was controlled released within 4 h, and 75% was released within 2 days. In the case of erythromycin-loaded PLGA nanoparticles, approximately 55% of drug was released within 4 h, while 75% was released within 3 days. These results evidenced that the antibiotic encapsulation into PLGA nanoparticles was a little bit slower than into PLA nanoparticles.

Surface morphologies (Figure 9) exhibited spherical, rigid, and ring-shaped nanoparticles. This panel revealed regular and isolated nanoparticles attributed to lactide content, which prevent aggregates formation.

The result of our investigation showed (Figure 10), in the case of Ery-PLA and Ery-PLGA samples, that the pure drug has an endothermic peak at around 150 °C, assigned to a melting point. The antibiotic melting peak disappeared for both antibiotic encapsulated PLA/PLGA nanoparticles thermographs, denoting the existence of erythromycin and indicating a possible uniform distribution of the drug throughout the host polymer at the molecular stage.

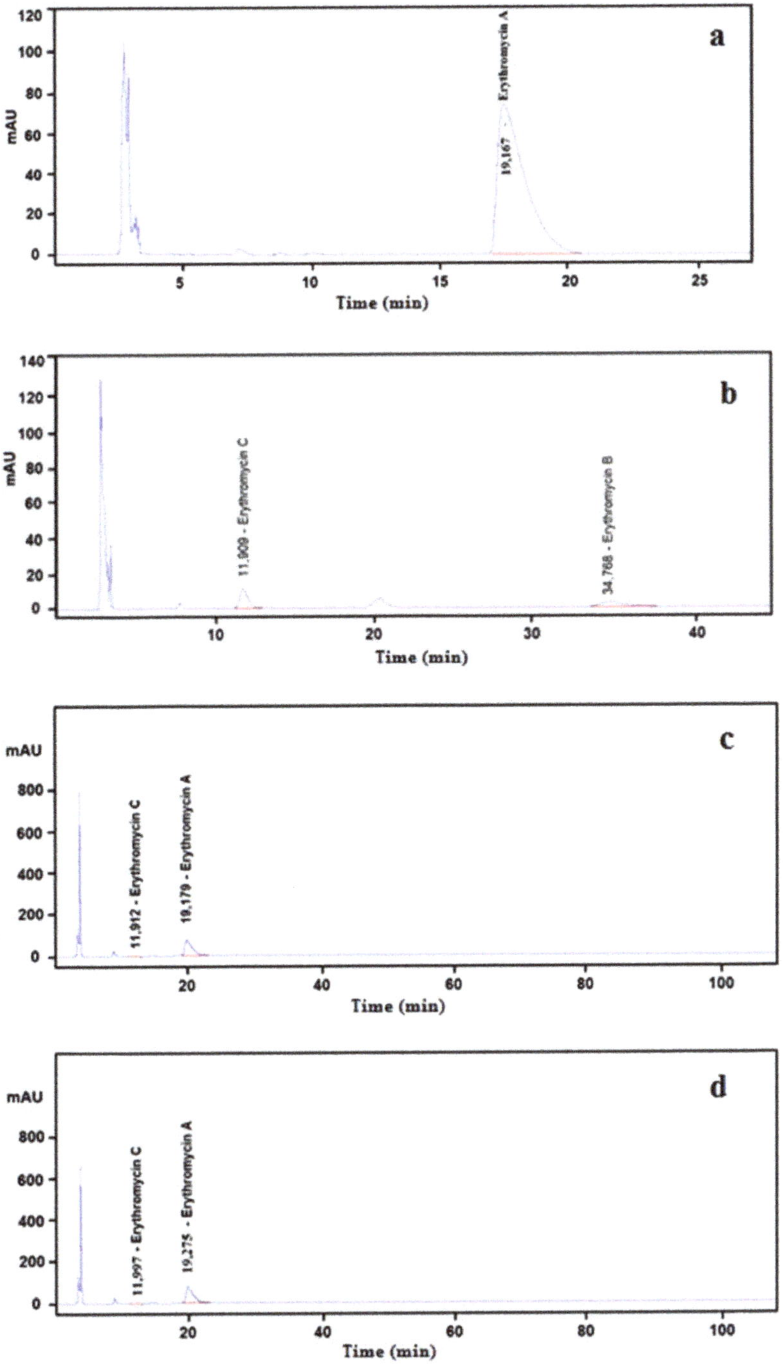

Figure 7. HPLC chromatogram for erythromycin references (**a**,**b**) and Ery-PLA/PLGA samples (**c**,**d**).

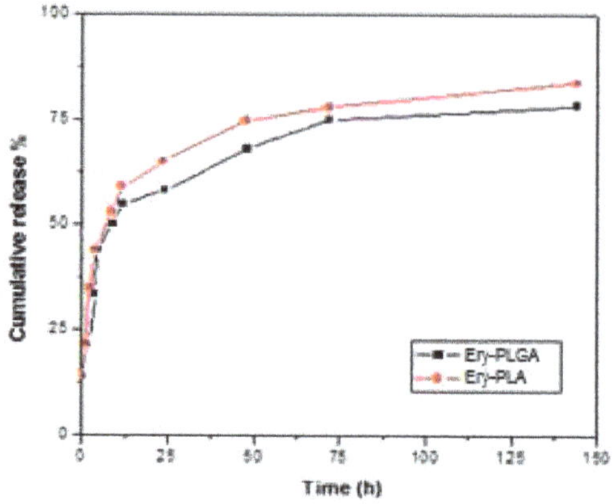

Figure 8. In vitro antibiotic release out of Ery-PLA/Ery-PLGA nanoparticles.

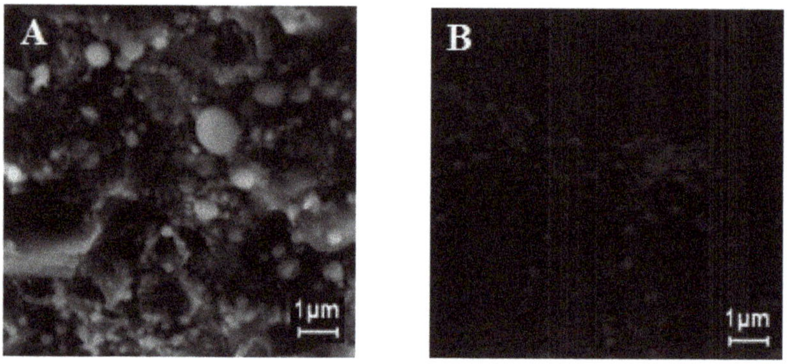

Figure 9. SEM images of (**A**) Ery-PLA and (**B**) Ery-PLGA.

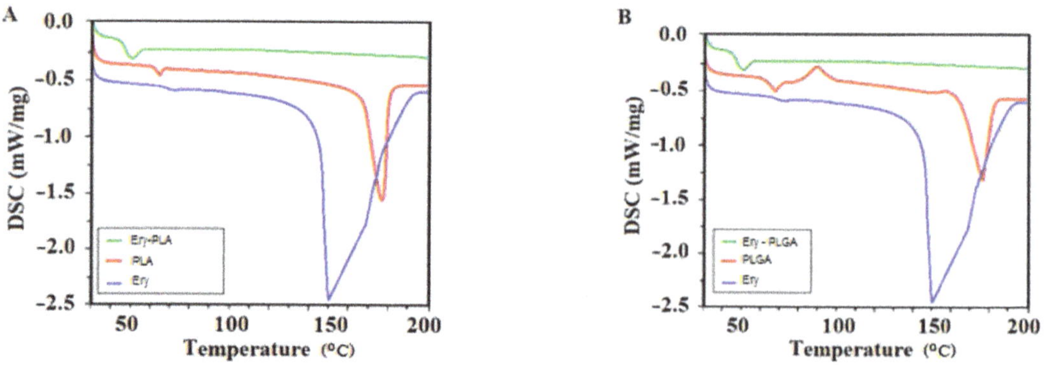

Figure 10. Thermal analysis graph of (**A**) erythromycin (Ery), PLA, and erythromycin-loaded PLA (Ery-PLA); (**B**) erythromycin (Ery), PLGA, and erythromycin-loaded PLGA (Ery-PLGA).

Furthermore, the thermographs also revealed a reduction in both PLA and PLGA transition temperature concerning nano-encapsulation of erythromycin due to increase of particles size, in the case of which glass transition temperature is inversely proportional to the amorphous polymers particle size.

FTIR spectra (Figure 11) of Ery-PLA and Ery-PLGA nano-architectures revealed chemical structure of drug-loaded polymer nanoparticles and peaks were almost similar.

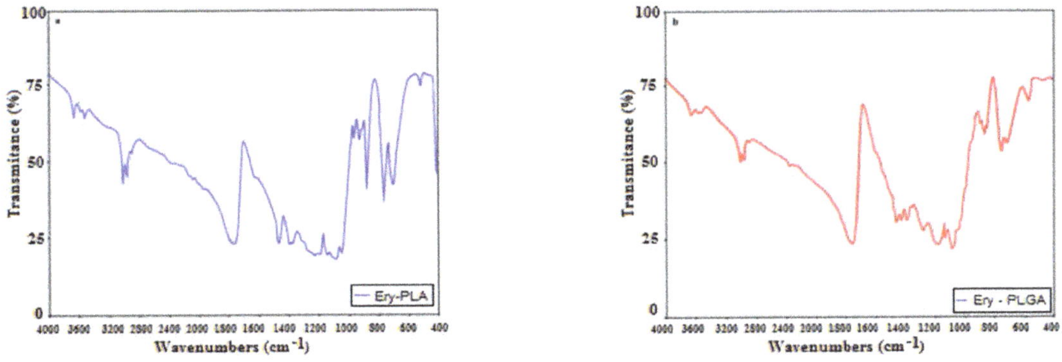

Figure 11. FTIR spectra of Ery-PLA and Ery-PLGA.

FTIR data revealed that Ery-PLGA nanoparticles' peaks were around 3650 cm^{-1} and 3500 cm^{-1} due to O-H stretching vibrations, at near 2950 cm^{-1} and 2990 cm^{-1} assigned to C-H stretch vibrations, 1760 cm^{-1} to C=O, 1460 cm^{-1} attributed to C-H binding, 1390 cm^{-1} to N=O link, 1180 cm^{-1} correlated to C-H stretching, 1080 cm^{-1} to C-O stretch vibrations, and 850 cm^{-1} and 760 cm^{-1} attributed to N-H shiver.

Furthermore, characteristic peaks of polymer were noticed, and there was no modification of characteristic peaks, denoting the lack of chemical interaction between erythromycin and polymer nanoparticles.

FTIR data revealed that Ery-PLGA nanoparticles' peaks at around 3650 cm^{-1} and 3000 cm^{-1}, attributed to O-H stretching vibrations, at 1760 cm^{-1} to C=O stretch, at 1460 cm^{-1} assigned to C-H link, at 1400 cm^{-1} to N=O binding, 1180 cm^{-1} and 1090 cm^{-1} to C-N stretch, and at 860 cm^{-1} and 750 cm^{-1} to N-H shiver. Some functional groups were less in number for simple erythromycin, confirming that the drug was not responsible for any interactions with the two polymers.

4. Discussion

In recent years, the administration of the drug after treatment of the infected area has become of major interest [33,34] because infections can recur. Recent studies show that poly (lactic acid) (PLA) and poly (lactic-co-glycolic acid) (PLGA) are biocompatible and biodegradable polymers, being a good host for the incorporation of a wide variety of antimicrobial drugs. Antibiotic-loaded chitosan nanoparticles obtained by ionic gelation technique using TPP as a crosslinking factor are of durable stability. Entrapment of erythromycin active molecules within the network of the polymer matrix implies more stable pharmaceutical formulation. This work was concerned with preparing a novel antibiotic carrier using PLA/PLGA nanoparticles loaded with erythromycin using the W/O/W double emulsion solvent evaporation method [35]. Erythromycin-loaded PLA exhibited 340 nm in size, while drug-loaded PLGA nanoparticles revealed approximately 270 nm in size. Particle size distribution results indicated that the nanoparticle sizes are related to the concentration of the antibiotic used to load PLA/PLGA polymer. Electro-kinetic potential (zeta potential) refers to the difference in electric potential between the electric charge at the surface of a solid nanoparticle in the dispersion and the charge of the diffuse electric layer, denoting the degree of rejection of similarly charged nanoparticles in dispersion. A higher zeta potential for the obtained nano-composites disclosed the stability of colloidal dispersions resisting nanoparticles aggregation. The amount of the erythromycin determined using HPLC analysis was expressed as sum of erythromycin A, erythromycin B, and erythromycin C, resulting in 99.7% of the initial amount used to prepare the nanocomposites [36]. Antibiotic release profile revealed that erythromycin release from Ery-PLA is faster than from Ery-PLGA nanocomposites. For the Ery-PLA sample, almost 50% of the antibiotic was rereleased within 4 h, and 75% was released within 2 days. For the Ery-PLGA sample, approximately 50% of drug was controlled released within 8 h, while 75% was released within 3 days. As a consequence of these results, antibiotic binding to PLGA was stronger than to PLA polymer, indicating that PLGA-drug can be a better delivery system. Morphological characterization of erythromycin encapsulated PLA and PLGA nanoparticles by scanning electron microscopy displayed spherical, rigid, and ring-shaped nanoparticles, as well as regular and isolated nanoparticles assigned to lactide content, which prevent aggregates formation. A correlation between the smooth surface and the lactide content which confer hydrophobicity to the polymer can be done. Moreover, due to the presence of lactide, the contact between particles becomes weaker, preventing agglomeration. A thermal analytical study indicated, for both samples, that pure drug has an endothermic peak at around 150 °C assigned to a melting point. The erythromycin melting peak disappeared for both antibiotic encapsulated PLA/PLGA nanoparticles thermographs. This evidenced the existence of erythromycin, showing a uniformly distribution of the drug throughout the host polymer. The thermographs also revealed a reduction in both polymers' transition temperature concerning nano-encapsulation of erythromycin. That was owed to the increase of particles size, in the case of which glass transition temperature is inversely proportional to amorphous polymer's particles' size. Thermal analysis of antibiotic-loaded polymer as a drug delivery system is of importance since the processes used to samples preparation are able to modify the organization of the polymer chains [21]. Vibration spectrum of encapsulated antibiotic involves the type of interaction occurring between the active molecule and polymer due to the vibrations of the atoms' interaction, which can be modified in frequency and intensity. From FTIR analysis, characteristic peaks of polymer was noticed, and there was no alteration of characteristic peaks by comparison with the spectra of erythromycin and the two neat polymers procured from an electronic spectra library (not shown, being used in another reference by some of the authors), demonstrating the absence of chemical interaction between erythromycin and PLA/PLGA nanoparticles. It can be concluded that the antibiotic is present with no alteration of its structure, without binding other functional groups by hydrogen bond. The two polymers and the antibiotic structure contain a C-O bond. FTIR spectra confirm the lack of interaction of the drug with the PLA and PLGA polymers [37]. These results suggest that the bactericidal effect of erythromycin-loaded PLA and PLGA nanoparticles would be significantly stronger than the antibiotic-active substance. Obtaining and characterizing antibiotic-loaded polymer nanoparticles was a first step in the development of a new drug-type formulations—PLA/PLGA nanoparticles. The authors also consider the stability study for one year, which will be performed every 3 months, as well as the microbial activity of the samples.

5. Conclusions

Many healthcare facilities worldwide are confronted with microbial pathogens that exhibit multi-drug resistance, which complicates the prevention or treatment of serious infections. Our study was based on obtaining and characterizing antibiotic-loaded polymeric nanoparticles, assuming that the therapeutic efficacy of erythromycin would be greatly increased by loading the drug into biocompatible polymer-based nanoparticles compared to the free antibiotic. This study showed that the structural and morphological aspects of the erythromycin-encapsulated polymer significantly improved the delivery of antibiotics. The treatment of infections refers to the destruction of the pathogenic factor, and the polymer loaded with erythromycin would be more effective due to the prolonged release and limitation of side effects. Moreover, the biocompatible and biodegradable polymer has been shown to be effective in encapsulating antibiotic molecules by acting as a carrier of drugs used to prevent and improve the treatment of bacterial infections.

Author Contributions: All authors have equal contribution to the work. Conceptualization, N.F., D.D.; project administration, G.C.; data curation, C.G. (Cristian Gutu); formal analysis, D.L.I., C.G. (Carmen Grierosu); investigation, L.D.D., M.G.D., L.E.; resources, M.I.C., E.R.B.G.; visualization, E.A.B.; validation, L.S.; investigation and writing—review & editing, N.F., C.G. (Ciolpan Gabriela); methodology, C.G. (Ciolpan Gabriela), G.M., C.M. All authors have read and agreed to the published version of the manuscript.

Funding: This research received no external funding.

Institutional Review Board Statement: Not applicable.

Informed Consent Statement: Not applicable.

Data Availability Statement: Not applicable.

Conflicts of Interest: The authors declare no conflict of interest.

References

1. Ranghar, S.; Sirohi, P.; Verma, P.; Agarwal, V. Nanoparticle-based Drug Delivery Systems: Promising Approaches Against Infections. *Braz. Arch. Biol. Technol.* **2014**, *57*, 209–222. [CrossRef]
2. Shen, S.; Wu, Y.; Liu, Y.; Wu, D. High drug-loading nanomedicines: Progress, current status, and prospect. *Int. J. Nanomed.* **2017**, *12*, 4085–4109. [CrossRef] [PubMed]
3. Amaral, L.; Viveiros, M.; Kristiansen, J.E. Non-antibiotics: Alternative therapy for the management of MDRTB and MRSA in economically disadvantaged countries. *Curr. Drug Targets* **2006**, *7*, 887–891. [CrossRef]
4. Mahapatro, A.; Singh, D.K. Biodegradable nanoparticles are excellent vehicle for site-directed in-vivo delivery of drugs and vaccines. *J. Nanobiotechnol.* **2011**, *9*, 55. [CrossRef] [PubMed]
5. Zhang, L.; Pornpattananangkul, D.; Hu, C.M.J.; Huang, C.M. Development of nanoparticle for antimicrobial drug delivery. *Curr. Med. Chem.* **2010**, *17*, 585–594. [CrossRef] [PubMed]
6. Cyphert, E.L.; Wallat, J.D.; Pokorski, J.K.; Von Recum, H.A. Erythromycin Modification That Improves Its Acidic Stability while Optimizing It for Local Drug Delivery. *Antibiotics* **2017**, *6*, 11. [CrossRef] [PubMed]
7. White, R.J. Why use erythromycin? *Thorax* **1994**, *49*, 944–945. [CrossRef]
8. Alvarez-Elcoro, S.; Enzler, M.J. The macrolides: Erythromycin, clarithromycin, and azithromycin. *Mayo Clin. Proc.* **1999**, *74*, 613–634. [CrossRef]
9. Pereira, J.M.; Mejia-Ariza, R.; Ilevbare, G.A.; McGettigan, H.E.; Sriranganathan, N.; Taylor, L.S.; Davis, R.M.; Edgar, K.J. Interplay of degradation, dissolution and stabilization of clarithromycin and its amorphous solid dispersions. *Mol. Pharm.* **2013**, *10*, 4640–4653. [CrossRef]
10. Neumann, K.; Lilienkampf, A.; Bradley, M. Responsive polymeric nanoparticles for controlled drug delivery. *Polym. Int.* **2017**, *66*, 1756–1764. [CrossRef]
11. Tran, L.T.C.; Gueutin, C.; Frebourg, G.; Burucoa, C.; Faivre, V. Erythromycin encapsulation in nanoemulsion-based delivery systems for treatment of Helicobacter pylori infection: Protection and *synergy*. *Biochem. Biophys. Res. Commun.* **2017**, *493*, 146–151. [CrossRef] [PubMed]
12. Brook, I.; Lewis, M.A.; Sándor, G.K.; Jeffcoat, M.; Samaranayakeet, L.P.; Rojas, J.V. Erythromycin in dentistry: More thanjust effective prophylaxis for endocarditis? *Oral Surg. Oral Med. Oral Pathol. Oral Radiol. Endod.* **2005**, *100*, 550–558. [CrossRef] [PubMed]
13. Annane, D.; Clair, B.; Salomon, J. Managing toxic shock syndrome with antibiotics. *Expert. Opin. Pharmacother.* **2004**, *5*, 1701–1710. [CrossRef] [PubMed]
14. Coyle, E.A. Targeting bacterial virulence: The role of protein synthesis inhibitors in severe infections. *Soc. Infect. Dis. Pharm. Pharmacother.* **2003**, *23*, 638–642. [CrossRef]
15. Gao, P.; Nie, X.; Zou, M.; Shi, Y.; Cheng, G. Recent advances in materials for extended-release antibiotic delivery system. *J. Antibiot.* **2011**, *64*, 625–634. [CrossRef]
16. Jong, W.H.D.; Borm, P.J.A. Drug delivery and nanoparticles: Applications and hazards. *Int. J. Nanomed.* **2008**, *3*, 133–149. [CrossRef]
17. Patel, A.; Patel, M.; Yang, X.; Mitra, A.K. Recent advances in protein and peptide drug delivery: A special emphasis on polymeric nanoparticles. *Protein Pept. Lett.* **2014**, *21*, 1102–1120. [CrossRef]
18. Diab, R.; Jaafar-Maalej, C.; Fessi, H.; Maincent, P. Engineered nanoparticulate drug delivery systems: The next frontier for oral administration? *AAPS J.* **2012**, *14*, 688–702. [CrossRef]
19. Hans, M.L.; Lowman, A.M. Biodegradable nanoparticles for drug delivery and targeting. *Curr. Opin. Sol. State Mater. Sci.* **2002**, *6*, 319–327. [CrossRef]
20. Davda, J.; Labhasetwar, V. Characterization of nanoparticle uptake by endothelial cells. *Int. J. Pharm.* **2002**, *233*, 51–59. [CrossRef]
21. Hariharan, M.; Price, J.C. Solvent, emulisifier and drug concentration factors in poly (D-L-Lacticacid) microspheres containing hexamethylmelamine. *J Microencapsul.* **2002**, *19*, 95–109. [CrossRef] [PubMed]

22. Buzea, C.; Pacheco, I.; Robbie, K. Nanomaterials and nanoparticles: Sources and toxicity. *Biointerphases* **2007**, *2*, MR17–MR71. [CrossRef] [PubMed]
23. McCarron, P.A.; Marouf, W.M.; Quinn, D.J.; Fay, F.; Burden, R.E.; Olwill, S.A.; Scott, C.J. Antibody targeting of camptothecin-loaded PLGA nanoparticles to tumorcells. *Bioconjug. Chem.* **2008**, *19*, 1561–1569. [CrossRef] [PubMed]
24. Nayak, B.; Panda, A.K.; Ray, P.; Ray, A.R. Formulation, characterization and evaluation of rotavirus encapsulated PLA and PLGA particles for oral vaccination. *J. Microencapsul.* **2009**, *26*, 154–156. [CrossRef]
25. Meera, S. Achrekar, Vedang Murthy, Sadhana Kanan, Rani Shetty, Mini Nair, and Navin Khattry, Introduction of Situation, Background, Assessment, Recommendation into Nursing Practice: A Prospective Study. *Asia Pac. J. Oncol. Nurs.* **2016**, *3*, 45–50.
26. Narayan, M.C. Using SBAR communications in efforts to prevent patient rehospitalizations. *Home Healthc. Now* **2013**, *31*, 504–515. [CrossRef]
27. Mundargi, R.C.; Babu, V.R.; Rangaswamy, V.; Patel, P.; Aminabhavi, T.M. Nano/micro technologies for delivering macromolecular therapeutic susing poly (D,L-lactide-co-glycolide) and its derivatives. *J. Control. Release* **2008**, *125*, 193–209. [CrossRef]
28. Ulery, B.D.; Nair, L.S.; Laurencin, C.T. Biomedical applications of biodegradablepolymers. *J. Polym. Sci. B Polym. Phys.* **2011**, *49*, 832–864. [CrossRef]
29. Vincent, D.; Salaar, K.; Alonzo, T. Applications of PLA in modern medicine. *Eng. Regen.* **2020**, *1*, 76–87. [CrossRef]
30. Zhao, X.; Hu, C.; Pan, G.; Cui, W. Pomegranate-structured electro sprayed microspheres for long-term controlled drug release. *Part. Part. Syst. Charact.* **2015**, *32*, 529–535. [CrossRef]
31. Tyler, B.; Gullotti, D.; Mangraviti, A.; Utsuki, T.; Brem, H. Polylactic acid (PLA) controlled delivery carriers for biomedical applications. *Adv. Drug Deliv. Rev.* **2016**, *107*, 163–175. [CrossRef] [PubMed]
32. Maji, R.; Dey, N.S.; Satapathy, B.S.; Mukherjee, B.; Mondal, S. Preparation and characterization of Tamoxifen citrate loaded nanoparticles for breast cancer therapy. *Int. J. Nanomed.* **2014**, *9*, 3107–3118.
33. Jiang, N.; Ma, Y.; Jiang, Y.; Zhao, X.; Xie, G.; Hu, Y.; Qin, C.; Yu, B. Clinical characteristics and treatment of extremity chronic osteomyelitis in Southern China. *Medicine* **2015**, *94*, e1874. [CrossRef] [PubMed]
34. Schroeder, J.E.; Girardi, F.P.; Sandhu, H.; Weinstein, J.; Cammisa, F.P.; Sama, A. The use of local vancomycin powder in degenerative spine surgery. *Eur. Spine J.* **2016**, *25*, 1029–1033.
35. Iqbal, M.; Zafar, N.; Fessi, H.; Elaissari, A. Double emulsion solvent evaporation techniques used for drug encapsulation. *Int. J. Pharm.* **2015**, *496*, 173–190. [CrossRef]
36. Mahmoudi, A.; Boukhechem, M.S. Simplified HPLC method for simultaneous determination of erythromycin and tretinoin in topical gel form. *Sep. Sci. Plus* **2020**, *3*, 86–93. [CrossRef]
37. Devi, T.S.R.; Gayathri, S. FTIR and FT-Raman spectral analysis of Paclitaxel drugs. *Int. J. Pharmaceut. Sci. Rev. Res.* **2010**, *2*, 106–110.

applied sciences

MDPI

Article

Effects of Dentifrices Containing Nanohydroxyapatite on Dentinal Tubule Occlusion—A Scanning Electron Microscopy and EDX Study

Emilia Bologa [1], Simona Stoleriu [1,*], Gianina Iovan [1], Cristina Angela Ghiorghe [1], Irina Nica [1], Sorin Andrian [1,*] and Oana Elena Amza [2]

[1] Faculty of Dental Medicine, Grigore T. Popa University of Medicine and Pharmacy, 16 Universitatii Str., 700115 Iasi, Romania; emilia.bologa@umfiasi.ro (E.B.); gianina.iovan@umfiasi.ro (G.I.); cristina.ghiorghe@umfiasi.ro (C.A.G.); irina.nica@umfiasi.ro (I.N.)

[2] Faculty of Dental Medicine, Carol Davila University of Medicine and Pharmacy Bucharest, 8 Eroii Sanitari Boulevard, 050474 Bucharest, Romania; oana_amza@yahoo.com

* Correspondence: simona.stoleriu@umfiasi.ro (S.S.); sorin.andrian@umfiasi.ro (S.A.); Tel.: +40-745-106-066 (S.S.); +40-745-213-742 (S.A.)

Received: 24 August 2020; Accepted: 14 September 2020; Published: 18 September 2020

Abstract: This in vitro study evaluated the effects of dentifrices containing nano-hydroxyapatite (n-HAp) on dentinal tubule occlusion and on mineral deposition. Dentin specimens of ten human teeth were submersed for 30 s in 40% citric acid and then randomly divided into four groups (three study groups and one control group). In the study groups, the dentin samples were exposed to three different n-HAp toothpastes: Karex (Dr. Kurt Wolff GmbH & Co. KG, Bielefeld, Germany), Biorepair Plus Sensitive (Coswell SpA, Bologna, Italy), and Dr. Wolff's Biorepair (Dr. Kurt Wolff GmbH & Co. KG, Bielefeld, Germany); in the control group no toothpaste was applied. All of the samples were evaluated using scanning electron microscopy (SEM) and energy-dispersive X-ray (EDX) analysis. In the control group all of the samples showed a frank and wide opening of the dentinal tubules, whereas in the study groups different degrees of tubule closure by mineral depositions were observed. Toothpastes containing n-HAp determined a significant occlusion of dentinal tubules and a significant increase of mineral deposition on the dentin surface. All three tested toothpastes showed similar results regarding the degree of dentinal tubule closure. Varying degrees of differences in calcium, phosphate, carbon, and oxygen ion concentrations among the three tested toothpastes were obtained.

Keywords: hydroxyapatite; dentifrices; dentin; tubule closure; dentin hypersensitivity

1. Introduction

Dentin hypersensitivity (DH) was described in 2009 by the FDI World Dental Federation as a "short sharp pain arising from exposed dentin most commonly at the tooth cervical area in response to stimuli (typically thermal, evaporative, tactile, osmotic or chemical), but which cannot be ascribed to any other dental defects, diseases or restorative treatments" [1]. Increased global average life expectancy and increased number of teeth because the population is now aware of the benefits of prevention and the importance of good oral hygiene [2,3] led to increased prevalence of DH, which commonly can range up to 74% [4,5].

Different theories were proposed throughout the years to explain the pathogenesis of DH, the hydrodynamic theory of Brännström being widely accepted nowadays. This theory postulates that external stimuli can rapidly dislocate the contents of dental tubules, distorting the nerve at pulp or predentin level or damaging the odontoblast cells and ultimately causing pain [2,6–9]. This is the explanation for two treatment strategies used in practice: dentinal tubule occlusion and nerve excitation prevention [10]. The agents in the first category determine the formation of an intratubular precipitate or the formation of an artificial smear layer over the dentin that blocks the entry into the tubules. Deposition of the thin coating layer determines the occlusion of the tubules, but also remineralization of the exposed dentin surface [11]. However, none of the agents in this category have been considered as a gold standard for dentinal tubule occlusion [12]. Various products are available on the market to be used either by the patients at home or by the dentists in the office, but the most common treatment of DH is the use of dentifrices [13,14].

Recommended toothpastes for DH contain different active ingredients such as: fluoride, calcium oxalate, calcium carbonate–phosphate, strontium acetate, arginine, and hydroxyapatite (HAp) [13,15–17]. HAp is the major inorganic component of natural teeth and bone and previous studies showed that nano-sized particles have similarities in morphology and structure to the tooth enamel apatite crystals [18,19]. Some studies and reviews have mentioned the capacity of this biofunctional material to remineralize enamel and dentin [19–22]. Nano-hydroxyapatite (n-HAp) was considered a promising active ingredient used for the treatment of DH due to high biocompatibility and bioactivity [23–25]. In toothpastes, HAp was included in the form of nanocrystals because they dissolve easier in this form [19,20,26]. Crystals of n-HAp included in dental products have a dimension of 50–1000 nm, which enables them to act like fillers [27]. These products can penetrate and block the exposed dentinal tubules which are responsible for DH. There is no common opinion in the literature regarding the efficiency of n-HAp based products used in DH treatment. Some studies have demonstrated higher desensitizing action and dentin remineralizing potential of n-HAp desensitizing agents [28,29], while other clinical studies have shown an equivalent result of n-HAp to other desensitizing agents [30].

This study aims to compare the effectiveness of three different commercial n-HAp toothpastes on dentinal tubule occlusion and on dentin mineral deposition by scanning electron microscopy (SEM) and energy-dispersive X-ray (EDX) analysis. The null hypothesis was that there is no significant difference regarding tubule occlusion between the three n-HAp dentifrices.

2. Materials and Methods

2.1. Sample Preparation

Ten extracted human permanent third molars were collected anonymously in the Oral Surgery Department/School of Dentistry at the University of Medicine and Pharmacy "Grigore T. Popa" Iași. A written informed consent form of biological materials used during the study was voluntarily obtained from the patients. The research was conducted in full accordance with the national research law 206/24.05.2004 and the Declaration of Helsinki, the protocol being approved by the Ethics Committee of the University (project no 3390). The teeth inclusion criteria were the presence of a complete crown, the absence of caries, cracks, or fillings. The teeth were cleaned to remove the soft tissues and stored in distilled water.

From each tooth, two dentin discs were obtained from mid-coronal part by cutting the tooth in two sections perpendicular to the long axis with a diamond disc (NTI-Kahla GmbH, Germany) at low speed under cooling water. Each dentin disc with a thickness of 1 mm was then cut in two half-disks in order to obtain four dentin pieces from each tooth. All the dentin samples were wet polished using 600-grit SiC papers. To open the dentinal tubules and to simulate the sensitive tooth model, all of the dentin half-discs were submersed for 30 s in 40% citric acid (Cerkamed, Poland). Then they were rinsed thoroughly with distilled water, followed by ultrasonic bathing for 10 min. After that the specimens were randomly and equally assigned to four groups. In groups 1–3, considered as study groups, three commercial n-HAp desensitizing toothpastes were applied on dentin disks and in group 4, considered as the control group, no toothpaste was applied. The toothpastes selected for the study groups were: Karex (Dr. Kurt Wolff GmbH & Co. KG, Bielefeld, Germany)—applied in group 1, Biorepair Plus Sensitive (Coswell SpA, Bologna, Italy)—applied in group 2, and Dr. Wolff's Biorepair

(Dr. Kurt Wolff GmbH & Co. KG, Bielefeld, Germany)—applied in group 3. Details regarding the chemical ingredients of the toothpastes are presented in Table 1.

Table 1. Chemical ingredients of the desensitizing toothpastes.

KAREX	Hydroxyapatite, Xylitol, Hydrated Silica, Tetrapotassium Pyrophosphate, Zinc Chloride, Cetylpyridinium Chloride, Sodium Methyl Cocoyl Taurate, Sodium Cocoyl Glycinate, Aqua, Glycerin, Phosphoric Acid, Silica, Hydrogenated Starch Hydrolysate, Cellulose Gum, Aroma
BIOREPAIR PLUS SENSITIVE COSWELL	Aqua, Zinc Hydroxyapatite, Glycerin, Sorbitol, Cellulose Gum, PEG-32, Silica, Sodium Myristoyl Sarcosinate, Sodium Methyl Cocoyl Taurate, Aroma, Sodium Saccharin, Citric Acid, Phenoxyethanol, Benzyl Alcohol, Sodium Benzoate
DR. WOLFF'S BIOREPAIR	Aqua, Zinc Hydroxyapatite, Hydrated Silica, Glycerin, Sorbitol, Silica, Aroma, Cellulose Gum, Sodium Myristoyl Sarcosinate, Sodium Methyl Cocoyl Taurate, Tetrapotassium Pyrophosphate, Zinc Pca, Sodium Saccharin, Phenoxyethanol, Benzyl Alcohol, Propylparaben, Methylparaben, Citric Acid, Sodium Benzoate.

The toothpastes were applied on the surface of the dentin disks using a brushing protocol which was described in some previous studies [31]. A tooth brushing machine designed to operate with back-and-forth movement at 1.5 Hz (250 g vertical load), with an amplitude of 30 mm (15 mm in each direction), and a frequency of 60 cycles/minute was used. The samples were brushed for 2 min, twice a day, for 14 days. The device was equipped with 4 stations to place the samples. Medium straight bristle toothbrushes (Classic Deep Clean, Colgate-Palmolive Company, New York, NY, USA) were placed in special attachments aligned parallel to the base. Abrasive slurries were prepared by mixing water and toothpastes (2:1 by volume). Between the tooth brushing sessions all the samples were stored in artificial saliva prepared according to AFNOR NF S90-701 standard procedure.

2.2. Scanning Electron Microscope (SEM) Analysis

Dentin samples were evaluated using scanning electron microscope VEGA II LSH (TESCAN, Brno, Czech Republic). Ten images at 2000× magnification were obtained of each dentin sample. The images were then assessed independently by two well-trained examiners, blinded to the tested toothpastes, to evaluate the degree of dentinal tubule occlusion on a five-grade scale, according to the tubule occlusion classification scoring system used in previous studies: 1 = occluded (100% of tubules occluded); 2 = mostly occluded (50–< 100% of tubules occluded); 3 = partially occluded (25–< 50% of tubules occluded); 4 = mostly unoccluded (<25% of tubules occluded); 5 = unoccluded (0%, no tubule occlusion) [26]. Each examiner counted the tubules in each of the 2000X images and established individually the dentinal tubule occlusion score on every image. If there was a disagreement in scoring between the two examiners, they re-examined the image until they arrived to a common opinion. For each sample, the average score of tubule occlusion from ten image evaluations was used for analysis.

2.3. Energy-Dispersive X-ray (EDX) Analysis

All of the specimens were examined by energy-dispersive X-ray microanalysis (QUANTAX QX2, BRUKER/ROENTEC, Berlin, Germany) for chemical determinations. Standard chemical composition determinations using PB-ZAF database on a selected area were used. Qualitative evaluation of the chemical elements revealed carbon (C), calcium (Ca), phosphorus (P), and oxygen (O) as dominant components. Quantitative evaluations of ion concentrations (wt%) were performed in ten different areas of each dentin sample. The ion concentrations on each sample were reported as the average value of ten determinations.

2.4. *Statistical Analysis*

The statistical software SPSS 20.0 (SPSS Inc., Chicago, IL, USA) was used for data analysis. Inter-examiner agreement on dentin tubule occlusion scores evaluated on SEM images was established using the Kappa test. The Kolmogorov–Smirnov normality test and nonparametric Mann–Whitney test were used to compare the dentin tubule occlusion values in groups. The Levene homogeneity of variance statistical test, the Kruskal–Wallis nonparametric test, and post hoc LSD were used to compare the percentages of chemical elements in the control and study groups.

3. Results

SEM images of some samples in the control and study groups are presented in Figure 1. In the control group, all of the samples showed a frank and wide opening of the dentinal tubules (Figure 1d), which confirmed the sensitive tooth model. In the study groups, obvious closure of the tubules by mineral depositions was observed, with different percentages of tubule occlusion being registered on the same sample or in different samples from the same group (Figure 1a–c). While most of the samples presented all the tubules partially occluded (as presented in Figure 1a,b), other samples presented completely unoccluded tubules beside partially occluded tubules. Mineral deposition on the dentin surface was also observed mostly on the samples in groups 1 and 2 (Figure 1a–c).

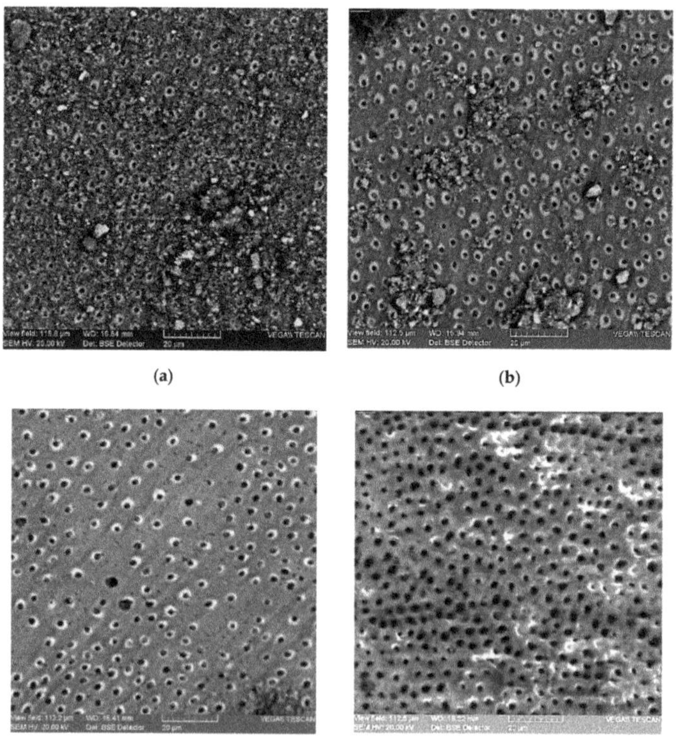

(a) (b) (c) (d)

Figure 1. SEM micrographs of dentin samples in study groups (**a**–**c**) and control group (**d**). (**a**) SEM image (2000×) of a dentin sample in group 1 (score 1). (**b**) SEM image (2000×) of a dentin sample in group 2 (score 1). (c) SEM image (2000×) of a dentin sample in group 3 (score 2). (**d**) SEM image (2000×) of a dentin sample in group 4 (score 5). SEM images on the dentin surface showing dentinal tubule occlusion with different scores in the study groups (group 1: Karex toothpaste, group 2: Biorepair Plus Sensitive toothpaste, and group 3: Dr. Wolff's Biorepair toothpaste) and in the control group (group 4).

Strong inter-examiner agreement on dentine tubule occlusion evaluated on SEM images was recorded (Kappa value of 0.961). The mean score values for dentin tubule occlusion and the number of samples specifically scored for tubule occlusion in each group are presented in Table 2. In group 3, the number of the samples with a score of 4 was higher than in groups 2 and 1. An equal number of samples with a score of 1 was recorded in groups 1 and 3 and an equal percentage of samples with a score of 3 was recorded in groups 2 and 3. The highest number of samples with a score of 2 was registered in group 1, followed by group 2 and 3, and the highest number of tubule closure with a score of 4 was obtained in group 3, followed by group 2 and group 1. Irrespective of the group, none of the samples in the study groups presented complete closure of dentinal tubules.

Table 2. The number of samples in the control and study groups scored with values between 1–5 for tubule occlusion.

	Number of Samples					Mean Score Value (± SD)
	Score 1	Score 2	Score 3	Score 4	Score 5	
Group 1	1	6	3	0	0	2.20 ± 0.632
Group 2	2	5	1	2	0	2.30 ± 1.059
Group 3	1	3	1	5	0	3.00 ± 1.155
Group 4	0	0	0	0	10	5.00 ± 0.000

The score values of dentin tubule occlusion in the study groups (group 1: Karex toothpaste, group 2: Biorepair Plus Sensitive toothpaste, and group 3: Dr. Wolff's Biorepair toothpaste) and in the control group (group 4).

The mean values of carbon, calcium, phosphorus, and oxygen ion concentration (wt%) are presented in Table 3. Carbon ion concentration in group 3 was significantly higher when compared to group 2 and group 1 ($p < 0.05$, Table 4). Statistically significant results were also recorded when comparing carbon ion concentration in the study groups and the control group ($p < 0.05$, Table 4). Significantly increased calcium ion concentration was recorded in the study groups when compared to the control group ($p < 0.05$, Table 4). In group 3, significantly lower calcium concentration was observed when compared to group 2 ($p < 0.05$, Table 4). Phosphorous ion concentration was significantly higher in the study groups when compared to the control group ($p < 0.05$, Table 4). Significantly lower phosphorous ion concentration was recorded in group 3 when compared to group 1 and group 2 ($p < 0.05$, Table 4). Oxygen ion concentration in the study groups was significantly higher when compared to the control group ($p < 0.05$, Table 4). Also, statistically significant differences were recorded when comparing the values of tubule closure in the study groups and the control group ($p < 0.05$, Table 4). There was no significant difference when comparing the values of dentin tubule occlusion in the study groups (p values > 0.05, Table 4).

Table 3. Mean values and standard deviation of carbon, calcium, phosphorus, and oxygen ion concentration (wt%) in control and study groups.

	Mean Value of Ion Concentrations (wt%) ± Standard Deviation			
	Carbon	Calcium	Phosphorus	Oxygen
Group 1	12.89 ± 7.64	22.40 ± 6.80	11.87 ± 2.76	48.32 ± 4.29
Group 2	10.58 ± 7.32	24.14 ± 3.86	12.96 ± 2.06	48.62 ±1.88
Group 3	23.73 ± 10.57	18.18 ± 6.50	9.58 ± 2.99	46.26 ± 3.71
Group 4	38.60 ± 6.600	13.15 ± 3.38	6.68 ± 1.87	41.54 ± 3.36

Carbon, calcium, phosphorous, and oxygen ion concentrations in the study groups (group 1: Karex toothpaste, group 2: Biorepair Plus Sensitive toothpaste, and group 3: Dr. Wolff's Biorepair toothpaste) and in the control group (group 4).

Table 4. Statistical test results when comparing the concentration of carbon, calcium, phosphorus, and oxygen ion concentration and the dentinal tubule occlusion score in groups.

Ion Concentration	Levene Test		Kruskal-Wallis Test		Post-Hoc LSD Test		
	Statistics	*p* Value	Chi-Square	*p* Value	Statistically Significant Results		*p* Value
Carbon	0.855	0.473	24.464	0.000 *	Group = 1	Group = 3	0.005 *
					Group = 1	Group = 4	0.000 *
					Group = 2	Group = 3	0.001 *
					Group = 2	Group = 4	0.000 *
					Group = 3	Group = 4	0.000 *
Calcium	1.843	0.157	8.333	0.000 *	Group = 1	Group = 4	0.000 *
					Group = 2	Group = 3	0.018 *
					Group = 2	Group = 4	0.000 *
					Group = 3	Group = 4	0.043 *
Phosphorous	1.364	0.269	12.665	0.000 *	Group = 1	Group = 3	0.045 *
					Group = 1	Group = 4	0.000 *
					Group = 2	Group = 3	0.004 *
					Group = 2	Group = 4	0.000 *
					Group = 3	Group = 4	0.012 *
Oxygen	1.169	0.335	9.055	0.000 *	Group = 1	Group = 4	0.000 *
					Group = 2	Group = 4	0.000 *
					Group = 3	Group = 4	0.004 *

Dentinal Tubules Occlusion Score	Kolmogorov Smirnov Test	Mann-Whitney Test		
	p Value	Statistical Results		*p* Value
	0.000	Group 1	Group 2	0.967
		Group 1	Group 3	0.092
		Group 1	Group 4	0.000 *
		Group 2	Group 3	0.173
		Group 2	Group 4	0.000 *
		Group 3	Group 4	0.000 *

Statistical test resuls of ion concentrations and dentinal tubule occlusion score comparison between the study and control groups. * statistically significant differences between groups.

4. Discussion

The current in vitro study aimed to establish the degree of dentinal tubule occlusion and mineral deposition on the dentin surface when using three different toothpastes containing n-HAp as an active ingredient. In our study none of the three n-HAp toothpastes determined complete occlusion of the dentinal tubules. Previous studies also showed partial closure of the dentinal tubules after an acid attack when using Biorepair Plus Sensitive toothpaste [32]. A study that tested n-HAp toothpaste Renamel (Sangi, Tokyo, Japan) had also demonstrated partial occlusion of dentinal tubules. Only one study reported total closure of dental tubules when an experimental pure n-HAp desensitizing toothpaste was tested [13].

Regarding the occlusion of dentinal tubules when comparing the toothpaste containing nano-HAp crystals to other products, the results of previous studies are contradictory. The toothpaste containing 1% n-HAp crystals led to a significantly higher percentage of tubule occlusion when compared to proargin technology [9]. Toothpastes containing 2 wt% n-HAp determined a reduction of DH in patients as a result of enamel surface restoration and dentinal tubule closure [21,33,34]. Some studies confirmed the efficacy of n-HAp dentifrices in achieving tubule occlusion [35]. On the contrary, in other studies there was no statistical difference between n-HAp products (Biorepair Plus) and Pronamel (Sensodyne) regarding the protection against an acid attack [29]. It was demonstrated that polishing the dentin surface with n-HAp led to partial tubular closure, while 30% S-PRG (surface pre-reacted glass-ionomer) filler-containing paste determined complete tubular closure [36].

In our study, SEM investigation of dentin samples treated by all three n-HAp toothpastes had shown mineral deposition that occluded the dentinal tubules and covered the intertubular dentin. Similar precipitate layer formation on the dentin surface was reported in previous studies that tested 10% and 15% n-HAp solutions [37]. Efficient dentinal tubule occlusion and mineral crystal growth on the dentin surface and into the dentinal tubules due to application of an n-HAp desensitizer was highlighted in a recent study [38]. Our study also reported increased calcium, phosphorous, carbon, and oxygen ion concentrations when n-HAp toothpastes had been applied on the dentin surface. In the remineralization process it was shown that n-HAp may act as a template in mineral crystal nucleation and growth to form a structure similar to dentin. Calcium and phosphate ions from the environment may be attracted on the dentin surface during the precipitation process and will fill the vacant places in the crystal structure [39]. Some previous studies have shown that toothpastes containing $ZnCO_3$/n-HAp or n-HAp have similar enamel and dentine remineralization capacity, and the remineralization potential was higher when compared to fluoride toothpastes [20]. On the contrary, another study that compared the remineralization capacity of HAp toothpaste to fluoride and bioactive glass toothpastes concluded that the effectiveness of the remineralization process was lower for HAp products than for fluoride toothpaste [33].

To our knowledge, there is no study in the scientific literature to evaluate Karex toothpaste efficacy on dentinal tubule occlusion and very few articles analyzed the efficacy of Biorepair Plus and Dr. Wolff's Biorepair toothpastes only in comparison with other desensitizing dentifrices. In our study there was no statistically significant difference in tubular occlusion between the study groups, so our null hypothesis was not rejected. This is in accordance with the studies of Tschoppe et al. [19] and Pei et al. [13] which have demonstrated that different n-HAp toothpastes have similar capacities in the remineralization process [13,19]. However, Tschopee et al. did not use the toothpastes tested in the present study and performed the experiment on bovine dentin specimens.

It is difficult to compare the results recorded in different studies due to different methodology of sample preparation and application technique. The present study simulates an acid attack using 40% citric acid in order to obtain the sensitive tooth dentin pattern. In our product application protocol, the dentin samples were brushed using a tooth brushing machine and this might preclude the probability of applying different forces on the manual toothbrush handle. One limitation of our study was that the dentine sample was obtained by cutting the tooth crown in the middle third. DH usually involves exposure of the cervical dentinal tubules, which have a number, diameter, and orientation that are different from those of the middle part of the crown. Therefore, the deposition pattern of the crystals might be different in vivo compared to our experiment. Another limitation was the evaluation of the toothpaste's action in the absence of any acidic attack between tooth brushing, which normally occurs in the oral cavity. Furthermore, the samples were stored in artificial saliva between tooth brushing sessions, but neither buffer effect of saliva nor enzyme activity were simulated. Further clinical studies are needed to evaluate the action of commercial n-HAp toothpaste on DH treatment.

5. Conclusions

Toothpastes containing n-HAp determined a significant closure of dentinal tubules and a significant increase of mineral deposition on the dentin surface. All three tested toothpastes showed similar results regarding the degree of tubule closure. There was a varying degree of differences in calcium, phosphate, carbon, and oxygen ion concentrations among the three tested toothpastes.

Author Contributions: Conceptualization, E.B. and S.A.; methodology, E.B., G.I., S.S. and I.N.; statistical analysis, O.E.A. and C.A.G.; writing—original draft preparation, S.S. and I.N.; writing—review and editing, G.I.; visualization, O.E.A. and E.B.; supervision, S.A. All authors have read and agreed to the published version of the manuscript.

Funding: This research received no external funding.

Conflicts of Interest: The authors declare no conflict of interest.

References

1. Dentin Hypersensitivity FDI World Dental Federation. Available online: https://www.fdiworlddental.org/resources/policy-statements-and-resolutions/dentin-hypersensitivity (accessed on 27 October 2019).
2. Baglar, S.; Erdem, U.; Dogan, M.; Turkoz, M. Dentinal tubule occluding capability of nano-hydroxyapatite; The in-vitro evaluation. *Microsc. Res. Tech.* **2018**, *81*, 843–854. [CrossRef]
3. Available online: https://www.who.int/gho/mortality_burden_disease/life_tables/situation_trends_text/en/ (accessed on 18 September 2020).
4. Rees, J.S.; Jin, L.J.; Lam, S.; Kudanowska, I.; Vowles, R. The prevalence of dentin hypersensitivity in a hospital clinic population in Hong Kong. *J. Dent.* **2003**, *31*, 453–461. [CrossRef]
5. Taani, D.Q.; Awartani, F. Prevalence and distribution of dentin hypersensitivity and plaque in a dental hospital population. *Quintessence Int.* **2001**, *32*, 372–376. [PubMed]
6. Hu, M.L.; Zheng, G.; Zhang, Y.D.; Yan, X.; Li, X.C.; Lin, H. Effect of desensitizing toothpastes on dentin hypersensitivity: A systematic review and meta-analysis. *J. Dent.* **2018**, *75*, 12–21. [CrossRef] [PubMed]
7. de Melo Alencar, C.; de Paula, B.L.F.; Guanipa Ortiz, M.I.; Baraúna Magno, M.; Martins Silva, C.; Cople Maia, L. Clinical efficacy of nano-hydroxyapatite in dentin hypersensitivity: A systematic review and meta-analysis. *J. Dent.* **2019**, *82*, 11–21. [CrossRef] [PubMed]
8. Iovan, G.; Andrian, S.; Devis, C. Mechanisms of dentin sensitivity: Theories and controversies. *J. Rom. Med. Dent.* **2008**, *12*, 6–11.
9. Kulal, R.; Jayanti, I.; Sambashivaiah, S.; Bilchodmath, S. An in-vitro comparison of nano hydroxyapatite, novamin and proargin desensitizing toothpastes—A SEM study. *J. Clin. Diagn. Res.* **2016**, *10*, ZC51-4. [CrossRef] [PubMed]
10. Gillam, D.; Talioti, E. The management of dentine hypersensitivity. In *Dentine Hypersensitivity: Developing a Person-Centred Approach to Oral Health*, 1st ed.; Robinson, P.G., Ed.; Academic Press: London, UK, 2015; pp. 45–76.
11. Cummins, D. Recent advances in dentin hypersensitivity: Clinically proven treatments for instant and lasting sensitivity relief. *Am. J. Dent.* **2010**, *23*, 3A–13A. [PubMed]
12. Amaechi, B.T.; Lemke, K.C.; Saha, S.; Gelfond, J. Clinical Efficacy in Relieving Dentin Hypersensitivity of Nanohydroxyapatite-containing Cream: A Randomized Controlled Trial. *Open Dent. J.* **2018**, *12*, 572–585. [CrossRef]
13. Pei, D.; Meng, Y.; Li, Y.; Liu, J.; Lu, Y. Influence of nano-hydroxyapatite containing desensitizing toothpastes on the sealing ability of dentinal tubules and bonding performance of self-etch adhesives. *J. Mech. Behav. Biomed. Mater.* **2019**, *91*, 38–44. [CrossRef]
14. Taha, S.T.; Han, H.; Chang, S.R.; Sovadinova, I.; Kuroda, K.; Langford, R.M.; Clarkson, B.H. Nano/micro fluorhydroxyapatite crystal pastes in the treatment of dentin hypersensitivity: An in vitro study. *Clin. Oral Investig.* **2015**, *19*, 1921–1930. [CrossRef] [PubMed]
15. Olley, R.C.; Pilecki, P.; Hughes, N.; Jeffery, P.; Austin, R.S.; Moazzez, R.; Bartlett, D. An in situ study investigating dentin tubule occlusion of dentifrices following acid challenge. *J. Dent.* **2012**, *40*, 585–593. [CrossRef] [PubMed]
16. Yuan, P.; Liu, S.; Lv, Y.; Liu, W.; Ma, W.; Xu, P. Effect of a dentifrice containing different particle sizes of hydroxyapatite on dentin tubule occlusion and aqueous Cr (VI) sorption. *Int. J. Nanomed.* **2019**, *14*, 5243–5256. [CrossRef] [PubMed]
17. Hiller, K.A.; Buchalla, W.; Grillmeier, I.; Neubauer, C.; Schmalz, G. In vitro effects of hydroxyapatite containing toothpastes on dentin permeability after multiple applications and ageing. *Sci. Rep.* **2018**, *8*, 1–13. [CrossRef] [PubMed]
18. Lin, X.; Xie, F.; Ma, X.; Hao, Y.; Qin, H.; Long, J. Fabrication and characterization of dendrimer-functionalized nano-hydroxyapatite and its application in dentin tubule occlusion. *J. Biomater. Sci. Polym. Ed.* **2017**, *28*, 846–863. [CrossRef]
19. Tschoppe, P.; Zandim, D.L.; Martus, P.; Kielbassa, A.M. Enamel and dentin remineralization by nano-hydroxyapatite toothpastes. *J. Dent.* **2011**, *39*, 430–437. [CrossRef]
20. Pepla, E.; Besharat, L.K.; Palaia, G.; Tenore, G.; Migliau, G. Nano-hydroxyapatite and its applications in preventive, restorative and regenerative dentistry: A review of literature. *Ann. Stomatal.* **2014**, *5*, 108–114. [CrossRef]

21. Schäfer, F.; Beasley, T.; Abraham, P. In vivo delivery of fluoride and calcium from toothpaste containing 2% hydroxyapatite. *Int. Dent. J.* **2009**, *59*, 321–324. [CrossRef]
22. Pajor, K.; Pajchel, L.; Kolmas, J. Hydroxyapatite and Fluorapatite in Conservative Dentistry and Oral Implantology—A Review. *Materials* **2019**, *12*, 2683. [CrossRef]
23. Coelho, C.C.; Grenho, L.; Gomes, P.S.; Quadros, P.A.; Fernandes, M.H. Nano-hydroxyapatite in oral care cosmetics: Characterization and cytotoxicity assessment. *Sci. Rep.* **2019**, *9*, 1–10. [CrossRef]
24. Besinis, A.; De Peralta, T.; Tredwin, C.J.; Handy, R.D. Review of Nanomaterials in Dentistry: Interactions with the Oral Microenvironment, Clinical Applications, Hazards, and Benefits. *ACS Nano* **2015**, *9*, 2255–2289. [CrossRef] [PubMed]
25. Singh, A.V.; Mehta, K.K. Top-Down Versus Bottom-Up Nanoengineering Routes to Design Advanced Oropharmacological Products. *Curr. Pharm. Dess.* **2016**, *22*, 1534–1545. [CrossRef] [PubMed]
26. Vano, M.; Derchi, G.; Barone, A.; Pinna, R.; Usai, P.; Covani, U. Reducing dentine hypersensitivity with nano-hydroxyapatite toothpaste: A double-blind randomized controlled trial. *Clin. Oral Investig.* **2018**, *22*, 313–320. [CrossRef] [PubMed]
27. Andrian, S.; Stoleriu, S.; Tărăboanţă, I.; Gamen, A.C.; Dîmbu, E.; Negraia, D. Remineralization of incipient enamel lesions using non-fluoride agents. A review. *Int. J. Med. Dent.* **2018**, *8*, 41–50.
28. Orsini, G.; Procaccini, M.; Manzoli, L.; Giuliodori, F.; Lorenzini, A.; Putignano, A. A double-blind randomized-controlled trial comparing the desensitizing efficacy of a new dentifrice containing carbonate/hydroxyapatite nanocrystals and a sodium fluoride/potassium nitrate dentifrice. *J. Clin. Periodontol.* **2010**, *37*, 510–517. [CrossRef] [PubMed]
29. Anand, S.; Rejula, F.; Sam, J.V.G.; Christaline, R.; Nair, M.G.; Dinakaran, S. Comparative evaluation of effect of nano-hydroxyapatite and 8% arginine containing toothpastes. *Acta Medica* **2017**, *60*, 114–119.
30. Wang, L.; Magalhães, A.C.; Francisconi-Dos-Rios, L.F.; Calabria, M.P.; Araújo, D.F.G.; Buzalaf, M.A.R.; Lauris, P.; Pereira, J.C. Treatment of dentin hypersensitivity using nano-hydroxyapatite pastes: A randomized three-month clinical trial. *Oper. Dent.* **2016**, *41*, E93–E101. [CrossRef]
31. Stoleriu, S.; Pancu, G.; Ghiorghe, A.; Sincar, D.C.; Solomon, S.; Andrian, S.; Iovan, G. Evaluation of dentinal changes following application of three different desensitizing agents. *Rev. Chim.* **2017**, *68*, 1573–1577. [CrossRef]
32. Poggio, C.; Lombardini, M.; Vigorelli, P.; Colombo, M.; Chiesa, M. The role of different toothpastes on preventing dentin erosion: An SEM and AFM study. *Scanning* **2014**, *36*, 301–310. [CrossRef]
33. Farooq, I.; Moheet, I.A.; AlShwaimi, E. In vitro dentin tubule occlusion and remineralization competence of various toothpastes. *Arch. Oral Biol.* **2015**, *60*, 1246–1253. [CrossRef]
34. Hill, R.G.; Chen, X.; Gillam, D.G. In vitro ability of a novel nanohydroxyapatite oral rinse to occlude dentine tubules. *Int. J. Dent.* **2015**, 153284. [CrossRef] [PubMed]
35. Jena, A.; Kala, S.; Shashirekha, G. Comparing the effectiveness of four desensitizing toothpastes on dentinal tubule occlusion: A scanning electron microscope analysis. *J. Conserv. Dent.* **2017**, *20*, 269–272. [CrossRef] [PubMed]
36. Iijima, M.; Ishikawa, R.; Kawaguchi, K.; Ito, S.; Saito, T.; Mizoguchi, I. Effects of pastes containing ion-releasing particles on dentin remineralization. *Dent. Mater. J.* **2019**, *38*, 271–277. [CrossRef] [PubMed]
37. Amaechi, B.T.; Mathews, S.M.; Ramalingam, K.; Mensinkai, P.K. Evaluation of nanohydroxyapatite-containing toothpaste for occluding dentin tubules. *Am. J. Dent.* **2015**, *28*, 33–39.
38. Luong, M.N.; Huang, L.; Chan, D.C.N.; Sadr, A. In Vitro Study on the Effect of a New Bioactive Desensitizer on Dentin Tubule Sealing and Bonding. *J. Funct. Biomater.* **2020**, *11*, 38. [CrossRef]
39. Zhang, M.; He, L.B.; Exterkate, R.A.M.; Cheng, L.; Li, J.; Ten Cate, J.M.; Crielaard, J.M.; Deng, D.M. Biofilm layers affect the treatment outcomes of NaF and nano-hydroxyapatite. *J. Dent. Res.* **2015**, *94*, 602–607. [CrossRef]

applied sciences

MDPI

Article

Facial Reconstruction: Anthropometric Studies Regarding the Morphology of the Nose for Romanian Adult Population I: Nose Width

Madalina Maria Diac [1], Kamel Earar [2,*], Simona Irina Damian [1,*], Anton Knieling [1,*], Tatiana Iov [3], Sarah Shrimpton [4], Maria Castaneyra-Ruiz [4], Caroline Wilkinson [4] and Diana Bulgaru Iliescu [1,3]

1 Faculty of Medicine, Department of Forensic Science, University of Medicine and Pharmacy "Grigore T. Popa", 700115 Iasi, Romania; madalina-maria.diac@umfiasi.ro (M.M.D.); diana.bulgaru@umfiasi.ro (D.B.I.)

2 University of Medicine and Pharmacy "Dunărea de Jos", 800201 Galati, Romania

3 Institute of Legal Medicine, 700450 Iasi, Romania; iovtatiana@yahoo.com

4 FaceLab–School of Arts and Design, Liverpool John Moores University, Liverpool L3 5TF, UK; s.l.shrimpton@ljmu.ac.uk (S.S.); m.castaneyraruiz@ljmu.ac.uk (M.C.-R.); c.m.wilkinson@ljmu.ac.uk (C.W.)

* Correspondence: erar_dr.kamel@yahoo.com (K.E.); simona.damian@umfiasi.ro (S.I.D.); anton.knieling@umfiasi.ro (A.K.)

Received: 27 August 2020; Accepted: 15 September 2020; Published: 17 September 2020

Abstract: Craniofacial reconstruction often represents a final step in medico-legal identification and is dependent on facial tissue thickness measurements and feature shape estimation. This study's aim is to create a reliable and readily reproductible method of predicting the maximum nose width (MNW) based on the maximum nasal aperture width (MAW) for a Romanian adult population. A sample of 55 computer tomography (CT) scans consisting of Romanian adult subjects was selected from the database of a neurosurgery hospital. The craniometrics measured consisted of a first measure of MAW and second one of the MNW using 3D systems Freeform Modelling Plus Software. Correlation analysis indicated a moderate link between the MAW and the MNW. Regression analysis showed that MAW and sex form a statistically significant regression pattern (R^2 = 0.340, SEE (Standard Error of Estimate) = 3.801). The preliminary results obtained provide reliable predictions of MNW for facial reconstruction based on MAW measured on the skull.

Keywords: facial reconstruction; nasal morphology; nose width; aperture width

1. Introduction

Human identification is one of the essential elements of forensic practice and investigation. This is optimally achieved through a close collaboration between forensic medicine, anthropology, and the judicial process; this collaboration needs to be materialized in laboratory methods (forensic, anthropological, genetic, craniofacial reconstruction) as well as in investigative methods (crime scene research, reconstitution, fingerprint, identification of personal objects, etc.). Human identification uses anthropological and laboratory methods in order to establish the identity of human remains, one of which is the reconstruction of the facial aspect from the skull.

In forensic practice, the skull helps perform several roles: establishing the subject's identity, reflecting the age of the subject, indicating sex as well as facial appearances [1]. Therefore, facial reconstruction depicts individual facial characteristics in order to help identify the person. In the facial reconstruction theory, all faces are unique, and the shape variation of the face is directly related to skeletal structure [2–5].

Human identification through facial appearance has preoccupied society for hundreds of years. The earliest example is the work of Alphonse Bertillon (1853–1914), a Parisian anthropologist who developed a system for recording detailed facial characteristics that could later be used for identification. Another pioneer was Mikhail Gerasimov (1907–1970), a Russian academic who carried out research relating to the morphology of the face and skull [4]. Facial analysis includes three methods: morphological—comparison of facial characteristics in terms of shape and facial type; proportional—comparison of facial characteristics in terms of size and distance between certain skull points; anthropometrical—comparison of measurements of the face and skull [6].

The anatomical relationship between soft and hard tissue can be established by means of radiological data—X-ray, computer tomography, nuclear magnetic resonance, etc. There are a number of recent publications that provide different imaging modalities for establishing the facial soft tissue thicknesses specific to different populations [7–10]. At an international level there is vast literature and research regarding this relationship as well as for the necessary parameters relating to craniofacial reconstruction. Thus, most countries (France, Brazil, Portugal, China, etc.) have conducted and are conducting numerous studies to develop methods for craniofacial reconstruction (based on computer-tomography images, cadavers, ultrasonography, etc.) and to establish the necessary correlations between bone and soft tissue specific to each population [11–15].

Facial reconstruction methods use two important facial elements (nose and mouth), which define the aspect of the face and contribute to a higher accuracy of the reconstructed physiognomy. Therefore, in reconstructing these elements, multiple studies which are related to multiple prediction methods, are necessary for correctly establishing the relation between the hard tissue and soft tissue. In this sense, as we mentioned before, a lot of countries are developing their own database with these correlations. One study, which developed such correlations for all populations, was done by Rynn and Wilkinson. The study revealed the relation between hard and soft tissue of the nose; in this sense the authors developed regression formulas based on aperture dimensions and nasal craniometric points [16].

Despite all these aspects, the Romanian population does not have any data on the relationship between soft and hard tissues in order to allow craniofacial reconstruction, as well as no data relating to the relationship between these tissues for the morphology of the nose. This data will allow, first of all, the alignment at an international standard of the forensic identification process, and secondly will help the development of research in forensics. Another important reason is the fact that Romania is a country with predisposition for disasters with multiple victims (earthquakes, flooding), but not only, situations where the identification process is primordial.

The main objective of this paper is to establish a correlation between the soft and hard tissues regarding the morphology of the nose specific to a Romanian adult population. In this direction, the authors endeavor to identify the maximum nose width (MNW) measured on the soft tissue based on the maximum width of nasal aperture (MNA) measured on the hard tissue.

2. Materials and Methods

The research is based on a retrospective study of 55 computer tomography (CT) images of the skull performed on living people aged 27–89 years (M = 60.95), SD = 16.53), of which 24 were females (43.6%) and 31 males (56.4%). The CT images were studied from the archives of a Neurosurgery Hospital in Iasi, Romania. Only the sex and age of the patients were considered in this research—no other details as to the identity of the person were targeted.

The practical procedures were carried out in accordance with the Declaration of Helsinki and the protocol was approved by the Ethical Committees of Grigore T. Popa Medicine and Pharmacy University and of the Neurosurgery Hospital (no. 9432/12 June 2020).

The inclusion criteria were represented by cases where the CT scans had over 60 slices (thus allowing the creation of 3D surfaces) and the images allowed the analysis and visualization of the whole nose region. Excluded from the present study were the cases where CT images showed fractures of the nasal bones or when the entire region of the nose could not be visualized, and it was not possible to correctly measure the nasal skull points.

The study was worked out at John Moores Liverpool University, Liverpool, United Kingdom for a period of two and a half months.

The resulting dataset consisted of 55 tomographic computer images that were worked on for the parameter maximum width of the nose. Images were processed using the InVesalius 3.0 software, creating separate surfaces (images) for both bone and soft tissue and exporting these images later as 3D images (obj. format). For analysis and reconstruction, the 3D images were imported into 3D Systems Free Form Modelling Plus Software using a Touch haptic feedback device. Freeform is a software with a high intelligence which can be applied in any sciences (medicine, engineering, etc.). Freeform features provide a high-level interface to a lot of representations. The purpose of freeform features is faster and more intuitive modelling with more guarantee for a high-quality design. Comparative with other ways of modelling, in freeform there is a better control over the 3D surface and it is easier to use, especially helping by touch haptic feedback device.

The above-mentioned steps allowed the import of all cases separately into FreeForm, with the skulls positioned in Frankfurt Horizontal Plane (FHP). This involves drawing a plane passing through the upper edge of the external acoustic meatus and the lower edge of the orbital rim, which being viewed from the interior, posterior, and lateral (right/left), must represent a straight line.

For the researched parameter, the maximum nose width (MNW) measurements were taken both on the skull and soft tissue. The measurements from the skull involved the maximum distance of the nasal aperture, for which the distance between the most lateral points of the aperture was taken into account (Figure 1). The measurements taken on the soft tissue involved the calculation of the maximum distance of the nasal alae, between the most lateral points of the nasal alae (Figure 2).

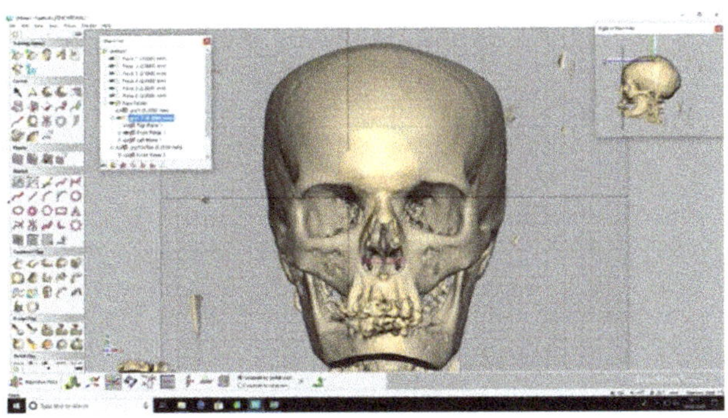

Figure 1. Distance between the most lateral points of the nasal aperture.

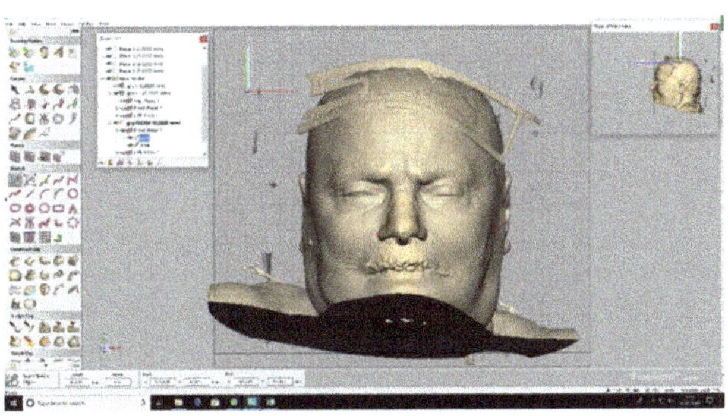

Figure 2. Distance between of the most lateral points of the nasal alae.

Thus, a series of measurements were made on hard tissue as well as soft tissue. For each parameter, we took two measurements and calculated an average of the two values; each measurement was made twice and rechecked. All the measurements used millimeters (mm) as measurement unit. Additionally, the intra-observer error (TEM) was calculated for a better evaluation of the two measurements which was performed by the same person. The value of TEM obtained following the methodology for TEM calculation is 0.341, which was classified as acceptable and shows that the variation between the two measurements suffers no influence of the systematic error.

The resulting database was analyzed using SPSS 3.0. Several statistical tests were performed in order to analyze the data. Kolmorov–Smirnov test was used to verify the normality of the statistical distribution. An independent samples *T*-test was performed to analyze sex difference regarding the maximum width of the nose and a repeated measures ANOVA was performed in order to compare the means for the maximum nose width calculated based on different regression formulas. Lastly, Pearson correlation indicated the relation between the variables and simple and multiple linear regression were conducted in order to provide linear regression equations.

3. Results

The Kolmogorov–Smirnov test indicated a normal distribution of variables: maximum width of the nose, maximum width of nasal aperture ($p = 0.200$).

The maximum width of the nose varies between 30.11 (mm) and 50.18 (mm), and the maximum width of the nasal aperture is between 17.84 and 28.16 (Table 1). The results are presented separately for male and female participants in Table 2.

Table 1. Descriptive statistics of the maximum width of the nose (MWN) and the maximum width of the nasal aperture (MAW).

	N	Min	Max	Mean	SD
MWN (mm)	55	30.11	50.18	39.14	4.25
MAW (mm)	55	17.84	28.16	23.58	1.94

Table 2. Descriptive statistics of the maximum width of the nose (MWN) and the maximum width of the nasal aperture (MAW) separately for males and females.

		N	Min	Max	Mean	SD
MWN	Women	24	32.22	44.84	36.95	3.02
	Men	31	30.11	50.18	40.84	4.32
MAW	Women	24	17.84	27.87	23.04	1.85
	Men	31	19.64	28.16	23.99	1.93

A *T*-test showed significant differences between females and males for the maximum width of the nose ($t\ (53) = -3.76, p < 0.001$). In this respect, the maximum width of the nose for males ($M = 40.84$, $SD = 4.32$) is significantly higher than that of females ($M = 36.95$, $SD = 3.02$) (Figure 3).

A *T*-test showed that there are no significant differences between males and females for the maximum width of the nasal aperture ($t\ (53) = -1.859, p = 0.069$) (Figure 3).

3.1. Correlational Analysis

Correlational analysis indicated a significantly moderate link between the maximum width of the nasal aperture and maximum width of the nose. The partial correlation between the maximum width of the nasal aperture and the maximum width of the nose, by controlling the sex of the participants, was carried out. In this respect, the greater the maximum width of the nasal aperture, the greater the maximum width of the nose (Table 3).

There was no significant link between the age of the participants and the maximum width of the nose ($r\ (53) = 0.101, p = 0.462$).

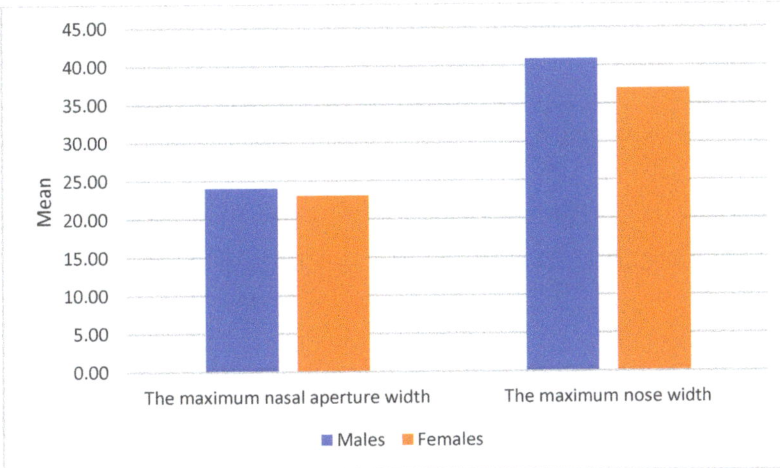

Figure 3. Differences between sexes for MAW and MNW.

Table 3. Correlations and partial correlations between the maximum width of the nasal aperture and the maximum width of the nose.

	MNW (Correlation)	MNW (Partial Correlation)
MAW	0.462 **	0.405 **

** Indicates significance at the 0.01 level.

3.2. Regression Analysis

Regression analysis showed that the maximum width of the nasal aperture and sex form a statistically significant regression pattern (F (2, 52) = 13.39, $p < 0.001$), which explains a proportion of 34% ($R^2 = 0.340$) of the variance of the maximum width of the nose. We found that sex predicts 21.1% of the variance of the maximum width of the nose (R2 = 0.211, $p < 0.001$) and the maximum width of the nasal aperture brings a significant explanatory addition of 12.9% (R2 change = 0.129, $p = 0.002$). (Table 4). SEE have small values indicating that the observations are closer to the fitted line.

Table 4. Regression model for maximum width of the nasal aperture and sex for maximum width of the nose.

Outcome (y)	Independent Variables (x)	B	Std. Error	Beta	R^2	SEE
Maximum width of the nose	Model 1					
	Sex	3.894	0.777	0.459 **	0.211 **	3.807
	Constant	36.948 **	1.035			
	Model 2					
	Sex	3.115	0.986	0.367 **	0.340 **	3.515
	Maximum width of the nasal aperture	0.813	0.255	0.371 **		
	Constant	18.220 **	5.915			

** Indicates significance at the 0.01 level.

Within the predictive model, the most important predictor of the maximum width of the nose is the maximum width of the nasal aperture (β = 0.371, p = 0. 002), followed by the sex of the participant (β = 0.367, p = 0.003).

The regression analysis showed that the maximum width of the nasal aperture statistically predicts the maximum width of the nose (F (1, 53) = 14.37, $p < 0.001$) and explains a proportion of 21.3% (R2 = 0.213) of the variance of the maximum width of the nose (Table 5) .

Table 5. Regression model for maximum width of the nasal aperture (without sex) for maximum width of the nose.

Outcome (y)	Independent Variables (x)	B	Std. Error	Beta	R^2	SEE
Maximum width of the nose	Maximum width of the nasal aperture	1.012	0.267	0.462 **	0.213 **	3.801
	Constant	15.280	6.316			

** Indicates significance at the 0.01 level.

The regression equation with sex: maximum width of the nose = 18.22 + 0.813* maximum width of the nasal aperture + 3.115*sex, for sex: male is represented by code 1, and female by code 0. (Table 6).

The simple regression equation (without sex) (Table 6):

maximum width of the nose = 15.28 + 1.012*maximum width of the nasal aperture,

All conditions of simple and multiple linear regression are met. Statistical analysis demonstrates the fact that the regression analysis is correct, with the homoscedasticity being present. Additionally, it can be observed that the errors are normally distributed, which means the regression equation is correct. All these prove that we can rely on both regression equations with precise results (Figure 4).

Table 6. Regression equations showing correlation coefficients (R) and linearity of relationship.

Predicted Dimension	Simplified Equation	R	R^2	<2.5 mm Error (%)	<5 mm Error (%)
1. MNW (mm)	15.28 + 1.012 * MAW	0.462 **	0.213	98.18%	100%
2. MAW (mm)	18.22 + 0.813 * MAW + 3.115*sex	0.583 **	0.340	98.18%	100%

* Indicates significance at the 0.05 level, ** Indicates significance at the 0.01 level.

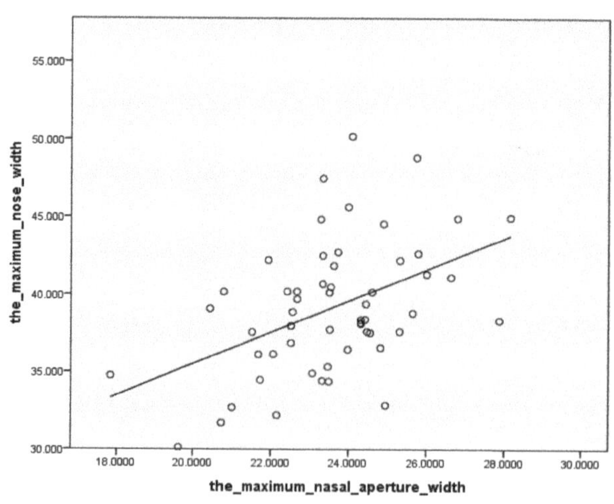

Figure 4. Regression plot.

3.3. Analysis of the Differences between the Real Values of the Maximum Width of the Nose and the Values Calculated Based on the Regression Formulas

Regarding the repeated measures ANOVA, for applying the test a GLM (general linear model) type was used and it shows that there are no statistically significant differences between the original values for maximum nose width measured on the Romanian population $F(3, 162) = 0.394$, $p = 0.758$) the values resulting from the current formulas (simple—without inclusion of sex; multiple—including the sex) and the values resulting from the regression formula developed by Rynn and Wilkinson (Table 7).

The *T*-test analysis for paired echantions shows that there are no statistically significant differences between the original values for maximum nose width measured on the Romanian population, the values resulting from the current formulas (simple—without inclusion of sex; multiple—including the sex), and the values resulting from the regression formula developed by Rynn and Wilkinson (Table 7).

For another line of evidence of the differences between the original values and values resulting from the regression formula developed by Rynn and Wilkinson, the Altman–Bland plot analysis was performed, which shows that there are no significant differences between the two values and that there are not many extreme cases, which shows that the regression is adequate (Figure 5).

Table 7. Descriptive statistics of maximum nose width calculated for each regression formula.

Maximum Width of the Nose	N	Min	Max	Mean	SD
Original values	55	30.11	50.18	39.14	4.25
Regression formula including the participants' sex	55	32.72	44.23	39.14	2.48
Simple regression formula	55	33.33	43.77	39.14	1.96
Regression formula determined by Rynn, Wilkinson, and Peters (2010)	55	31.21	46.93	39.49	3.22

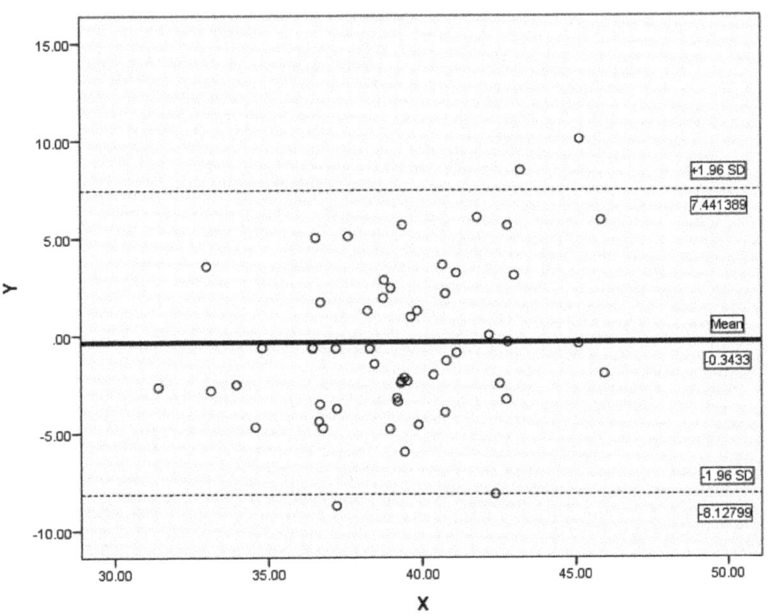

Figure 5. Bland and Altman plot with the representation of the limits of agreement (dotted line), from −1.96 s to +1.96 s, where Y = PNW-RMNW, X = (PNW + RMNW)/2.

Our analysis found a threshold of marginal significance approaching the standard threshold ($p = 0.05$), which may indicate possible differences between the two types of values (the ones resulting from the formula developed by Rynn and colleagues being higher) (Figure 6).

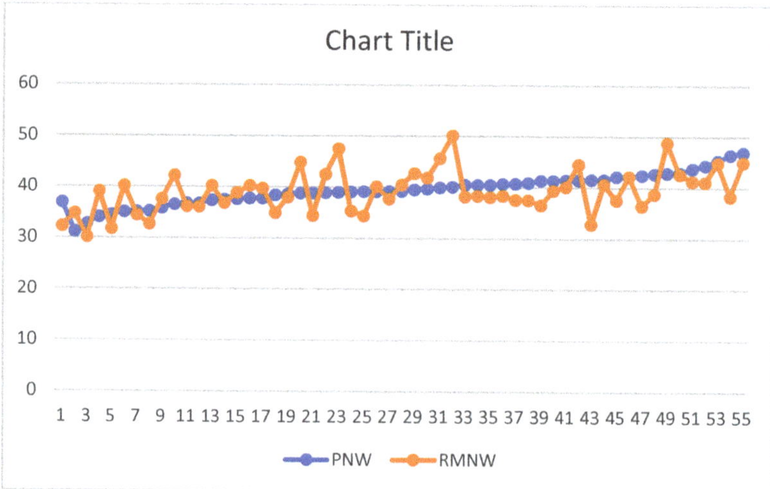

Figure 6. Differences between the real values of the maximum nose width and the predicted nose width following the Rynn and Wilkinson formula.

4. Discussion

Craniofacial reconstruction is currently the elective method in identifying victims, based on examination of the skull. The purpose of craniofacial reconstruction is therefore to create an image as close as possible to that of the individual before death, thus allowing skeletal identification in the absence of other means of identification or helping to guide it.

In order to be able to perform the technique of craniofacial reconstruction and, implicitly, to obtain an image as reliable and as close to reality as possible, it is necessary to know the thickness of soft tissues in specific locations (cheeks, chin, nose, lips, eyes, forehead). This parameter requires a wealth of research carried out on specific populations [8–14]. In this respect, the objective of the present study was to analyze the relationship between hard tissue and soft tissue with regard to the maximum width of the nose, a first parameter in the construction of the morphological aspect of the nose [17].

The objective pursued in this paper was achieved, the results providing both a normal distribution of variables and a significant relationship between the hard tissue and soft tissue; thus, the thickness and morphology of soft tissues are dependent on skeletal structure. This helps the craniofacial reconstruction process.

The interest of this research work in facial reconstruction is given primarily by the originality of the work at national level, in Romania, the craniofacial reconstruction being a completely new field in legal medicine in this country. As a result of the above, it goes without saying that there are currently no valid data on the correlation between soft and hard tissue specific to the Romanian population. Thus, the present paper represents an important contribution in the field of the identification process in legal medicine by introducing craniofacial reconstruction methods as well as presenting data specific to the adult population of Romania. At the same time, the craniofacial reconstruction process is an important aspect for the progress of knowledge and research in the field of identification and, implicitly, forensic anthropology. Even if craniofacial reconstruction is a separate work area, or can be considered a separate area, it could represent, in legal medicine, a final complementary examination for anthropological expertise on deceased persons (skeleton, skull, bone remains, etc.) [18].

The literature mentions that the anatomical relations between soft and skeletal tissues can be obtained using imaging methods: X-ray, computer tomography, magnetic resonance, or ultrasonography [19–21]. In the present study, the authors chose the cranio-cerebral tomographic computer images as working material to obtain accurate data for the relationship between soft tissue and bone for the morphology of the nose. The advantages of using the computer tomography images are as follows: (i) they offer a better view of the bone compared to MRI or ultrasonography; (ii) are much easier to interpret and do not require exhaustive knowledge in radiology; (iii) the available database is larger [22,23].

In the study conducted by Rynn and Wilkinson, the authors addressed this topic by developing regression formulas for the morphology of the nose (maximum width of the nose, length, height, projection, and more) valid for the white Caucasoid population. In the aforementioned study, the participating subjects consisted of Northern Brits and white Americans. Rynn and Wilkinson noted that the maximum width of the nose (predicted nose width) represents the maximum width of the nasal aperture/3 × 5, without taking into account the sex or age of the subjects included in the study [16].

Initially, we compared the results of the research carried out by Rynn and colleagues on a new adult Romanian population. According to the results of the present study, there are differences between the actual value of the maximum width of the nose and the maximum width of the nose predicted by the regression equation developed by Rynn and Wilkinson (Table 7, Figure 6). Therefore, the authors developed a regression formula to predict the aforementioned parameter specific to the adult population of Romania. This research needs to be continued and validated on a larger number of participants.

The results confirm the close relationship between the maximum width of the nasal aperture and the maximum width of the nose, the bone being a significant predictor for the morphology of soft tissue (Table 3). At the same time the statistical analysis carried out shows that in the case of the adult population of Romania, the sex of the individual is also an important predictor for estimating the maximum width of the nose starting from the maximum distance of the nasal aperture. Figure 6 highlights the differences between the real values of the maximum nose width of the Romanian adult population and the values resulting from the Rynn and colleague's formula (the predicted nose width).

Figure 6 also shows that the values predicted according to the regression equation that includes the sex of the participants developed in this research are the closest, one can even say similar to the actual values of the maximum nose width of the adult population of Romania. The second set of values, those predicted as a result of the simple regression equation without taking into account the sex of the participants is slightly larger, but also close to the actual values. These may be important in cases where sex cannot be estimated accurately and we can use only the simple equation, without expecting significant errors.

The literature mentions that the nasal morphology interacts statistically with age of the participants, especially for the parameters which outline the nasal structure (the anterior projection, the height and length of the nose) [16,19]. One study shows a small increasing of the piriform aperture and of the nasal width for each 10 years [7,24]. In our study, this difference is not applicable because the age is not a predictable factor for maximum width of the nose or for maximum width of the aperture, as resulting from the statistical analysis. Regarding sex as a factor that has been suggested to affect the morphology of the maximum width of the nose, our study confirms the results from the literature [23,24].

In this research, the authors developed two types of regression formulas to estimate the maximum width of the nose: a regression formula that includes the sex of the participants as well as a regression formula without including the sex of the participants. Both regression formulas have the same mean (Table 7) and slightly different standard deviations, and Figure 6 shows that the results of the two regression formulas have similar values.

The regression formula generated from this database, specific to the adult population of Romania regarding the maximum width of the nose, has the advantage that it is easy to use as measurements can be made either using a computer tomographic image or even directly on the available skull. It can also be used for 2D and 3D facial reconstruction.

We would like to underline the most important aspect of the study: the authors developed two regression equations for estimating the maximum nose width for Romanian adult population (one including the sex of the person, another one without it). This is an essential aspect for the future of

Romanian anthropological research in general, and facial reconstruction in particular. Future research should be carried out based on the measurements of more landmarks and of more anatomical components of the facial skull and from other radiological data.

5. Conclusions

The results of the study show that each population is different. Additionally, the intense populational migration in recent years creates an increase in mixed populations in various parts of the globe. This proves the need for intensive studies and the development of new research for each type of population. Of course, new studies on the adult population of Romania are needed on a much larger number of cases, both for the morphology of the nose and for other anatomical components of the facial mass. The data used in this paper provides reliable results with a large applicability in the future for estimating the maximum nose width in facial reconstruction for Romanian adult populations. This paper is a good start in the development of the forensic facial reconstruction studies for Romanian adult populations.

Author Contributions: Conceptualization and design, C.W., M.M.D. and D.B.I.; methodology, M.M.D., S.S. and A.K.; software, M.M.D., S.S. and M.C.-R.; statistical analysis, M.M.D., S.S., M.C.-R.; data curation, M.M.D., S.I.D. and T.I.; writing—original draft preparation, M.M.D.; writing—review and editing, M.M.D., S.S., M.C.-R., K.E., C.W. and D.B.I.; supervision, C.W. and D.B.I. All authors have read and agreed to the published version of the manuscript.

Funding: This research was partially supported by COST, within MULTI-modal Imaging of FOREnsic SciEnce Evidence-tools for Forensic Science program, Action Number CA16101.

Acknowledgments: The authors thank the FaceLab team, John Moores University, Liverpool, UK for invaluable support and research facilities.

Conflicts of Interest: The authors declare no conflict of interest.

References

1. Wilkinson, C. *Forensic Facial Reconstruction;* Cambridge University Press: New York, NY, USA, 2004; pp. 6–27.
2. Wilkinson, C. Computerized Forensic Facial Reconstruction a Review of Current Systems. *Forensic Sci. Med. Pathol.* **2005**, *1*, 173–177. [CrossRef]
3. Gupta, S.; Gupta, V.; Vij, H.; Vij, R.; Tyagi, N. Forensic Facial Reconstruction: The final frontier. *J. Clin. Diagn. Res.* **2015**, *9*, 26–28. [CrossRef] [PubMed]
4. Drgacova, A.; Dupej, J.; Veleminska, J. Facial soft tissue thicknesses in the present Czech population. *Forensic Sci. Int.* **2016**, *260*, 106.e1–106.e7.
5. Guyomarch, P.; Santos, F.; Dutailly, B.; Coqueugniot, H. Facial soft tissue depths in French adults: Variability, specificity and estimation. *Forensic Sci. Int.* **2013**, *231*, 411.e1–411.e10. [CrossRef] [PubMed]
6. Verze, L. History of facial reconstruction. *Acta Biomed* **2009**, *80*, 5–12.
7. Chen, F.; Chen, Y.; Yu, Y.; Qiang, Y.; Liu, M.; Fulton, D.; Chen, T. Age and sex related measurement of craniofacial soft tissue thickness and nasal profile in the Chinese population. *Forensic Sci. Int.* **2018**, *212*, 272.e1–272.e6. [CrossRef]
8. Stephan, C.N.; Sievwright, E. Facial soft tissue thickness (FSTT) estimation models-and the strength of correlations between craniometric dimensions and FSTTs. *Forensic Sci. Int.* **2018**, *286*, 128–140. [CrossRef]
9. De Greef, S.; Claes, P.; Vandermeulen, D.; Mollemans, W.; Suetens, P.; Willems, G. Large-scale in-vivo Caucasian facial soft tissue thickness database for craniofacial reconstruction. *Forensic Sci. Int.* **2006**, *1595*, S126–S146. [CrossRef]
10. El-Mehallawi, I.H.; Soliman, E.M. Ultrasonic assessment of facial soft tissue thicknesses in adult Egyptians. *Forensic Sci. Int.* **2001**, *117*, 99–107. [CrossRef]
11. Claes, P.; Vandermeulen, D.; De Greef, S.; Willems, G.; Clement, J.G.; Suetens, P. Computerized craniofacial reconstruction: Conceptual framework and review. *Forensic Sci. Int.* **2010**, *201*, 138–145. [CrossRef]
12. Ridel, A.F.; Demeter, F.; Liebenberg, J.; L'Abbe, E.N.; Vadermeulen, D.; Oettle, A.C. Skeletal dimensions as predictors for the shape of the nose in a South African sample: A cone-beam computed tomography (CBCT) study. *Forensic Sci. Int.* **2018**, *289*, 18–26. [CrossRef] [PubMed]

13. Chung, J.H.; Chen, H.T.; Wan-Yi, H.; Huang, G.S.; Shaw, K.P. A CT-scan database for the facial soft tissue thickness of Taiwan adults. *Forensic Sci. Int.* **2015**, *253*, 132.e1–132.e11. [CrossRef] [PubMed]
14. Bulut, O.; Sipahioglu, S.; Hekimoglu, B. Facial soft tissue thicknesses database for craniofacial reconstruction in the Turkish adult population. *Forensic Sci. Int.* **2014**, *242*, 44–61. [PubMed]
15. Sipahioglu, S.; Uluay, H.; Diren, H.B. Midline facial soft tissue thickness database of Turkish population: MRI study. *Forensic Sci. Int.* **2012**, *219*, 282.e1–282.e8. [CrossRef] [PubMed]
16. Rynn, C.; Peters, H.L.; Wilkinson, C. Prediction of nasal morphology from the skull. *Forensic Sci. Med. Pathol.* **2010**, *6*, 20–34. [CrossRef]
17. Tedeschi-Oliveira, S.V.; Haltenhoff Melani, R.F.; Haddad de Almeida, N.; Saavedra de Paiva, L.A. Facial soft tissue thickness of Brazilian adults. *Forensic Sci. Int.* **2009**, *193*, 127.e1–127.e7. [CrossRef]
18. Wilkinson, C.; Motwani, M.; Chiang, E. The relationship between the soft tissues and the skeletal detail of the mouth. *J. Forensic Sci.* **2003**, *48*, 1–5. [CrossRef]
19. Paim Strapasson, R.A.; Herrera, L.M.; Haltenhoff Melani, R.F. Forensic facial reconstruction: Relationship between the alar cartilage and piriform aperture. *J. Forensic Sci.* **2017**, *10*, 1–6. [CrossRef]
20. Haddad de Almeida, N.; Michel-Crosato, E.; Saavedra de Paiva, L.A.; Haye Biazevic, M.G. Facial soft tissue thickness in the Brazilian population: New reference data and anatomical landmarks. *Forensic Sci. Int.* **2013**, *231*, 404.e1–404.e7. [CrossRef]
21. Stephan, C.N.; Preisler, R. In vivo facial soft tissue thicknesses of adult Australians. *Forensic Sci. Int.* **2018**, *282*, 220.e1–220.e12. [CrossRef]
22. Panenkova, P.; Benus, R.; Masnicova, S.; Obertova, Z.; Grunt, J. Facial soft tissue thicknesses of the mid-face for Slovak population. *Forensic Sci. Int.* **2012**, *220*, 293.e1–293.e6. [CrossRef] [PubMed]
23. Perlaza Ruiz, N.A. Facial soft tissue thickness of Colombian adults. *Forensic Sci. Int.* **2013**, *229*, 160.e1–160.e9. [CrossRef] [PubMed]
24. Starbuck, J.M.; Ward, R.E. The affect of tissue depth variation on craniofacial reconstructions. *Forensic Sci. Int.* **2007**, *172*, 130–136. [CrossRef] [PubMed]

applied sciences

MDPI

Article

The Interdisciplinary Approach of Some Middle Bronze Age Pottery from Eastern Romania

Ana Drob [1,2], Viorica Vasilache [1,*] and Neculai Bolohan [3]

1 Institute of Interdisciplinary Research-Interdisciplinary Sciences Department, ARHEOINVEST Center, "Alexandru Ioan Cuza", University of Iași, Bld Carol I no. 11, 700505 Iași, Romania; ana.drob@uaic.ro
2 Doctoral School of History, Faculty of History, "Alexandru Ioan Cuza", University of Iași, Bld Carol I no. 11, 700505 Iași, Romania
3 Faculty of History, "Alexandru Ioan Cuza", University of Iași, Bld Carol I no. 11, 700505 Iași, Romania; neculai.bolohan@uaic.ro
* Correspondence: viorica_18v@yahoo.com

Abstract: Prehistoric pottery is the most abundant material discovered in archaeological sites and represents the main element of knowledge about human communities from the past. This study presents a model of interdisciplinary investigation of pottery through several types of analyses, enabling the scientific study of this category of artifacts. The analyses were performed on 11 ceramic fragments from the Middle Bronze Age settlement of Piatra Neamț–*Lutărie*, Eastern Romania, considering information about the color, production technique, type, size, functionality and category of the vessel, but also data related to ceramic paste inclusions. The samples were studied by optical microscopy (OM), scanning electron microscopy (SEM) coupled with energy-dispersive X-ray analysis (EDX) and micro-Fourier-transform infrared spectroscopy (μFTIR). The results obtained provide important information regarding pottery manufacturing technologies, such as sources of the raw materials and firing temperatures, and revealed the functionality of various vessel categories within a prehistoric settlement.

Keywords: middle bronze age; eastern Romania; pottery; macroscopic analysis; OM; SEM-EDX; μFTIR

Citation: Drob, A.; Vasilache, V.; Bolohan, N. The Interdisciplinary Approach of Some Middle Bronze Age Pottery from Eastern Romania. *Appl. Sci.* **2021**, *11*, 4885. https://doi.org/10.3390/app11114885

Academic Editor: Filomena De Leo

Received: 10 May 2021
Accepted: 25 May 2021
Published: 26 May 2021

1. Introduction

Using the archaeological methodology, we know a repertoire of seven vessel categories for Costișa and six vessel categories for Monteoru. Some of these vessels are also found in the group of fragments from Piatra Neamț–*Lutărie*. This approach is built on a case study that includes archaeological knowledge and aims to identify the physical and chemical characteristics specific to the pottery of a site in an area of interaction. In the same context, the present study proposes a scientific recovery and interdisciplinary analysis of a group of ceramic fragments to enlighten the history of a site, which we do not have much data.

The ceramic fragments analyzed belong to the Middle Bronze Age (1955/1879–1630/1614 BC) [1–3]. From an archaeological point of view, these materials are characteristic of the local Costișa community, with the possibility of some southern Monteoru pottery. These two Bronze Age communities were overlapping in the Subcarpathian area of Moldavia, where they coexisted both in individual settlements but also within the same one. Although some scholars have tried to highlight the existence of conflicts between these communities [4,5], recent research has disproved this hypothesis [6–8]. Thus, in this paper, the pottery of these communities is studied to identify their behavior in terms of similarities or differences concerning the pottery technologies they used, which could indicate the nature of the relationship between these two communities. Moreover, there is relatively little information about the everyday life of these groups, as no details are known about producing pottery, the process of manufacturing and pyrotechnology, aspects that will be studied in this work.

This paper presents the interdisciplinary study of 11 ceramic fragments from the Piatra Neamț–*Lutărie* site through several types of analyses. Thus, they were first investigated macroscopically, taking into account a series of information, such as color, production technique, type, size, category of the vessel and functionality, but also data related to inclusions (type, size, frequency, shape). Regarding the chemical composition and other microelements, several analytical techniques were used, such as OM, SEM-EDX and μFTIR. The results obtained from interdisciplinary analyses provide data on the nature of the raw material, firing technologies, and data on the functionality of the vessels.

2. Materials and Methods

2.1. Site Information

The site from Piatra Neamț–*Lutărie* [9,10] is located in Eastern Romania (Figure 1), in the hydrographic basin of the Bistrița river (GPS: N 46°56′08.13″; E 26°23′03.06″), which represents the main prehistoric communication route in this area. This settlement was excavated in 1957, leading to the discovery of archeological artifacts from the Eneolithic, Bronze Age and Iron Age periods and the II–III centuries AD. Information on the Costișa habitation is scarce, with no mention of living structures or firing installations assigned to this community. The settlement dimensions are unclear because the site was affected by landslides and the rescue archaeological excavation area was approximately 200 m^2. From this area were recovered only ceramic fragments. Other types of objects attributed to the middle bronze age period are not mentioned. Moreover, in the 1959 report, it is specified that some archaeological complexes were affected by the "flooded soil," and the archaeological material was mixed [9].

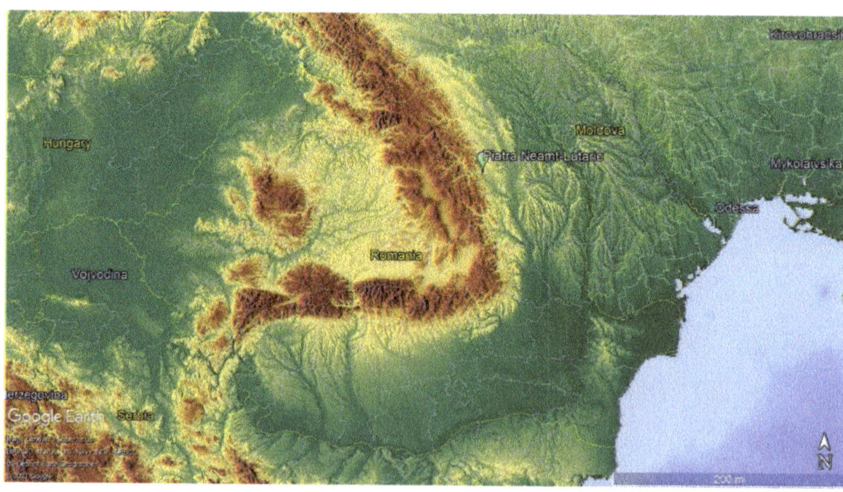

Figure 1. Geographical location of the Piatra Neamț–*Lutărie* settlement.

Geologically, the site area is located on Paleogene deposits (Oligocene), consisting of clays, breccias, sandstone and conglomerates. In the western part of the site, represented by the Pietricica peak, Paleogene deposits are composed of sandstone-shale flysch, bituminous flysch with Kliwa sandstones, and conglomerates. To the east, the area is made up of Quaternary formations (Pleistocene) consisting of sands, gravels, boulders and loess deposits [11]. Pedologically, in the settlement area and in its vicinity, there are only chernozems [12]. Therefore, the settlement from Piatra Neamț–*Lutărie* is located in a geological area rich in raw clay materials that could be used in pottery making and fertile soils used in agricultural activities.

2.2. Methodology

The pottery fragments were discovered in 1957 recovered from depths between −0.55 m and −1.50 m. The materials are in the repositories of the Museum of History and Archeology Piatra-Neamț. In this regard, some of the methodological limitations of this study should be mentioned. First of all, these 11 fragments studied are the only existing ceramic materials from this site, which did not allow the analysis of a more consistent database that would enable more comprehensive analysis. The small number of ceramic fragments is most likely caused by excavation or storage deficiencies because, in the other known Costișa sites, the appreciable amount of pottery is completed by other objects, such as those made of bone, stone or bronze. Second, the small number of preserved ceramic fragments also determined the size of the samples taken so that they were no larger than 2–3 cm, which unfortunately did not permit a complex petrographic study. As an extensive petrographic study could not be performed, the information on mineralogy will be limited to microscopic observations, which are supplemented by the results of chemical analyses, in particular the data provided by the FT-IR analysis. In addition, to these issues can be added an extremely small number of archaeometric studies for the Middle Bronze Age pottery in the eastern Carpathian area, which could have been used as a reference basis for the present study. However, even in the case of these limitations, the ceramic materials from the old excavations can be capitalized through such archaeometric studies, which contribute to the understanding of certain technological aspects involved in the processes of making prehistoric pottery in the eastern Carpathian area.

The 11 pottery fragments analyzed were marked with the letter L (Lutărie) and with numbers from 1–10. They all belong to Costișa ceramic group. They were fragments of liquid transport or storage vessels (L1), cooking (L2, L3, L4, L8), storage (L5, L6, L7) and serving vessels (L9, L10) (Figure 2). The Monteoru ceramic group is represented by a single fragment of liquid transport or storage vessel, noted L11 (Figure 2/L11). All the fragments were studied macro- and microscopically, but also by SEM-EDX and μFTIR analyses.

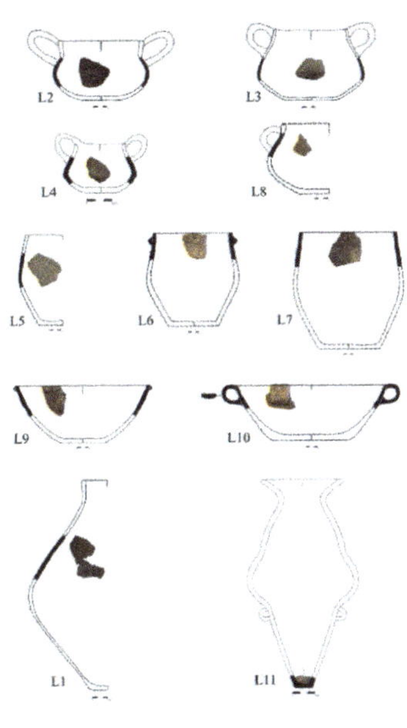

Figure 2. The analyzed pottery samples from the Piatra Neamț–*Lutărie*.

2.2.1. Macroscopic Analyses

The analysis criteria used are a standard for macroscopic investigations. Thus, for this type of approach, the following elements were considered: the color of fragments, shape/type of vessel, modeling technique, wall thickness, functionality, hardness, surface, surface treatment and firing atmosphere [13].

The color is determined through the Munsell Soil Color Charts, which allows a neutral rendering of nuances [14].

The shape and type of the vessels are important for identifying the technological features of different ceramic categories. Moreover, there were attempts to identify the production techniques used to manufacture the vessels [15] by studying the edges and surfaces of the ceramic fragment [16].

Measuring the vessel wall thickness is an important step because it is an indicator of its functionality and the technical skills of prehistoric potters [13,17].

Regarding the functionality of the vessels, relevant publications were considered trying to determine whether the functional criteria known through ethnoarchaeology [18,19] and experimental archeology are applicable in our study [20,21]

The hardness of the ceramic fragments and their surface provide important clues about the quality of the vessels, but also about technical skills and functionality [22]. These data were supplemented by information on surface treatment [23], which involved smoothing and/or burnishing [14,22,24] and ceramic slips [23].

The firing atmosphere can be oxidizing, reducing or mixed, and it has a significant effect on both the color of the paste and the resulting surfaces [22,25]. The aspects related to pyrotechnology provided important data about the quality of the vessels, but also about the technological knowledge involved in the production process [23].

In the paste analyses, it is necessary to identify the component inclusions [26]. They can be ceramoclasts (crushed shards), organic matter, lithoclasts (crushed stones) or clay inclusions (small pieces of unhomogenized clay) [27]. Next, it is necessary to determine their size, frequency, shape and sorting [22]. The shape of the inclusions may indicate if they are part of the raw material. In the case of an intentional addition, their shape will be angular, caused by the crushing of the materials used [13]. The sorting and shape of the inclusions indicate the homogeneity of the paste.

The homogeneity of the paste is determined by the quality of mixing and the amount of inclusions [22]. Mixing also influences the porosity of the paste. There are two types of pores: primary and secondary. The main pores are open spaces left after the clay modeling, and the secondary pores appear as a result of the burning organic matter, appearing in the form of black spots [28].

2.2.2. OM Analyses

Microscopic analysis of the fragments was performed with a Zeiss Imager.a1M microscope with a built-in AXIOCAM camera, which uses AxionVisionRelease 4.7.1 software. The images were obtained at 50× and 100× magnifications in the dark field for a clear view of the inclusions. Samples were sanded with a Stroers LaboPol device using discs with different granulations.

2.2.3. SEM-EDX Analyses

In the current analyses was used an electron microscope with SEM scan, model VEGA II LSH, produced in the Czech Republic by TESCAN, coupled with an EDX detector type QUANTAX QX2, produced in Germany by BRUKER/ROENTEC. This microscope has a tungsten electron gun that can achieve a resolution of 3 nm at 30 KV, with a magnification of between 30 and 1,000,000 × in "resolution" mode, the acceleration voltage between 200 V and 30 kV, scan speed between 200 ns and 10 ms per pixel. The working pressure is less than 1×10^{-2} Pa. The EDX detector is used for microanalysis, which allows quantitative measurements without using specific calibration standards. It has an active area of 10 mm^2, analyzing sanded or irregular surfaces. The SEM micrographs obtained consisted of backscattered electrons (BSE) at 200× and 500× magnifications for ceramic paste and surfaces without being covered with metal or graphite.

2.2.4. μFTIR Analyses

The spectra were recorded with an FTIR spectrophotometer coupled with an HYPERION 1000 microscope, both equipment from Bruker Optic, Germany. The FTIR spectrophotometer is of the TENSOR 27 type, which is predominantly suitable for measurements in close IR. The standard detector is DLaTGS covers the spectral range 7500–370 cm^{-1}, and works at room temperature. The resolution is usually 4 cm^{-1} but can also reach 1 cm^{-1}. TENSOR 27 is equipped with a He-Ne laser that emits at 633 nm and a power of 1 mW and has a ROCKSOLID alignment of the interferometer. The signal/noise ratio of this device is very good. The TENSIONER is completely controlled by the OPUS software. The HYPERION 1000 microscope is an accessory that can be paired with almost any Bruker FTIR spectrophotometer. For completely nondestructive measurements, the TENSOR 27 spectrophotometer is connected to the HYPERION 1000 microscope, and, usually, the solid samples are analyzed in reflection mode. The software is OPUS/VIDEO for interactive video data acquisition. The microscope can work in both transmission and reflection, being equipped with a 15× lens. The detector is of the MCT type cooled with liquid nitrogen (−196 °C). The spectral range is 600–7500 cm^{-1}, and the measured area is optimized to a diameter of 250 μm with the possibility of reaching a minimum of 20 μm.

3. Results and Discussions

3.1. *Macroscopic Results*

From the category of vessels for storage and transport of liquids, there is a single pot (L1) fired in a reducing atmosphere, with a finely burnished surface. The wall thickness is 7.87 mm. In the paste, medium-size ceramoclasts and lithoclasts, sub-angular with a distribution of 10–15%, were identified. These inclusions are well mixed and integrated into the clay matrix. Lithoclasts were isolated presences and were represented by three rounded pebbles, most likely from the raw material.

The cooking vessels (L2, L3, L4, L8) were fired in a reducing atmosphere, having fine (L2, L3) and semi-fine surfaces (L4, L8). Two vessels had an outer ceramic slip as surface treatment (L4, L8), and two other pots had their exterior surface burnished with a hard object (L2, L3). The vessel L2 had a burning spot with fine cracks on the outer surface, below the line of maximum diameter, both caused by extended heat exposure. The wall thickness is between 6.97 and 7.83 mm, and the identified inclusions in the ceramic paste are small, subrounded ceramoclasts, with a distribution of 5–10% (L3) and medium-size ones, with a distribution of 10–15% (L2, L4). Large sub-angular inclusions were also observed, with a frequency of 10–15% (L8). In the paste of cooking vessels, mainly the secondary pores were visible, indicating a good mixing of the components.

For the category of serving vessels (L9, L10), the firing took place in a mixed environment (L9) and in an incomplete oxidizing one (L10). The L9 vessel had a ceramic slip applied on the inner surface, and the L10 has both surfaces very well smoothed. On the exterior of both containers, there were visible secondary burning spots resulted from heat exposure. The thickness of the vessel wall is between 10.56 and 11.04 mm, and the paste of both pots have medium, sub-angular ceramoclasts with a distribution of 15–20% (L9) and 20–25% (L10).

Storage vessels (L5, L6, L7) were fired in a reducing (L5) and mixed (L6, L7) atmosphere and had a semi-fine (L5) and semi-coarse (L6, L7) surface. Two of the vessels have their outer surface coated with a ceramic slip (L5, L6), while another vessel had its inner surface burnished with a hard tool (L7). The wall thickness is between 9.84 and 11.74 mm, and the main inclusions are the ceramoclasts, but the lithoclasts are also present in small numbers, resulting from the raw materials. The identified ceramoclasts are large, sub-angular and have a frequency of 10–15% in the clay matrix, and the paste mixing is fair, with primary and secondary pores. On the inner surface of two vessels (L6 and L7), there are visible traces of smudging (Figure 3) caused by using these containers in the preparation of hot food, information that will be verified through chemical analyses.

Figure 3. Storage vessels with traces of smudging (**a**) L6, (**b**) L7.

The Monteoru vessel intended for storing or transporting liquid (L11) is brown, both on the inside and the outside. It was fired in a reducing atmosphere. The wall thickness is 9.16 mm. The fragment has a semi-fine surface and an exterior ceramic slip. The inclusions identified in the clay matrix are large-sized, subrounded, with a frequency of 10–15%. On the base of the vessel, secondary burning traces are visible and a series of large pores indicating negligent mixing. In the vessel's interior, a black layer is visible that covers the entire inner surface of the fragment (Figure 4). Those residues resulted from using the vessel will be of particular interest in chemical analyses.

Figure 4. Sample L11 (**top**) interior of the vessel; (**bottom**) outside of the vessel.

All vessels were made by the coiling technique, as indicated by the traces of joining or by noticeable bumps, especially on the inner surface of the vessels. The auxiliary elements (handles, grips) were made by modeling. Traces of joining and finishing of these components are also visible.

Surface treatments consisted mainly of applying an external ceramic slip, noticeable in cooking and storage vessels. The same treatment is applied to the interior for the serving vessels. The outer luster was intended to waterproof cooking and liquid transporting/storing vessels, and the inner one is present in the case of a storage vessel.

The presence of ceramoclasts is an important aspect related to the manufacture of vessels that facilitate certain functionalities. This type of inclusion is easier to break than other materials. They are already burned, which makes them more stable during the firing of the pots, and they have the same properties as the material in which they were incorporated. Therefore, it can be assumed that ceramoclasts were preferred because when they are burned, they have the same thermal characteristics and the same rate of expansion as clay, which demonstrates the prehistoric potters knew that aspect. Thus, in the paste of all the studied fragments, there are ceramoclasts present, which, depending on the functional category, have different dimensions, shapes and distributions. Based on the dimensions, shape and distribution of the main intentional inclusions, which are the ceramoclasts, three paste categories were identified. The first one, marked CP1 (Paste Category 1), is used only for cooking vessels, the inclusions being of various sizes and shapes, but the distribution in ceramic paste does not exceed 10–15%. This category is also used for liquid transport or storage vessels. The second category, CP2 (Paste Category 2), is used for storage vessels, where the ceramoclasts are large, sub-angular and have a frequency of 10–15%. The Monteoru pot was included in this category (L11). The Monteoru vessel does not present major differences in manufacturing and finishing, compared to the Costișa fragments, wherein the same inclusions with the same size, shape and frequency have been identified. The last category, CP3 (Paste Category 3), is found in serving vessels, where the ceramoclasts have a medium size, sub-angular shape and distribution of 15–25%. They were found in a larger quantity than in the case of the other functional categories.

3.2. Interdisciplinary Analyses

OM, SEM-EDX and μFTIR analyses were performed for all fragments, keeping the notation for the ceramic paste to differentiate the inner surface by marking it with I and the outer surface marked with E.

3.2.1. OM Analyses

Through optical microscopy (Figure 5), it was observed that all the analyzed ceramic fragments contain quartz, mica and iron oxides, all representing natural inclusions from the raw material.

The presence of ceramoclasts was also detected macroscopically, but microscopically reused ceramoclasts were identified. The reused ceramoclasts result from crushing other pottery fragments, which, in their turn, also contain ceramoclasts added in the paste of the new vessels. They are present only in the case of cooking (L2, L3, L8), storage (L5, L6, L7) and serving vessels (L10).

In some ceramic fragments (L2, L3, L6, L7, L9, L10), the presence of possible plant materials as small pores and dark lamellae was observed. Visible organic matter may be the result of intentional addition or may be part of the clay used. Moreover, in all samples, the iron oxides from the raw material were identified, an important clue for identifying the type o of clay used in pottery manufacturing.

In the case of the Monteoru vessel, the black layer had multiple fine cracks, resembling a viscous, dry substance (Figure 6). Its nature will be verified, as precisely as possible, through EDX and μFTIR analysis.

Reused ceramoclasts identified by optical microscopy indicate a concern for reusing damaged vessels and their "recycling" as inclusions of the new ones. The organic matter visible in most of the ceramic fragments could indicate the firing temperatures, which did not reach high values. The minerals identified microscopically were quartz, mica and iron oxides, which were part of the raw material clay used to manufacture pottery from Piatra Neamț–*Lutărie*.

Figure 5. OM for the Costișa ceramic fragments (L1–L10) from Piatra Neamț–*Lutărie* (C) ceramoclasts; (RC) reused ceramoclasts).

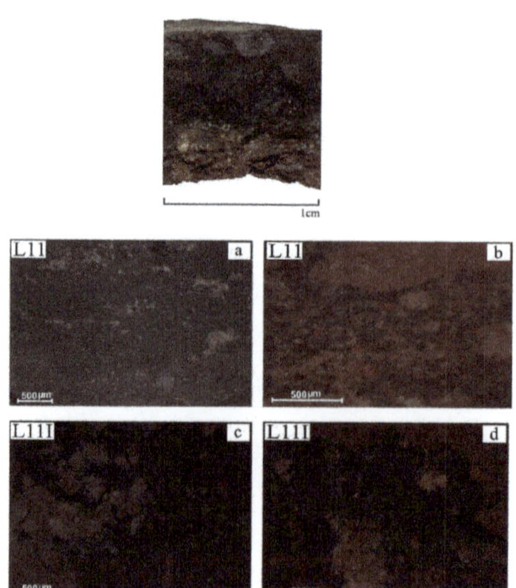

Figure 6. OM for the L11 sample **(a,b)** core; **(c,d)** inner surface with residues.

3.2.2. SEM-EDX Analyses

The SEM-EDX analyses were performed both on the core and on the surfaces, keeping the notations from the OM analyses. The SEM micrographs for the core of the Costișa samples (Figures 7 and 8) show a good homogeneity of the microstructural elements, which are well incorporated in the clay matrix (L1, L4). A compact lamellar structure was also visible (L8).

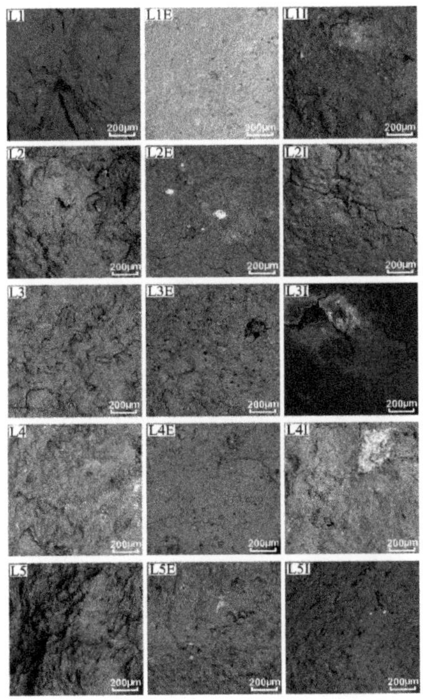

Figure 7. SEM micrographs for the Costișa ceramic fragments (L1–L5) from Piatra Neamț-*Lutărie*.

In most samples (L2, L3, L5, L6, L7, L9, L10), individualized mineral particles are visible, with a low homogeneity, and in other samples, traces of vegetal fibers were noticed (L7, L10), aspects that suggest relatively low firing temperatures. The SEM images also show elongated and flattened pores, present in five samples (L1, L5, L8, L9, L10), which indicate using the coiling technique in making the vessels [29].

SEM analyses of the surfaces revealed the presence of luster (Figures 7 and 8). This type of treatment closes ceramic pores, making the vessel walls smooth and waterproof, as noticed in the case of vessels L1, L2, L3 for the exterior and L7 for the interior. The rest of the ceramic fragments have well-smoothed surfaces, with partially covered pores and individualized mineral particles. In the case of samples L2E+I, L3I, L4E+I, L7E+I and L8E, fine cracks are visible on the surfaces, which may have resulted from using the vessel. Such traces that represent indicators of functionality are also the carbon deposits [30] on the outer surfaces of vessels L3 and L7.

The SEM micrographs for Monteoru sample L11 (Figure 9) indicate a good homogeneity, with mineral elements well incorporated in the clay mass. Carbon deposits are visible on the outer surface of the fragment, and the black layer was observed on the inside.

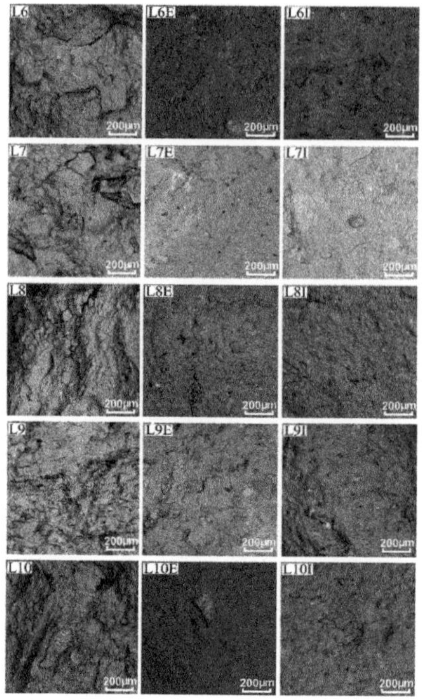

Figure 8. SEM micrographs for the Costișa ceramic fragments (L6–L10) from Piatra Neamț-*Lutărie*.

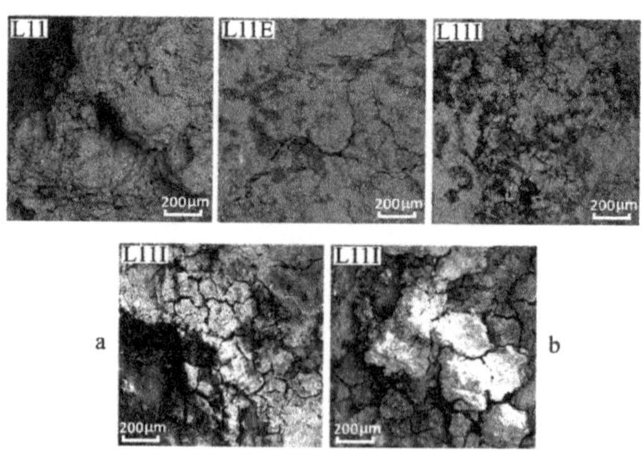

Figure 9. SEM micrographs for Monteoru sample L11 (**a**,**b**) details for the inner residues.

SEM analyses of the ceramic paste indicated that most of the samples have a low homogeneity, and the component microstructures were well individualized. The vitrification process was not observed in any sample, an indication that the firing temperatures have not exceeded 800–850 °C [31–33]. For five ceramic fragments, it was confirmed by SEM using the coiling technique in vessel manufacturing, a fact observed macroscopically for all vessels.

EDX analyses for ceramic paste were performed for all 11 samples taken from the Piatra Neamț-*Lutărie*. The chemical composition of the studied fragments contains the common elements found in clays, such as Si, Al, P, Mg, Ca, K, Na, Fe, Ti, O or C [34–39]. The elements that are of interest in archaeometric studies are iron, calcium, phosphorus and carbon (Table 1). In pottery studies, these elements are considered significant because their presence in certain concentrations allows arguments about the type of clay (Fe > 4% ferruginous clay, Ca > 5% calcareous clay), the firing temperature (C is found in samples up to 750 °C, and its presence in high concentration on the inner surface of the vessel may be due to the nature of the goods stored in the vessel) and the functionality of the vessels (the presence of P> 2% demonstrates that the vessels were used in food preparation).

Table 1. Elemental composition of the ceramic fragments from the Piatra Neamț-*Lutărie*.

Sample	Elemental Composition in Weight Percent (%)										
	Si	Al	P	Mg	Ca	K	Na	Fe	Ti	C	O
L1	24.65	9.42	1.30	0.84	1.83	3.76	0.34	4.63	1.43	-	51.80
L2	22.30	9.96	2.10	1.46	1.86	2.50	0.60	3.42	0.72	0.64	54.44
L3	26.22	9.94	2.44	1.14	1.47	2.68	0.37	4.96	0.56	0.55	49.67
L4	26.24	9.81	2.55	1.25	1.46	2.41	0.72	5.47	0.71	-	49.38
L5	23.44	9.10	1.37	1.21	2.70	3.07	0.70	4.27	0.85	-	53.29
L6	22.76	8.26	2.51	0.80	2.43	3.17	0.64	4.45	0.92	0.27	53.79
L7	22.18	9.94	2.30	1.33	2.13	2.97	0.53	5.11	0.82	0.23	52.46
L8	23.95	10.80	3.04	1.17	1.43	2.23	0.78	4.91	0.66	-	51.03
L9	18.27	7.14	2.10	0.70	2.60	2.15	0.52	3.79	0.40	1.19	61.14
L10	22.57	10.10	4.75	0.95	2.37	2.17	0.90	5.15	0.84	0.22	49.98

The presence of iron and/or calcium indicates the nature of the clay used, which can be ferruginous (4%) [36–39] or calcareous (5%) [31,40]. The analyzed samples have iron values from 3.42% to 5.97%, which suggests using ferruginous clay in vessel manufacturing. Calcium is present in all samples, with concentrations not exceeding 2.70%, so the presence of a calcareous clay is not detected. The Monteoru fragment (L11) has a higher calcium content (4.13%), indicating a possible low calcareous clay.

Phosphorus values higher than 2% result from using a vessel for boiling or for storing liquids rich in phosphorus [41–43]. In the analyzed group, phosphorus has concentrations between 2.09 and 3.55% for cooking vessels samples L2, L3, L4, L8, 2.30–2.51% for storage vessels (L6, L7) and 2.10–4.70% for serving vessels (L9, L10).

Carbon is present in the chemical composition of six samples, two for cooking (L2, L3), two for storage (L6, L7) and two for food serving (L9, L10). Concentrations are between 0.22 and 1.18%, indicating that the firing temperature for these vessels did not exceed 700 °C [33,44].

3.2.3. µFTIR Analyses

The µFTIR analyses performed for all 11 samples showed similarities in terms of chemical compounds present in the raw clay used for pottery making. The appearance of the obtained spectra is the same, except for the L11 fragment, which is a feature that will be discussed at length (Figure 10).

The range 4000–3000 cm^{-1} is attributed to water [45], and the absorbed water [46,47] from the samples was also detected through obvious peaks from ~3368 cm^{-1} and ~1630 cm^{-1}, corresponding to the -OH deformations, most likely as a result of deposition processes.

Kaolinite ($Al_2Si_2O_5(OH)_4$) was identified in almost all samples by peak position in the region 3500–3750 cm^{-1} [46,48,49]. All fragments were fired at temperatures above 500 °C, as indicated by the absence of the 915 cm^{-1} doublets [46,50,51] of kaolinite, except for the L11 sample.

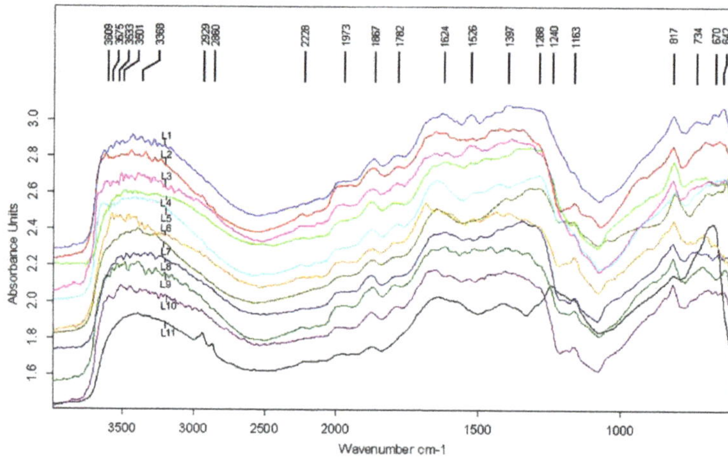

Figure 10. FT-IR spectra of the ceramic fragments from Piatra Neamț–*Lutărie*.

In all analyzed samples, calcite ($CaCO_3$) was identified by peaks at ~2970 cm^{-1}, ~2936 cm^{-1}, ~2519 cm^{-1}, ~737 cm^{-1}. In addition, the presence of carbonates [51,52] is visible from the broad band at 1300–1500 cm^{-1}, which indicates that the firing temperatures of the vessels did not exceed 700–750 °C [51]. At ~706 cm^{-1}, aragonite was identified [53], the anhydrous variant of calcium carbonate, which was present in five samples (L2, L4, L6, L8, L9).

Silicates are represented by the Si-H stretching of quartz [45] in the region 1900–1870 cm^{-1}, the representative peaks from 1973 cm^{-1} and 1870 cm^{-1} being visible in all samples. Moreover, quartz [48–50,54,55] was also identified at ~2231 cm^{-1}, ~1162 cm^{-1} and ~692 cm^{-1}. Wollastonite [49] ($CaSiO_3$) was identified at 925 cm^{-1} in sample L7 and at 1055 cm^{-1} in the L8. In the same sample, the presence of diopside [49] ($CaMg[Si_2O_6]$) at 628 cm^{-1} was highlighted so that both identified minerals come from the raw clay. In addition, muscovite [56,57] ($KAl_3Si_2O_{10}(OH)_{2)}$ is present in all samples at 1270–1272 cm^{-1} and ~818 cm^{-1}.

Feldspars are present at ~1772 cm^{-1}, establishing their alkaline nature by identifying the anorthoclase [51,56] ($(NaK)[AlSi_3O_8]$) at ~667 cm^{-1} and albite [49,51,58] ($NaAlSi_3O_8$) at ~645 cm^{-1}. Iron oxides [51] were detected in all samples at ~656 cm^{-1}.

The presence of feldspars is related to the aluminosilicates, potassium and magnesium from the EDX analysis. The presence of iron oxides identified by macro- and microscopy is also found in the μFTIR spectra, which supports the ferruginous nature of clay established by EDX. The high phosphorus concentration identified by the EDX analysis and the lack of phosphates in the μFTIR spectra supports the hypothesis on the origin of this element from vessel usage. Although no carbon was identified in some fragments, the correlation of EDX results with μFTIR spectra indicated the presence of carbonates and calcite in all samples so that the firing temperatures did not exceed the decomposition limit of carbonates.

Thus, the EDX and μFTIR analyses illustrate using a local clay that has the same mineralogical and physicochemical characteristics, and the firing temperatures of the vessels were below 500 °C for the L11 sample and between 550 and 600 °C and 700–750 °C for the rest of the samples.

3.3. L11 and the Presence of Organic Residues

The Monteoru fragment (L11) has a higher iron content (5.97%) that suggests using the same ferruginous clay, but also the presence of an appreciable calcium content could indicate a low calcareous clay. The phosphorus concentration of this vessel is 5.23%, most likely determined by the storage of substances with a high phosphorus content.

The presence of a dark-colored layer identified macro- and microscopically is an indication in this regard.

In an attempt to determine the nature of the substance identified inside the vessel, several EDX analyses were performed, both on the outside (L11E), and in particular on the black layer (L11I), including at the interface with the vessel wall (L11I2) to eliminate the possible effects of contamination (Table 2).

Table 2. Elemental composition of the Monteoru fragment (L11) for ceramic paste and surfaces.

Sample	Elemental Composition in Weight Percent (%)										
	Si	Al	C	P	Mg	Ca	K	Na	Fe	Ti	O
L11	12.14	7.95	7.07	5.23	1.14	4.13	1.68	0.80	5.97	0.49	53.40
L11 E	7.55	2.83	16.96	0.37	0.26	3.18	1.55	0.10	3.96	0.47	62.77
L11 I	7.85	2.95	22.21	1.47	0.35	5.32	1.56	0.12	4.49	0.40	53.28
L11 I 2	5.68	2.53	22.91	1.43	0.40	6.63	1.77	0.28	2.77	0.54	55.06

The compositional results show, in addition to the elements specific to the raw material, the presence of a very high concentration of carbon, which at the interface with the vessel wall is up to 22.9%. This value certainly indicates that this element is organic in nature. Moreover, the carbon appears to have migrated from the inside to the outside, penetrating the vessel wall, which precludes possible contamination. To clarify the SEM-EDX results, an analysis was performed by infrared spectroscopy for the ceramic paste and surfaces (Figure 11).

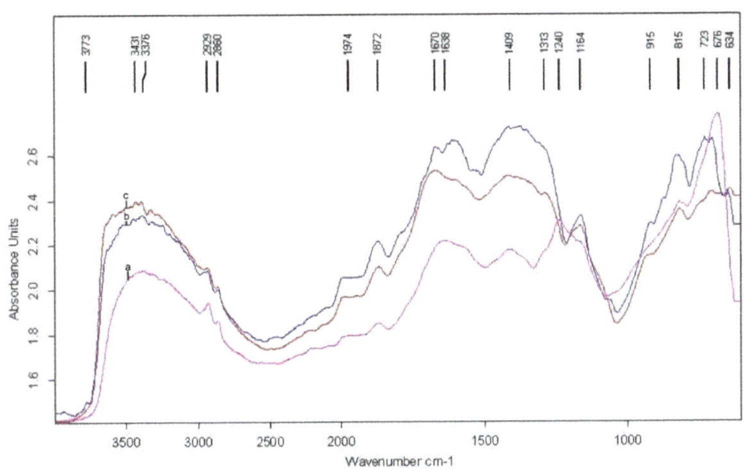

Figure 11. FT-IR spectra for sample L11 (a) ceramic paste; (b) outer surface; (c) inner surface.

Thus, for the paste of the L11 sample, montmorillonite [51] ($(Na,Ca)_{0.33}(Al,Mg)_2$ $(Si4O10)(OH)_2{\cdot}nH_2O$) were identified mainly by the presence of the central peak at 3376 cm^{-1}, from the specific range 3300–3500 cm^{-1}. In addition, the inner surface spectrum shows the presence of kaolinite [46,48,49] at 3562 cm^{-1}. The presence of the doublets at 915 cm^{-1}, attributed to OH deformations of kaolinite, is visible up to temperatures of 500 °C, after which it disappears [46,50,51]. It is important to mention that the vessel has a ceramic slip applied on the exterior, which is made of the same type of clay as the vessel. There is also the possibility of an inner slip that was not noticed due to the black layer covering the entire inner surface of the vessel. The intense peaks from 2929 cm^{-1} and 2860 cm^{-1} are attributed to the organic carbon [47,50,55,59–61], present on the surfaces but also in the paste, this being the result of the blackish substance identified on the vessel. Metal carboxylates at ~1409^{-1} and oxalates ~1313 cm^{-1} may derive from the degradation of some organic compounds. In this regard, it is worth noting that the stretching of C–H contributions appearing in the same spectrum at ~2860 cm^{-1} and ~2929 cm^{-1} support the presence of a protein material subjected to degradation processes [62,63]. The other minerals identified by the μFTIR analysis of the L11 sample are the same as in the case of Costișa samples and will not be repeated.

The results of the chemical analyses highlighted using the coiling technique in the process of pottery manufacturing, as well as the effect of the luster on the surfaces, a technique of closing the ceramic pores, while the ceramic slip or only finished surfaces have visible pores and mineral elements partially integrated into the clay.

All vessels were made of ferruginous clay whose mineralogy indicates using a local source. The containers were fired at temperatures between 550 and 600 °C and 700–750 °C. The presence of phosphorus confirmed the functionality of cooking vessels and indicated that storage and serving vessels were also used for cooking or storage of phosphorus-rich substances [64]. The Monteoru vessel (L11) was also manufactured from ferruginous clay, being fired at temperatures that did not exceed 500 °C. The chemical composition of the Monteoru fragment indicates using two types of clay, one for making the vessel and another for the ceramic slip.

4. Conclusions

The present archaeometric study generated a series of general data for the pottery from Piatra Neamț–*Lutărie* settlement. Thus, all the pots were made of local kaolinitic clay, by the coiling technique, with well-finished and smoothed surfaces, burnished or covered with a ceramic slip, treatments adapted according to the functionality of the vessels. The physicochemical analysis indicated that the two communities used the same source of raw material located in the vicinity of the settlement, which shows that the potters preferred the immediate and effortless source. In this sense, the positioning of the settlement was made taking into account several factors, including the proximity of water and clay reserves necessary for the manufacture of pottery.

The inclusions used in the manufacture of vessels are ceramoclasts, which, depending on their size, shape and frequency, led to identifying three types of paste correlated with certain functional categories. The use of ceramoclasts demonstrates the knowledge of their thermal properties by the Costișa and Monteoru potters. Moreover, using reused ceramoclasts could have a dual role, one of a practical nature, which involves integrating damaged containers into the paste of new vessels and would constitute a "recycling" system and a special one, which could represent the conservation of cultural ideas and identities.

Pyrotechnology information, both at the macroscopic level and through the interdisciplinary study, suggests an uncontrolled firing, with temperatures between 550 and 700 °C. The identification of relatively low-temperature ranges, together with the variety of colors of the vessels and the uneven atmosphere, indicates that the pottery was fired in pits or above ground.

To conclude, using the methods applied in this paper, we found no notable differences in the pottery of these two communities. The manufacturing techniques, including using the same source of clay, finishing treatments and firing pottery technologies, are the same, suggesting ordinary contacts between Costișa and Monteoru ceramic groups in the interference zone of Eastern Romania. The typology of pottery, details of the chemical composition, physical characteristics, macro- and microscopic observations suggest the existence of a stable community, which has the same technological skills for producing the usual pottery of a Middle Bronze Age community.

Author Contributions: Conceptualization, A.D., V.V. and N.B.; methodology, A.D., N.B.; validation, V.V. and A.D.; investigation, V.V., A.D.; writing—original draft preparation, A.D., V.V., N.B.; All authors have read and agreed to the published version of the manuscript.

Funding: This work was co-funded by the European Social Fund, through Operational Programme Human Capital 2014–2020, project number POCU/380/6/13/123623, project title "PhD Students and Postdoctoral Researchers Prepared for the Labor Market!".

Institutional Review Board Statement: Not applicable.

Informed Consent Statement: Not applicable.

Data Availability Statement: All required data provided in the manuscript.

Acknowledgments: We would like to thank Ciprian-Dorin NECULA, the director of the Neamț County Museum Complex, for the access to the archaeological material.

Conflicts of Interest: The authors declare no conflict of interest.

References

1. Bolohan, N. "All in one". Issues of Methodology, Paradigms and Radiocarbon Datings Concerning the Outer Eastern Carpathian Are. In *Signa Praehistorica. Studia in Honorem Magistri Attila László Septuagesimo Anno. Honoraria, 9*; Bolohan, N., Mățău, F., Tencariu, A.F., Eds.; Editura Universității „Alexandru Ioan Cuza" din Iași: Iași, Romania, 2010; pp. 229–244.
2. Popescu, A.; Băjenaru, R. From the Middle Danube to Anatolia: Contacts During the Second Millennium BC. A Case Study. In *SOMA 2011, Proceedings of the 15th Symposium on Mediterranean Archaeology, University of Catania, Catania, Italy, 3–5 March 2011*; Militello, P.M., Öniz, H., Eds.; BAR IS 2695 (I): Oxford, UK, 2015; pp. 35–43.
3. Bolohan, N.; Gafincu, A.; Stoleriu, I. Middle Bronze Age Chronology East of the Carpathian Area. A Bayesian Model. In *Bronze Age Chronology in the Carpathian Basin, Proceedings of the International Colloquium from Târgu Mureș, Romania, 2–4 October 2014*; Németh, R.E., Rezi, B., Eds.; Editura Mega: Târgu Mureș, Romania, 2015; pp. 131–146.
4. Vulpe, A. Voice Costișa. In *Enciclopedia Arheologiei și Istoriei Vechi a României 1 (A–C)*; Preda, C., Ed.; Editura Enciclopedică: București, Romania, 1994; p. 362.
5. Vulpe, A. Perioada Mijlocie a Epocii Bronzului. In *Istoria Românilor I, Moștenirea Timpurilor Îndepărtate*, 2nd ed.; revised; Petrescu-Dîmbovița, M., Vulpe, A., Eds.; Editura Enciclopedică: București, Romania, 2010; pp. 244–265.
6. Bolohan, N. Recent Discoveries Belonging to Early/Middle Bronze Age in Central Moldavia. *Arh. Mold.* **2003**, *26*, 195–206.
7. Bolohan, N.; Crețu, C. Recent Discoveries Belonging to Early/Middle Bronze Age in Central Moldova. In *Thracians and Circumpontic World*; Niculiță, I., Zanoci, A., Băț, M., Eds.; Cartdidact: Chișinău, Moldavia, 2004; pp. 55–76.
8. Popescu, A.-D.; Băjenaru, R. Rivalries and conflicts in the bronze age: Two contemporary communities in the same space. *Dacia NS* **2009**, *52*, 5–22.
9. Nițu, A.; Zamoșteanu, I.; Zamoșteanu, M. Sondajele de la Piatra-Neamț. *MCA* **1959**, *6*, 359–374.
10. Diaconu, V. *Repertoriul Descoperirilor Atribuite Epocii Bronzului din Județul Neamț*; Editura „Constantin Matasă": Piatra-Neamț, Romania, 2016.
11. *Geological Map of Romania, Scale 1:200000*; Editura Institutului Geologic al României: București, Romania, 1968.
12. *Soil Map of Romania, Scale 1:1000000*; Institutul de Geodezie, Fotogrammetrie, Cartografie și Organizarea Teritoriului: București, Romania, 1978.
13. Prehistoric Ceramics Research Group. *The Study of Prehistoric Pottery: General Policies and Guidelines for Analysis and Publication*, 3rd ed.; Wessex Archaeology Ltd.: Salisbury, UK, 2010.
14. Rice, P.M. *Pottery Analysis: A Sourcebook*; University of Chicago Press: Chicago, IL, USA, 1987.
15. Vuković, J. Archaeological Evidence of Pottery Forming Sequence: Traces of Manufacture in Late Neolithic Vinća Assemblage. In *Archaeotechnology: Studying Technology from Prehistory to the Middle Ages*; Vitezović, S., Antonović, D., Eds.; DC Grafički Centar: Belgrade, Serbia, 2014; pp. 177–199.
16. Roux, V. La Technique du Tournage: Définition et Reconnaissance par les Macrotraces. In *Terre Cuite et Société. La Céramique, Document Technique, Économique et Culturel. XIVe Rencontres Internationales d'Archéologie et d'Histoire d'Antibes*; Audouze, F., Binder, D., Eds.; Editions APDCA: Juan-les-Pins, France, 1994; pp. 45–58.
17. Joukowsky, M. *A Complete Manual of Field Archaeology, Tools and Techniques of Field Work for Archaeologists*; Prentice-Hall INC: New Jersey, NJ, USA, 1980.
18. Henrickson, E.; McDonald, M.M.A. Ceramic form and Function: An Ethnographic Search and an archaeological Application. *Am. Anthropol.* **1983**, *85*, 630–643. [CrossRef]
19. Bodi, G.; Solcan, L. Considerații asupra unei posibile tipologii funcționale a formelor ceramice cucuteniene specifice fazei A. *Arh. Mold.* **2010**, *33*, 315–323.
20. Forte, V.; Cesaro, S.N.; Medeghini, L. Cooking traces on Copper Age pottery from central Italy: An integrated approach comparising use wear analysis, spectroscopic analysis and experimental archaeology. *JAS Rep.* **2018**, *18*, 121–138. [CrossRef]
21. Vitelli, K.D. Were pots first made for foods? Doubts from Franchthi. *World Archaeol.* **1989**, *21*, 17–29. [CrossRef]
22. Orton, C.; Tyres, P.; Vince, A. *Pottery in Archaeology*; Cambridge University Press: London, UK, 1993.
23. Rye, O.S. *Pottery Technology, Principles and Reconstruction (Manuals on Archaeology)*; Taraxacum: Washington, WA, USA, 1981.

24. Ionescu, C.; Hoeck, V. Ceramic technology. How to investigate surface finishing. *Archaeol. Anthrop. Sci.* **2020**, *12*, 204–220. [CrossRef]
25. Tite, M.S. Ceramic production, provenance and use–a review. *Archaeometry* **2008**, *50*, 216–231. [CrossRef]
26. Ionescu, C.; Ghergari, L. Mic glosar de termeni geologici utilizați în studiul ceramicii arheologice. *CA* **2006**, *33*, 451–460. [CrossRef]
27. Kreiter, A. *Technological Choices and Material Meanings in Early and Middle Bronze Age Hungary: Understanding the Active Role of Material Culture through Ceramic Analysis*; BAR IS 1604, Archaeopress: Oxford, UK, 2007.
28. Skibo, J.M.; Schiffer, M.B.; Reid, K.C. Organic-tempered pottery: An experimental study. *Am. Antiq.* **1989**, *54*, 122–146. [CrossRef]
29. Palanivel, R.; Meyvel, S. Microstructural and microanalytical study–(SEM) of archaeological pottery artifacts. *Rom. J. Phys.* **2010**, *55*, 333–341.
30. Łaciaka, D.; Borowskib, M.P.; Łydżba-Kopczyńskac, B.; Barona, J.; Furmaneka, M. Archaeometric characterization and origin of black coatings on prehistoric pottery. *Geochem. Int.* **2019**, *79*, 453–466. [CrossRef]
31. Maniatis, Y.; Tite, M.S. Technological Examination of Neolithic–Bronze Age Pottery from Central and Southeast Europe and from the Near East. *J. Archaeol. Sci.* **1981**, *8*, 59–76. [CrossRef]
32. Karapukaitytė, A.; Pakutinskienė, I.; Tautkus, S.; Kareiva, A. SEM and EDX characterization of ancient pottery. *Lith. J. Phys.* **2006**, *46*, 383–388. [CrossRef]
33. Amicone, S.; Radivojević, M.; Quinn, P.S.; Berthold, C.; Rehren, T. Pyrotechnological connections? Re-investigating the link between pottery firing technology and the origins of metallurgy in the Vinča Culture, Serbia. *J. Archaeol. Sci.* **2020**, *11*, 1–19. [CrossRef]
34. Sandu, I.; Vasilache, V.; Tencariu, F.A.; Cotiugă, V. *Conservarea Științifică a Artefactelor din Ceramică*; Editura Universității „Alexandru Ioan Cuza": Iași, România, 2010.
35. Sandu, I.; Cotiugă, V.; Sandu, A.V.; Ciocan, A.C.; Olteanu, G.I.; Vasilache, V. New archaeometric characteristics for ancient pottery identification. *Int. J. Conserv. Sci.* **2010**, *1*, 75–82.
36. Vasilache, V.; Sandu, I.; Enea, S.C.; Sandu, I.G. Determinări ceramografice pe loturi din siturile Costești și Giurgești. In *Comunitățile Cucuteniene din Zona Târgului Frumos: Cercetări Interdisciplinare în Siturile de la Costești și Giurgești*; Boghian, D., Enea, S.C., Ignătescu, S., Stanc, S.M., Eds.; Editura Universității „Alexandru Ioan Cuza": Iași, Romania, 2014; pp. 138–147.
37. Crupi, V.; Kasztovszky, Z.; Khalilli, F.; La Russa, M.F.; Macchia, A.; Majolino, D.; Rossi, B.; Rovella, N.; Ruffolo, S.A.; Venuti, V. Evaluation of complementary methodologies applied to a preliminary archaeometric study of glazed pottery from Agsu (Azerbaijan). *Int. J. Conserv. Sci.* **2016**, *7*, 901–912.
38. Sharmin, D. Technological Interpretation of Traditional Ceramic Materials on Thin-Section Petrography. *Int. J. Conserv. Sci.* **2020**, *11*, 539–556.
39. Sandu, I.G.; Vasilache, V.; Cotiugă, V. Revaluation of archaeological ceramics assets under generally accepted principles of integrated scientific conservation. *Int. J. Conserv. Sci.* **2010**, *1*, 241–247.
40. Ravisankar, R.; Kiruba, S.; Shamira, C.; Naseerutheen, A.; Balaji, P.D.; Seran, M. Spectroscopic techniques applied to characterization of recently excavated ancient potteries from Thruverkadu Tamilandu, India. *Microchem. J.* **2011**, *99*, 370–375. [CrossRef]
41. Duma, G. Phosphate Content of Ancient Pots as Indication of Use. *Curr. Anthropol.* **1972**, *13*, 127–130. [CrossRef]
42. Béarat, H.; Dufournier, D. Quelques expériences sur la fixation du phosphore par les céramiques. *Revue d'Archéométrie* **1972**, *18*, 65–73. [CrossRef]
43. Santos Rodrigues, S.F.; Lima da Costa, M. Phosphorus in archaeological ceramics as evidence of the use of pots for cooking food. *Appl. Clay Sci.* **2016**, *123*, 224–231. [CrossRef]
44. Bong, W.S.K.; Matsumura, K.; Nakai, I. Firing Technologies and Raw Materials of Typical Early and Middle Bronze Age Pottery from Kaman-Kalehöyük: A Statistical and Chemical Analysis. *Anatol. Stud.* **2008**, *17*, 295–311.
45. Aroke, U.O.; Abdulkarim, A.; Ogubunka, R.O. Fourier-transform Infrared Characterization of Kaolin, Granite, Bentonite and Barite. *ATBU J. Environ. Technol.* **2013**, *6*, 42–53.
46. Frost, R.L.; Vassallo, A.M. The dehydroxylation of the kaolinite clay minerals using infrared emission spectroscopy. *Clays Clay Miner.* **1996**, *44*, 635–651. [CrossRef]
47. Damjanović, L.; Holclajtner-Antunović, I.; Mioć, U.B.; Bikić, V.; Milovanović, D.; Radosavlkević Evans, I. Archaeometric study of medieval pottery excavated at Stari (old) Ras, Serbia. *J. Archaeol. Sci.* **2011**, *38*, 818–828. [CrossRef]
48. Chen, Y.; Zou, C.; Mastalerz, M.; Hu, S.; Gasaway, C.; Tao, X. Applications of Micro-Fourier Transform Infrared Spectroscopy (FTIR) in the Geological Sciences—A Review. *Int. J. Mol. Sci.* **2015**, *16*, 30223–30250. [CrossRef]
49. Oancea, A.V.; Bodi, G.; Nica, V.; Ursu, L.E.; Drobota, M.; Cotofana, C.; Vasiliu, A.L.; Simionescu, B.C.; Olaru, M. Multi-analytical characterization of Cucuteni pottery. *J. Eur. Ceram.* **2017**, *37*, 5079–5098. [CrossRef]
50. Velraj, G.; Janaki, K.; Mustafa, A.M.; Palanivel, R. Estimation of firing temperatures of some archaeological pottery sherds excavated recently in Tamilandu, India. *Spectrochim. Acta A Mol. Biomol. Spectrosc.* **2009**, *72*, 730–733. [CrossRef]
51. Ravisankar, R.; Kiruba, S.; Eswaran, P.; Senthilkumar, G.; Chandrasekaran, A. Mineralogical Characterization Studies of Ancient Potteries of Tamilnadu, India by FT-IR Spectroscopic Technique. *E-J. Chem.* **2010**, *7*, S185–S190. [CrossRef]
52. Nodari, L.; Marcuz, E.; Maritan, L.; Mazzoli, C.; Russo, U. Hematite nucleation and growth in the firing of carbonate-rich clay for pottery production. *J. Eur. Ceram.* **2007**, *27*, 4665–4673. [CrossRef]

53. Gasaway, C.; Mastalerz, M.; Krause, F.; Clarkson, C.; Debuhr, C. Applicability of Micro-FTIR in Detecting Shale Heterogenity. *Microscopy* **2017**, *265*, 60–72. [CrossRef] [PubMed]
54. Palanivel, R.; Kumar, U.R. Thermal and spectroscopic analysis of ancient potteries. *Rom. J. Phys.* **2011**, *56*, 195–208.
55. Kumar, R.S.; Rajkumar, P. Characterization of minerals in air dust particles in the state of Tamilandu, India through FTIR Spectroscopy. *Infrared Phys. Tchenol.* **2014**, *67*, 30–41. [CrossRef]
56. Barilaro, D.; Barone, G.; Crupi, V.; Majolino, D.; Mazzoleni, P.; Tigano, G.; Venuti, V. FT-IR absorbance spectroscopy to study Sicilian "proto-majolica" pottery. *Vib. Spectrosc.* **2008**, *48*, 269–275. [CrossRef]
57. Vasilache, V.; Kavruk, V.; Tencariu, F.A. OM, SEM-EDX, and micro-FTIR analysis of the Bronze Age pottery from the Băile Figa salt production site (Transylvania, Romania). *Microsc. Res. Tech.* **2020**, *83*, 604–617. [CrossRef]
58. Akyuza, S.; Guliyev, F.; Celik, S.; Ozel, A.E.; Alakbarov, V. Investigations of the Neolithic potteries of 6th millennium BC from Goytepe-Azerbaijan by vibrational spectroscopy and chemometric techniques. *Vib. Spectrosc.* **2019**, *105*, 102980. [CrossRef]
59. Columbini, M.P.; Giachi, G.; Modugno, F.; Ribechini, E. Characterization of organic residues in pottery vessels of Roman age from Antinoe (Egypt). *Microchem. J.* **2005**, *79*, 83–90. [CrossRef]
60. Maritan, L.; Mazzoli, C.; Nodari, L.; Russo, U. Second Iron Age grey pottery from Este (northeastern Italy): Study of provenance and technology. *Appl. Clay Sci.* **2005**, *29*, 31–44. [CrossRef]
61. Costa, T.G.; de Correia, M.D.M.; Reis, L.B.; dos Santos, S.S.; Machado, J.S.; Bueno, L.; da Silva Müller, I. Spectroscopic characterization of recently excavated archaeological potsherds of Taquaea/Itararé tradition from Tobias Wagner site (Santa Catarina-Brazil). *J. Archaeol. Sci. Rep.* **2017**, *12*, 561–568.
62. Torrisi, L.; Venuti, V.; Crupi, V.; Silipigni, L.; Cutroneo, M.; Paladini, G.; Torrisi, A.; Havránek, V.; Macková, A.; La Russa, M.F.; et al. RBS, PIXE, Ion-Microbeam and SR-FTIR Analyses of Pottery Fragments from Azerbaijan. *Heritage* **2019**, *2*, 1852–1873. [CrossRef]
63. Sotiropoulou, S.; Papliaka, Z.E.; Vaccari, L. Micro FTIR imaging for the investigation of deteriorated organic binders in wall painting stratigraphies of diferent techniques and periods. *Microchem. J.* **2016**, *124*, 559–567. [CrossRef]
64. Bolohan, N. On Clay and Pots in the Middle Bronze Age. A case Study from Siliștea-Pe Cetățuie, Neamț County. In *Bronze Age Crafts and Craftsmen in the Carpathian Basin, Proceedings of the International Colloquium from Târgu Mureș, Romania, 5–7 October 2012*; Rezi, B., Németh, R.E., Berecki, S., Eds.; Editura Mega: Târgu Mureș, Romania, 2013; pp. 33–56.

applied sciences

MDPI

Article

The Archaeometallurgical Evaluation of Three Bronze Socketed Axes, Discovered in Eastern Romania

Viorica Vasilache [1], Vasile Diaconu [2], Otilia Mircea [3], Ana Drob [1,4,*] and Ion Sandu [5,6,*]

1 ARHEOINVEST Center, Interdisciplinary Sciences Department, Institute of Interdisciplinary Research, "Alexandru Ioan Cuza" University of Iași, Bld Carol I no. 11, 700505 Iași, Romania; viorica_18v@yahoo.com
2 Museum of History and Ethnography Târgu Neamț, Neamț National Museum Complex, 37 Ștefan cel Mare Str., 615200 Târgu Neamț, Romania; diavas_n82@yahoo.com
3 Museum of History Roman, Neamț National Museum Complex, 16 Cuza Vodă Str., 611008 Roman, Romania; otiliamircea@yahoo.com
4 Faculty of History, Doctoral School of History, "Alexandru Ioan Cuza" University of Iași, Bld Carol I no. 11, 700505 Iași, Romania
5 Romanian Inventors Forum, Str. Sf. P. Movila 3, 700089 Iasi, Romania
6 Academy of Romanian Scientists (AOSR), 54 Splaiul Independentei St., Sect. 5, 050094 Bucharest, Romania
* Correspondence: anadrob1@gmail.com (A.D.); ion.sandu@uaic.ro (I.S.)

Abstract: This paper presents the study of three bronze socketed axes discovered in Neamţ County, Romania. The surface structures and those from the interface of the corrosion layer with the metal core of the basic alloy were analyzed, in order to elucidate the nature of the materials used and the manufacturing processes. The analyzes by optical microscopy (OM) and electron microscopy (SEM), coupled with X-ray spectrometry (EDX), revealed the type of their degradation during the depositional period, as a result of the processes of chemical alteration and physical damage. A series of metallurgical techniques used were also established, as well as the identification of some finishing and decoration processes that led to the establishment of the objects' functionality.

Keywords: OM; SEM-EDX; socketed axes; bronze; corrosion

Citation: Vasilache, V.; Diaconu, V.; Mircea, O.; Drob, A.; Sandu, I. The Archaeometallurgical Evaluation of Three Bronze Socketed Axes, Discovered in Eastern Romania. *Appl. Sci.* **2021**, *11*, 1811. https://doi.org/10.3390/app11041811

Academic Editor: Karl Ulrich Kainer
Received: 21 January 2021
Accepted: 16 February 2021
Published: 18 February 2021

1. Introduction

Researchers in the field of scientific conservation of cultural heritage pay special attention to each archaeological discovery, and in the case of an archaeological artefact, first and foremost is considered the context of the discovery, but also other aspects, such as its manufacture, use and disposal. The context of the discovery provides a series of very important data in determining the patrimonial features and functions of the artifact [1].

The evolution of the conservation state is strongly influenced by the depositional period in the archaeological site. From the beginning, the primary patina is formed. It is based on chemical redox processes (oxides, sulfides, etc.), when continuous and uniform films are formed. Due to these layers with their preservative role, it is also called the noble patina. Before the abandonment and immediately after (the final phase of the period of use and continuing with the incipient phase after the abandonment), the secondary patina, also called poor patina, is reformed over the noble one. It occurs due to the aggression of the exogenous factors, when as a result of redox electrochemical processes, assisted by acido-basic, ionic exchange and hydrolysis processes, the following are formed: oxyhydroxides, hydroxy salts, halogenates, carbonates, sulfates, phosphates, etc., which may be anhydrous or hydrated in the form of crusts, nodes and moles, concentrated in or differentiated across certain areas. For certain artifacts, in this phase they may suffer thermal effects (calcination, recrystallization, etc.) following incineration and anthropic or natural fires. In the archaeological site, over the secondary patina, tertiary or contamination patina forms, under the influence of pedological, chemical and microbiological processes (segregation, diffusion, osmosis, monolithization, fossilization, mineralization, hydration/dehydration, structural reforming, etc.). The three types of structures are often difficult to identify, on artifacts from both disturbed and undisturbed sites, due to variable/oscillating aggression of exogenous factors [2–17].

It is known that the authentication of an artifact means much more than dating and identifying the metal worker and the place of manufacture. In fact, attention is paid to a number of attributes related to the origin of the raw material, the elaboration of the basic material, the manufacturing technology, the craftsman/workshop, the owner/holder, the distribution area, the way of use and the course taken until the abandonment. These aspects contribute to the determination of patrimonial elements and functions.

The old collections of some museums are composed from a series of objects that deserve a revaluation and a proper scientific valorification through modern methods, as is the case of the three socketed axes discovered in Neamț County, Romania. These objects represent single finds, accidental discoveries without a clear context; but, as far as the metal objects are rare presences, these can be reevaluated and exploited from a scientific point of view through interdisciplinary methods, in order to gain more data. These axes were analyzed using noninvasive OM and SEM-EDX techniques to evaluate their archaeometallurgical characteristics, based on the chemical composition, the morphology of the corrosion crusts and the basic alloy. The experimental data allowed to indicate the artistic techniques and the manufacturing technology involved.

The analogies of the pieces and their chronological frame indicate different cultural frameworks in three consecutive historical periods, so the main objective of this paper is to try to identify the manufacturing techniques that were used in order to follow the evolution and changes of these technologies in time in this specific area. Another objective is to investigate the depositional environment of the artifacts and its impact on their preservation trough the study of corrosion products and patina to find optimal conservation solutions.

2. Materials and Methods

2.1. Description of the Artifacts

The three bronze socketed axes (Figure 1) analyzed were selected from the collections of the History and Ethnography Museum in Târgu Neamţ, presenting specific differences in shape, size, ornamentation and site aggression/conservation status. The artifacts were discovered in the eastern part of Romania, namely in the sub-Carpathian area of Neamț county (Figure 2), a rich area in salt springs (brine) that attracted prehistoric communities, exploiting the area's resources since the Chalcolithic and which continue to be processed even today [18–21]. Thus, these bronze artifacts could represent goods obtained after the exchange with salt and may indicate a chief that oversaw and protected the area and the resources. Even if these objects represent accidental discoveries, without a clear context, the archaeometallurgical study could lead to new information about their use and symbolism.

Figure 1. Images of the three socketed axes and the analyzed areas (**A**) inventory no. 4649 (**B**) inventory no. 2113, (**C**) inventory no. 4650.

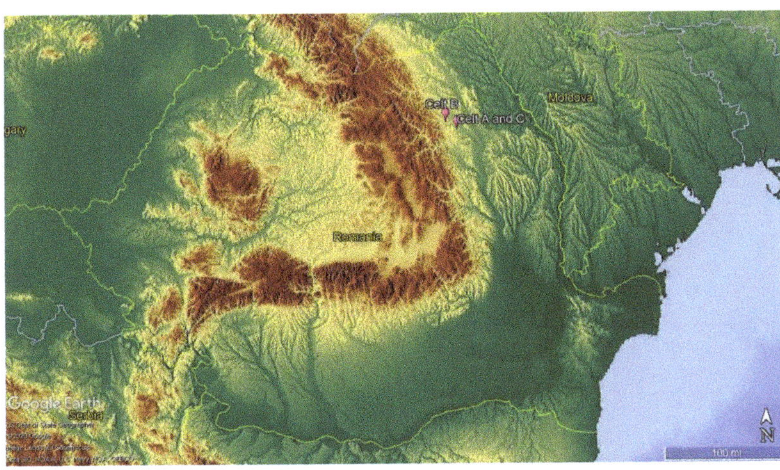

Figure 2. Geographic location of the discoveries, Neamţ county, Romania: **A** and **C**—Ţibucani; **B**—Tg. Neamţ (Google Earth 2020).

Socketed axe A (Figure 1A), inventory no. 4649, was discovered in 1965 on the territory of Ţibucani, Neamţ County, Romania. The piece was discovered on the bank of a creek, within a settlement of the Noua Culture [22,23]. The piece has a length of 124 mm, 0.413 kg weight, a blade width of 45 mm, the depth of the fixing orifice is 76 mm, the diameter of the fixing orifice is 35 mm $\times$ 23 mm, the length of the attachment loop is 37 mm and its diameter is 6 mm $\times$ 5 mm. The item has a massive body; the upper edge is thick, and it is oval in plane and is provided with a small loop for attaching. The blade is slightly curved, and on the sides the burrs of casting in the bivalve pattern are observed. On the main faces, below the upper edge, there is a perforated alveoli and colorizations/traces of use. The entire surface of the item is covered by a consistent patina, colored in malachite green, dispersed under tenorite layers and sprinkled with small azurite points, in deepened surface areas. From a typological point of view, this artifact can be included, according to some specialists, in the eastern version of the Transylvanian Socketed Axe [24] or in the C3 version (according to a broader classification based on morphological criteria of the socketed axes types from the Late Bronze and Early Iron Age) [25]. More recently, these socketed axes were included in the Râșești type [22]. Many analogies can be found for this item, especially in the eastern area of Romania [22]. These socketed axes are specific to the Late Bronze Age and were created by the Noua Culture communities. From a chronological point of view, they belong to the Bronzezeit D (Br D) period (according to Reineke chronological system, 14th–12th c. BCE). The Reineke chronological system is a Central European periodization system based on metallic discoveries, and according to the synchronisms of these objects form a uniformization from a chronological point of view.

Socketed axe B (Figure 1B), inventory no. 2113, was discovered by chance in 1957 on the territory of Târgu Neamţ, Romania [26]. The item is small in size and has a slightly trapezoidal body. It is equipped with an attaching loop. The piece has a length of 70 mm, 0.142 kg weight, a blade width of 40 mm, the diameter of the fixing orifice is 22 mm × 27 mm, the depth of the fixing orifice is 40 mm, the length of the attachment loop is 43 mm and its diameter is 5 mm × 5 mm. The upper edge is straight, thick and oval in plane, and it preserves the burrs from casting. Below the edge there is a cast ornament, consisting of six oblique ribs, and on the body of the item, inside the trapezoidal facets, there are four others vertically arranged ribs, which confer certain aesthetics, with a rank attribute [27]. The cutting edge is slightly arched and shows, like the other socketed axe, visible use wear marks. The entire surface of the item is covered with a blue-azure patina, with small white areas of stannite, below malachite green and tenorite black layers, over which there are grey nantokite crusts. From a typological point of view, this artifact was attributed to the Cozia-Saharna type [28], and according to the scheme proposed by V. Dergačev it has correspondences in the group of socketed axes with trapezoidal facets and vertical ribs [22]. The best analogies for this socketed axe are found at Brza Palanca, Serbia [22]; Jupalnic, Mehedinţi county, Romania [22,29]; and Pietrosu, Buzău county, Romania [22,29]. Considering the fact that similar specimens have been identified in the middle part of the Danube, it is supposed that this item also originated in that area and it could have reached the submontane area of Neamţ, either through the mountain passes towards Transylvania or through the curve of the Carpathians. On the basis of the abovementioned analogies, but also in the absence of a clear archaeological context, for the item from Târgu Neamţ we can admit a dating as from the beginning of the Early Iron Age, at the Ha A2–Ha B2 level (Hallstattzeit); from a cultural point of view, it must be associated with the Corlăteni-Chișinău Culture (11th–9th c. BCE) [30].

Socketed axe C (Figure 1C), inventory no. 4650, was accidentally discovered on the territory of the village of Ţibucani, Neamţ county, Romania [23]. It is a small specimen with an elongated body. The piece has a length of 78 mm, 0.54 kg weight, a blade width of 37 mm, the diameter of the fixing orifice is 20 mm × 14 mm, the depth of the fixing orifice is 50 mm, the length of the attachment loop is 21 mm and its diameter is 2 mm. The upper edge is straight, slightly thickened and oval in plan. The item was provided with an attachment loop, broken during the period of use. Below the upper edge there are three horizontal ribs. The lower part of the item is flared, and the cutting edge is slightly arched. The surface of the artifact is covered with a black tenorite layer, uniform in thickness, which contains superficially dispersed small spots of azurite and malachite, over which there are brown-gray crusts of nantokite. The main analogies for this artifact are found in bronze deposits and singular discoveries, such as those from Rădeni, Neamţ county, Romania [23]; Floreni, Vaslui county, Romania [31]; Sâmbăta Nouă, Tulcea county, Romania [29]; Techirghiol, Constanţa county, Romania [29]; Căpușu de Câmpie, Mureș county, Romania [29]; Rusi, Sibiu county, Romania [32]; Iara II, Cluj county, Romania [29]; Boldești, Prahova county, Romania [29]; Ciunga, Alba county, Romania [33]; Obreja, Alba county, Romania [34]; Valea Mare, Vaslui county, Romania [34]; and Dupliska, Ukraine [35]. Similar examples are known at the level of the Gyermely and Hajdúböszörmény bronze deposits horizons, as shown by specimens from Szendrőlád, Hungary [36]; Jászkarajenő, Hungary [36]; Balmazújváros, Hungary [37]; Nádudvar-Bojárhollós, Hungary [37]; and Polgár, Hungary [37]. On the basis of the abovementioned analogies, the item from

Ţibucani can be framed chronologically in the Early Iron Age, between Ha B1 and Ha B2, or even in Ha B3 (Hallstattzeit), in the 10th–8th c. BCE.

2.2. Methods of Analysis

The three artifacts were studied by surface optic microscopy (OM) without any sampling to obtain information about the resulting crusts from alterations of the base alloy and those from incorporation and monolithization in the ground (identifying similarities or differences in appearance and color and homogeneity of the chemical compounds from the primary, secondary or tertiary patina). A ZEISS-type optical microscope with video camera and computer co-assistance was used for the analyses. The samples were analyzed by reflection at different magnification orders (×50 and ×100) in dark field.

For elemental and microstructural analyses, the samples were collected non-destructively from the fixing orifice and from areas with destruction, being studied the areas with corrosion and the metal alloy (samples from the metal interface). All samples were studied through scanning electron microscopy (SEM) coupled with energy-dispersive X-ray spectroscopy (EDX). A SEM microscope, model VEGA II LSH, produced by TESCAN Czech Republic, coupled with an EDX detector, type QUANTAX QX2, manufactured by BRUKER/ROENTEC Germany, was used in the analyses. The microscope, controlled by a computer, has a tungsten filament electron cannon that can achieve a resolution of 3 nm at 30 kV, with a magnification between ×30 and ×1,000,000 in the "resolution" operating mode, an acceleration voltage between 200 V and 30 kV and scanning speed between 200 ns and 10 ms per pixel. The working pressure was less than 1×10^{-2} Pa. Quantax QX2 is an EDX detector used for the qualitative and quantitative micro analyses. The EDX detector was a third-generation, X-flash-type detector. The images obtained for the analyzed samples were made up of secondary electrons (SE) or backscattered electrons (BSE) at magnifications between ×200 and ×1000.

2.3. Areas of Analyses

Figure 1 shows four analyzed areas for each socketed axe, one for the identification of the processed alloy and three representatives for the conservation status, which allow to highlight the effects of the deterioration and degradation resulting from the use and abandonment periods, in particular the morpho-structural components of the three patinas (noble, poor and contamination).

3. Results and Discussions

From the microphotographs obtained by OM for the representative areas, each of the three analyzed socketed axes show interesting characteristics in terms of corrosion and use.

Specifically, socketed axe A has a relatively uniform, thin layer of corrosion on the surface, completely covering the metal core (Figure 3a). Near the blade, in the outer layer, there were identified traces of deep scratches (Figure 3b), attributed to the use of the item before deposition, and at the interface of the metal core were found chemical compounds resulting from the corrosion processes occurring on the base alloy, namely, copper oxides—red—and the basic copper carbonates—green (Figure 3c).

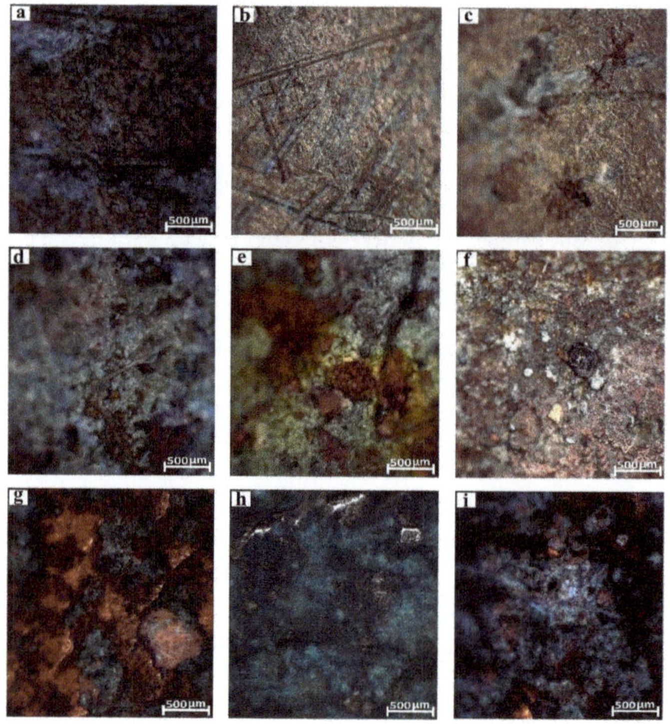

Figure 3. OM images of the three representative areas of the surface of the socketed axes: (**a**) uniformity of the corrosion layer for axe A; (**b**) scratches caused by use before deposition for axe A; (**c**) the oxides and basic carbonates of copper for axe A; (**d**) disposal of the corrosion products in the patina for axe B; (**e**) the basic carbonates and copper oxides for axe B; (**f**) copper oxides for axe B; (**g–i**) the corrosion layer with the disposal of the chemical compounds of the patina for axe C.

Socketed axe B presents a corrosion layer distributed very heterogeneously over the entire surface of the item (Figure 3d), in which deposits of the basic carbonates and copper oxides were randomly located on the micro-zones in the form of crusts (Figure 3e), and at the interface with the metal core were found thin layers of copper oxides, also disposed unevenly (Figure 3f).

Socketed axe C presents a discontinuous, thin layer of corrosion with unevenly arranged deposits of oxides (cuprite/tenorite), chlorides (nantochite) and copper basic carbonates (malachite/azurite), cassiterite and silver veins, unevenly distributed in the form of crusts or thin layers (Figure 3g–i).

SEM-EDX analyses were performed on both the surface corrosion layer and the stratigraphic structure at the interface with the base alloy to highlight the correlation between the chemical composition of the artifact, the nature of the corrosion products and the elements taken from the soil (contamination).

From the SEM micrographs of the corrosion layer of socketed axe A (Figure 4S1-1, S1-2 and S1-3) was identified a structure with a non-homogeneous distribution, characterized by the presence of cracks, microcavities and microphases differentiated as superficial structures in the form of irregularly shaped granules.

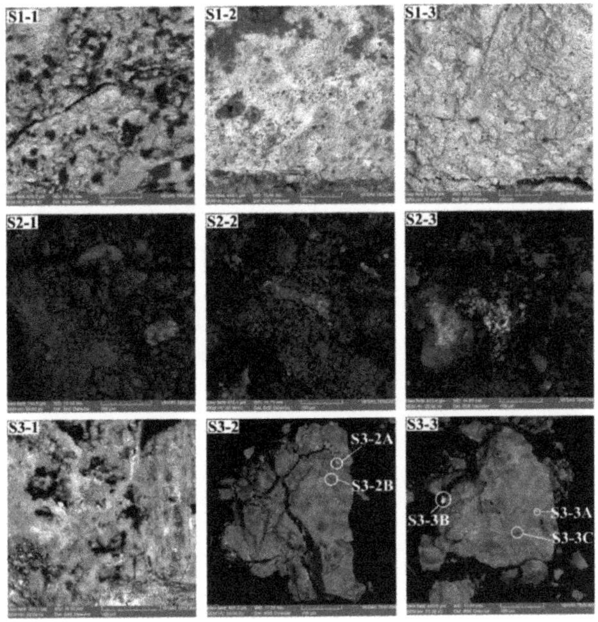

Figure 4. SEM images of the corrosion layers of socketed axes A, B and C.

The EDX analyses of these surfaces identified the chemical elements corresponding to the basic alloy (Cu and Sn) and the ore impurities (Pb and Fe), as well as the elements of the corrosion products (C, O and Cl) and of the contamination from the soil, such as Si, Al, Mg, P and S (Table 1). This composition shows that the artifact has undergone corrosion and segregation processes under the influence of an environment with fluctuating humidity and temperature, rich in oxygen and carbon dioxide, in the presence of alkaline and alkaline-earth cations, phosphates, carbonates, aluminosilicates and silicon dioxide (quartz micro-crystallites).

The SEM micrographs of the corrosion layer of socketed axe B, also performed on three representative areas, are shown in Figure 4(S2-1,S2-2,S2-3). Through the analyses, the surface structures, with a non-homogeneous granulometric distribution of the corrosion crusts and metallic micro-phases with uneven geometric profiles, can be observed.

The elemental EDX analyzes of the corrosion layer of socketed axe B, performed on the three areas (S2-1, S2-2 and S2-3), corresponding to the SEM micrographs in Figure 4, is shown in Table 1.

Thus, in samples S2-1 and S2-3 were highlighted the chemical elements corresponding to the basic alloy (Cu and Sn), and the micro-elements from the ore were based on Ni, Pb and Fe. Instead, the sample S2-2 contains corrosion products of the copper, in the form of basic carbonates and oxides, but also iron oxides, possibly from contamination, because the Fe concentration is quite high (14.23%).

Table 1. Chemical composition of the surface structures for socketed axes A (S1), B (S2) and C (S3).

Samples	Elemental Composition—Weight Percent (%)														
	Cu	Sn	Pb	Fe	Ag	Ni	Si	Al	Ca	Mg	K	O	S	C	P	Cl
S1-1	17.07	21.69	0.14	1.69	-	-	1.84	0.40	-	0.12	-	40.61	0.15	14.30	1.34	0.63
S1-2	16.03	24.59	0.09	2.66	-	-	3.99	1.70	-	0.31	-	44.37	0.13	4.93	1.07	0.13
S1-3	21.11	23.45	0.11	1.49	-	-	1.19	0.10	-	0.35	-	45.25	0.07	4.88	1.91	0.09
S2-1	64.59	3.11	0.14	1.71	-	1.38	0.01	0.01	-	0.04	-	26.51	0.01	2.34	0.15	-
S2-2	23.39	-	-	14.23	-	-	0.45	0.23	0.75	-	0.47	42.71	0.27	17.00	0.17	0.33
S2-3	68.43	1.57	0.14	0.83	-	1.36	0.02	0.03	-	-	-	25.69	0.02	1.89	0.02	-
S3-1	32.31	17.38	1.09	0.72	-	-	1.98	0.52	-	0.67	-	37.30	-	6.35	1.30	0.38
S3-2A	34.99	11.07	1.09	-	5.65	-	0.82	0.09	-	0.72	-	39.75	0.66	4.61	0.55	-
S3-2B	46.17	11.81	0.49	-	-	-	0.91	0.02	-	0.40	-	36.44	0.02	3.17	0.57	-
S3-3A	33.35	16.48	0.29	-	4.44	-	1.14	0.14	-	0.78	-	40.12	0.43	1.87	0.96	-
S3-3B	9.24	-	-	-	35.41	-	0.42	0.09	-	0.31	-	28.28	4.63	21.62	-	-
S3-3C	36.12	12.95	1.15	-	-	-	1.44	0.27	-	0.89	-	40.59	-	3.55	1.09	-

For socketed axe C, analyses were also performed on three representative areas, corresponding to the SEM micrographs in Figure 4, namely, the areas S3-1, S3-2 and S3-3. The three SEM micrographs of the corrosion layer of socketed axe C, shown in Figure 4, have completely different surface structures. The micrograph S3-1 shows the presence of cavities formed by dissolution in the presence of groundwater and crusts with coarse granulometry, incorporating contamination inclusions. Crusts S3-2 and S3-3, taken from two representative areas on socketed axe C, have very similar structural profiles, with cracks resulting from contraction and the conglomerate incorporates, in addition to corrosion products and ground contamination components. In addition, silver has been identified, and that seems to have come from the silvering process because it appears concentrated on certain areas only on the surface and it is not present in the composition of the alloy [38].

In all three samples, the chemical elements of the base alloy—Cu and Sn (Table 1)—were identified alongside the microelements originating in the ore and soil compounds from depositional processes. In sample S3-1, besides Pb and Fe as micro-elements from the ore, many contamination elements from the soil are also present (Si, Al, Mg, P, C and O). Instead, Ag was identified in the S3-2 and S3-3 samples. For this reason, several analyses were made and it was found that Ag originates from the silvering of the socketed axe. Moreover, the results highlighted the compounds resulting from the alteration processes (redox, acido-base and complexation) of the base alloy during the deposition, namely, the oxides and copper carbonates, tin, lead, iron and nickel, together with the superficial contamination structures corresponding to the elements Si, Al, Mg, P, S, C and O.

From the SEM micrographs (Figure 5) of the three socketed axes, performed on clean areas, without corrosion or contamination products, for socketed axe A one can see a homogeneous structure of the basic alloy (Figure 5S1) containing Cu (88.84%) and Sn (5.92%), as alloying elements from the ore, and Pb (0.73 %) and Fe (0.75%) as alloying microelements, also from ore. The oxygen (40.62%) comes from the newly formed oxides. Instead, socketed axe B and C show voids, microcracks and microphases that are differentiated, with a complex granulometry, randomly distributed.

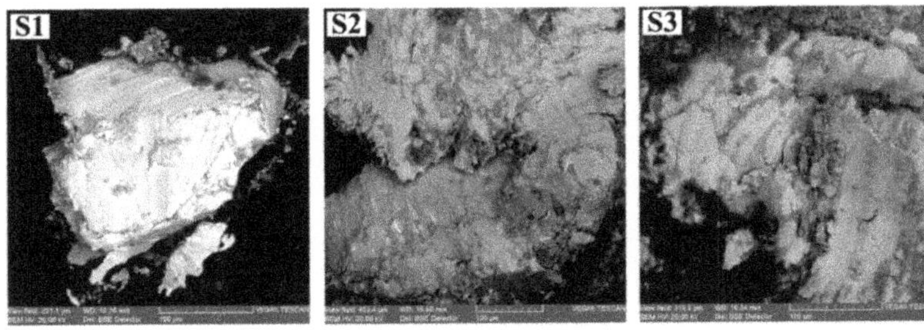

Figure 5. SEM images of the samples from the metallic core: **S1**—socketed axe A; **S2**—socketed axe B; and **S3**—socketed axe C.

The chemical composition of socketed axe B indicates the presence of Cu (91.29%), Sn (1.14%,) Pb (0.66%) Fe (2.32%), O and Ni (1.00%), and socketed axe C have 88.22% Cu, 5.76% Sn, 1.21% Pb, 1.01% Ni and 3.80% O.

The correlation between the data obtained for the three socketed axes indicates that they come from different production areas and the raw materials are, most likely, from different sources. Their manufacturing technologies are close (making the alloy by smelting ores, pouring into bivalve moulds, finishing, ornamentation through polishing and/or gilding), with socketed axe C being special because it was ornamented by silvering. From an archaeological point of view, this is a very important aspect that illustrates the function of the artifact, namely the prestigious one. This aspect is also supported by the lack of traces of use wear on the surface of the artifact, but which are present in socketed axes A and B.

The three socketed axes also differ constructively and functionally, in terms of ornamentation and size, and in terms of conservation status; they can be discussed in relation to the composition of the base alloy, the aggressiveness of the site, the age of the artifact and the depositional period.

Based on the results obtained, it can be emphasized that the corrosion products were formed by redox, acido-basic and complexation reactions, which have an important role in hydric processes (hydration/drying). These occurred at the surface of the alloy and in the fixing orifices. Over time, due to cyclic variations in humidity and temperature, corrosion layers or crusts have undergone contraction, segregation and diffusion processes. These processes have been accelerated by the differences in chemical load of the groundwater and the presence of hydrogel coatings of Sn (II), Pb (II) and Fe (II), whose instability did not lead to the Liesegang phenomenon (this phenomenon is characterized by the appearance of concentric formations of corrosion products resulting from structural reform) [10,11,13,15,27,39] but to coarse structures similar to crusts and moles [6,7], with uneven distributions. Moreover, under the influence of the chemical and electrochemical potentials at the interfaces, they led to processes of segregation and diffusion of some components from the volume phase into the corrosion structures, respectively, between them and the depositional environment, and vice versa. These lead to cracks, voids, cavities and contamination by incorporating the microstructures from the archaeological site during the deposition.

The high content of Sn in the corrosion crusts of socketed axes A and C highlights a high rate of segregation and diffusion from the volume phase of the alloy to the surface, under the influence of electrochemical potentials activated by the chemical loading of the groundwater, differentiated according to variations and differences of the cryptoclimate (internal parameters of the depositional environment) and microclimate parameters (external parameters from the soil surface). These potentials occur in high-porosity sites of sandstone systems. In contrast, for socketed axe B, these potentials are greatly diminished, coming from a site semi-protected by waterproof structures.

Compared to other bronze age artifacts, the high content of the base alloy in Sn and Pb demonstrates that soil aggression has been greatly differentiated in the case of the three socketed axes. The strongest aggressiveness of the laying environment is in the case of socketed axe C, which has resulted in the total destruction of the silver coating and the formation of a thin layer of cuprite and malachite. Then there is socketed axe A, in the case of which the aggression has led to a high rate of segregation and diffusion of Fe from the volume phase to the surface (found in the corrosion compounds). Socketed axe B has laid in a sandstone site, with high porosity, where superficial altering processes of the base alloy were very strong, leading to thick corrosion layers and crusts over the passive malachite layer. Porosity also allowed a strong contamination explained by the high concentration of Fe (14.23%) in the corrosion products.

4. Conclusions

A series of historical and chemical conclusions can be drawn based on the results obtained by corroborating the OM and SEM-EDX analyses of the basic alloy and the compounds from the corrosion films and crusts of the three socketed axes.

The data obtained for the three socketed axes show that they come from different areas and periods, and in their manufacturing process, probably, copper ores from different sources were used. They were made by casting in bivalve molds made of siliceous stone or clay, and the presence of burrs and holes proves some casting defects. Important differences in the manufacturing technology were not observed between the three objects across the three chronological periods, indicating the same metallurgical techniques.

Corrosion products were formed by redox, acido-base and complexation reactions, and the presence of phosphate anions and oxyhydroxides led to the formation of films over the primary and secondary patina structures, with a semi-membrane effect, which allowed the formation of crusts, moles and thin coarse structures, overlayed and/or interleaved with the contamination structures. The aggressiveness of the depositional environment was different for each artifact; the most aggressive was that of socketed axe C, followed by socketed axe A and socketed axe B.

The socketed axe C is special because it was decorated by silvering, which illustrates its functionality as a prestigious object; its function was also supported by the presence of the decoration. Decoration is found also on socketed axe B, an aspect that could indicate the same symbolic functionality, but the presence of the wear traces also indicates its use as a tool or weapon. Socketed axe A has only use wear traces, indicating a practical use rather than a symbolic one.

The shape of the artifacts and the composition of the alloy demonstrate that the three socketed axes come from different areas and distinct chronological stages. Socketed axe A is specific to the Late Bronze Age, being framed in the Noua Culture; socketed axe B can be classified in chronological stage Ha A2–Ha B2, and from a cultural point of view it is associated with the Corlăteni-Chișinău Culture; and socketed axe C can be framed chronologically between Ha B1 and Ha B2, or even Ha B3 (10th–8th c. BCE). This cultural and chronological variety could support the hypothesis of salt exchanges and the existence of an elite that controlled these resources in the eastern sub-Carpathians area of Romania.

Author Contributions: Supervision and methodology, V.V. and I.S.; formal analysis and investigation, A.D. and O.M.; writing—review and editing and data curation, V.D.; writing—original draft and final correction, A.D. and V.V.; data curation and validation, V.V. and I.S. All authors have read and agree to the published version of the manuscript.

Funding: This work was co-funded by the European Social Fund, through Operational Programme Human Capital 2014–2020, project number POCU/380/6/13/123623, project title "PhD Students and Postdoctoral Researchers Prepared for the Labour Market!"

Institutional Review Board Statement: Not applicable.

Informed Consent Statement: Not applicable.

Data Availability Statement: All required data provided in the manuscript.

Conflicts of Interest: The authors declare no conflict of interest.

References

1. Sandu, I. *Deteriorarea și Degradarea Bunurilor de Patrimoniu Cultural*; Universitatii "Al. I. Cuza": Iași, Romania, 2008.
2. Domenech-Carbo, A. Tracing, authenticating and dating archaeological metal using the voltammetry of microparticles. *Anal. Methods* **2011**, *3*, 2181–2188. [CrossRef]
3. Doménech-Carbó, A.; Doménech-Carbó, M.T.; Peiró-Ronda, M.A. Dating archaeological lead artifacts from measurement of the corrosion content using the voltammetry of microparticles. *Anal. Chem.* **2011**, *83*, 5639–5644. [CrossRef] [PubMed]
4. Doménech-Carbó, A.; Doménech-Carbó, M.T.; Peiró-Ronda, M.A.; Osete-Cortina, L. Electrochemistry and Authentication of Archaeological Lead Using Voltammetry of Microparticles: Application to the Tossal de Sant Miquel Iberian Plate. *Archaeometry* **2011**, *53*, 1193–1211. [CrossRef]
5. Goffer, Z. *Archaeological Chemistry*, 2nd ed.; Wiley-Interscience a John Wiley & Sons: Hoboken, NJ, USA, 2007.
6. Robbiola, L.; Blengino, J.M.; Fiaud, C. Morphology and mechanisms formation of natural patinas on archaeological Cu-Sn alloys. *Corros Sci.* **1998**, *40*, 2083–2111. [CrossRef]
7. Robbiola, L.; Portier, R. A global approach to the authentication of ancient bronzes based on the characterization of the alloy—Patina—Environment system. *J. Cult. Heritage* **2006**, *7*, 1–12. [CrossRef]
8. Scott, D.A. *Copper and Bronze in Art*; Getty Publications: Los Angeles, CA, USA, 2002.
9. Mircea, O.; Sandu, I.; Sarghie, I.; Sandu, A.V. The Identified Effects of Degradation in Archaeological Artifacts with Overlapped Metals Used in Authentication. *Int. J. Conserv. Sci.* **2011**, *1*, 27–40.
10. Mircea, O.; Sandu, I.; Vasilache, V.; Sandu, A.V. Study of the Atypical Formations in the Corrosion Bulks of an Ancient Bronze Shield, by Optical and Electron Microscopy. *Microsc. Res. Tech.* **2012**, *75*, 1467–1474. [CrossRef] [PubMed]
11. Mircea, O.; Sandu, I.; Vasilache, V.; Sandu, A.V. Research on Atypical Formations from Corrosion Bulks of an Ancient Bronze. *Rev. Chim.* **2012**, *63*, 893–899.
12. Sandu, I.; Ursulescu, N.; Sandu, I.G.; Bounegru, O.; Sandu, I.C.A.; Alexianu, A. The pedological stratification effect of corrosion and contamination products on byzantine bronze artefacts. *Corros. Eng. Sci. Technol.* **2008**, *43*, 256–266. [CrossRef]
13. Sandu, I.; Mircea, O.; Sandu, I.G.; Vasilache, V.; Sandu, A.V. Study of the Liesegang Chemical Effects in Antique Bronze Artefacts During Their Stay Within an Archaeological Site. *Rev. Chim.* **2014**, *65*, 311–319.
14. Sandu, I.G.; Sandu, I.; Dima, A. *Restaurarea şi Conservarea Obiectelor din Metal*; Corson: Iași, Romania, 2002.
15. Sandu, I.G.; Mircea, O.; Vasilache, V.; Sandu, I. Influence of the Archaeological Environment on Ancient Copper Alloy Artifacts. *Microsc. Res. Tech.* **2012**, *75*, 1646–1652. [CrossRef] [PubMed]
16. Vasilache, V.; Aparaschivei, D.; Sandu, I. A Scientific Investigation of the Ancient Jewels Found in the Ibida Site, Romania. *Int. J. Conserv. Sci.* **2011**, *2*, 117–126.
17. Vasilache, V.; Sandu, I.; Lazanu, C.C.; Sandu, I.G. Archaeometallurgical evaluation of two spearheads from the bronze age. *Int. J. Conserv. Sci.* **2015**, *6*, 633–642.
18. Monah, D.; Dumitroaia, G.; Weller, O.; Chapman, J. *L'exploitation du sel a Travers le Temps (Actes du Colloque International, Piatra Neamț, Roumanie, Oct. 2004)*; BMA XVIII, C; Matasă: Piatra Neamț, Romania, 2007.
19. Cavruc, V.; Harding, A. Prehistoric production and exchange of saltin the Carpathian-Danube Region. In *Salz und Gold: Die Rolle des Salzes im Prähistorischen Europa/Saltand Gold: The Role of Salt in Prehistoric Europe*; Nikolov, V., Bacvarov, K., Eds.; Provadia & Veliko: Tarnovo, The Republic of Bulgaria, 2012; pp. 173–200.
20. Brigand, R.; Weller, O. Neolithic and Chalcolithic settlement pattern in central Moldavia (Romania). *Doc. Praehist.* **2013**, *40*, 195–207. [CrossRef]
21. Weller, O. Firt salt making in Europe: An overview from Neolithic times. *Doc. Praehist.* **2015**, *XLII*, 185–196.
22. Dergačev, V. *Topory-kel'ty pozdnej bronzy Karpato-Podynav'ja, Vyšysk II, Kel'ty i serpy nižnego Podynav'ja*; Čentral'naja: Moldova, 2010.
23. Dumitroaia, G. Obiecte de aramă şi bronz descoperite pe teritoriul judeţului Neamţ. *Mem. Ant.* **1985**, *IX–XI*, 465–482.
24. Petrescu-Dîmboviţa, M. Contribuţii la problema sfârşitului epocii bronzului şi începutului epocii fierului în Moldova. *SCIV* **1953**, *4*, 443–486.
25. Rusu, M. Depozitul de bronzuri de la Balşa. *Sargetia* **1966**, *4*, 17–40.
26. Petrescu-Dîmboviţa, M. Date noi relativ la descoperirile de obiecte de bronz de la sfârşitul epocii bronzului şi începutul Hallstattului în Moldova. *ArhMold* **1964**, *II–III*, 251–272.
27. Sandu, I.G.; Tencariu, F.A.; Vornicu, D.M.; Sandu, A.V.; Vornicu, A.; Vasilache, V.; Sandu, I. Establishing the archaeo-metallurgic ornamentation process of an axe from the bronze age by OM, SEM-EDX and micro-FTIR. *Microsc. Res. Tech.* **2014**, *77*, 918–927. [CrossRef] [PubMed]
28. Uşurelu, E. Consideraţii cu Privire la Celturile de Tip Cozia-Saharna. In *Interferenţe Cultural-Cronologice în Spaţiul Nord-Pontic*; Sava, E., Ed.; Centrul Editorial al Universității de Stat Moldova: Chişinău: Moldova, 2003; pp. 211–219.
29. Petrescu-Dîmboviţa, M. *Depozitele de Bronzuri din România*; Academia Română: Bucureşti, Romania, 1977.
30. László, A. *Începuturile Epocii Fierului la est de Carpaţi: Culturile Gava-Holihrady și Corlăteni Chișinău pe Teritoriul României, B.THR. VI*; Publirom: Bucureşti, Romania, 1994.
31. Udrescu, T. Descoperiri arheologice în jumătatea sudică a Moldovei cu privire la cultura Noua. *Carpica* **1974**, *VI 1973–1974*, 17–42.

32. Ciugudean, H.; Luca, S.A.; Georgescu, A. *Depozite de Bronzuri de Bronzuri Preistorice din Colecţia BRUKENTHAL (I)*; ALTIP: Sibiu, Romania, 2008.
33. Ciugudean, H. Noi descoperiri arheologice pe teritoriul judeţului Alba (I). *Apulum* **1978**, *16*, 39–53.
34. Soroceanu, T.; Lakó, E. Der zwiete Depotfund von Dragu, Kr. Sălaj. Zu den Tüllenbeiledepotfunde in Rumänien. In *Bronzefunde aus Rumänien*; Soroceanu (Hrsg.), T., Ed.; PAS: Berlin, Germany, 1995; pp. 187–195.
35. Żurowski, K. Zabytki brązove z młodszej epoki brązu i wczesnego okresu żelaza z dorzecza górnego Dniestru. *PrzArch* **1949**, *8*, 155–284.
36. Mozsolics, A. *Bronzefunde aus Ungarn. Depotfundhorizonte von Aranyos, Kurd und Gyermely*; Akadémiai Kiadó: Budapesta, Hungary, 1985.
37. Mozsolics, A. *Bronzefunde aus Ungarn. Depotfundhorizonte Hajdúböszörmény*; Románd und Bükkszentlászló: Verlag Oetket/Voges Kiel, Germany, 2000.
38. Figueiredo, E.; Silva, R.J.C.; Fátima Araújo, M.; Senns-Martinez, J.C. Identification of ancient gilding technology and Late Bronze Age metallurgy by EDXRF, Micro-EDXRF, SEM-EDS and metallographic techniques. *Microchim Acta* **2010**, *168*, 283–291. [CrossRef]
39. Sandu, I.; Mircea, O.; Sandu, A.V.; Sarghie, I.; Sandu, I.G.; Vasilache, V. Non-invasive Techniques in the Analysis of Corrosion Crusts formed on Archaeological Metal Objects. *Rev. Chim* **2010**, *61*, 1054–1058.

MDPI
St. Alban-Anlage 66
4052 Basel
Switzerland
Tel. +41 61 683 77 34
Fax +41 61 302 89 18
www.mdpi.com

Applied Sciences Editorial Office
E-mail: applsci@mdpi.com
www.mdpi.com/journal/applsci

www.ingramcontent.com/pod-product-compliance
Lightning Source LLC
Chambersburg PA
CBHW061509210926
51850CB00008B/1864
* 9 7 8 3 0 3 6 5 8 2 4 3 6 *